Low-Power Computer Vision

T0313066

Low-Power Computer Vision

Improve the Efficiency of Artificial Intelligence

Edited by

George K. Thiruvathukal
Yung-Hsiang Lu
Jaeyoun Kim
Yiran Chen
Bo Chen

CRC Press
Taylor & Francis Group
Boca Raton London New York

CRC Press is an imprint of the
Taylor & Francis Group, an **Informa** business

A CHAPMAN & HALL BOOK

First edition published 2022
by CRC Press
6000 Broken Sound Parkway NW, Suite 300, Boca Raton, FL 33487-2742

and by CRC Press
4 Park Square, Milton Park, Abingdon, Oxon OX14 4RN

CRC Press is an imprint of Taylor & Francis Group, an Informa business

© 2022 selection and editorial matter, George K. Thiruvathukal, Yung-Hsiang Lu, Jaeyoun Kim, Yiran Chen, Bo Chen; individual chapters, the contributors

Reasonable efforts have been made to publish reliable data and information, but the author and publisher cannot assume responsibility for the validity of all materials or the consequences of their use. The authors and publishers have attempted to trace the copyright holders of all material reproduced in this publication and apologize to copyright holders if permission to publish in this form has not been obtained. If any copyright material has not been acknowledged please write and let us know so we may rectify in any future reprint.

Except as permitted under U.S. Copyright Law, no part of this book may be reprinted, reproduced, transmitted, or utilized in any form by any electronic, mechanical, or other means, now known or hereafter invented, including photocopying, microfilming, and recording, or in any information storage or retrieval system, without written permission from the publishers.

Trademark notice: Product or corporate names may be trademarks or registered trademarks and are used only for identification and explanation without intent to infringe.

Library of Congress Cataloging-in-Publication Data
Names: Thiruvathukal, George K. (George Kuriakose), editor.
Title: Low-power computer vision : improve the efficiency of artificial intelligence / edited by George K. Thiruvathukal, Yung-Hsiang Lu, Jaeyoun Kim, Yiran Chen, Bo Chen.
Description: First edition.	Boca Raton : CRC Press, [2022]	Includes bibliographical references and index.
Identifiers: LCCN 2021042753	ISBN 9780367744700 (hbk)	ISBN 9780367755287 (pbk)	ISBN 9781003162810 (ebk)
Subjects: LCSH: Computer vision.	Low voltage systems.
Classification: LCC TA1634 .L69 2022	DDC 006.3/7--dc23/eng/20211028
LC record available at https://lccn.loc.gov/2021042753

ISBN: 978-0-367-74470-0 (hbk)
ISBN: 978-0-367-75528-7 (pbk)
ISBN: 978-1-003-16281-0 (ebk)

DOI: 10.1201/9781003162810

Publisher's note: This book has been prepared from camera-ready copy provided by the authors.

Typeset in LM Roman
by KnowledgeWorks Global Ltd.

Contents

YUNG-HSIANG LU AND XIAO HU, YIRAN CHEN, JOE SPISAK, GAURAV AGGARWAL, AND MIKE ZHENG SHOU, AND GEORGE K. THIRUVATHUKAL

ABHINAV GOEL, CALEB TUNG, XIAO HU, HAOBO WANG, AND YUNG-HSIANG LU AND GEORGE K. THIRUVATHUKAL

SECTION II   Competition Winners

CHAPTER   4 ■ Hardware Design and Software Practices for Efficient Neural Network Inference   55

YU WANG, XUEFEI NING, SHULIN ZENG, YI CAI, KAIYUAN GUO, AND HANBO SUN, CHANGCHENG TANG, TIANYI LU, AND SHUANG LIANG, AND TIANCHEN ZHAO

CHAPTER   5 ▪ Progressive Automatic Design of Search Space
              for One-Shot Neural Architecture Search   91

XIN XIA, XUEFENG XIAO, AND XING WANG

CHAPTER   6 ▪ Fast Adjustable Threshold for Uniform Neural
              Network Quantization   111

ALEXANDER GONCHARENKO, ANDREY DENISOV, AND SERGEY ALYAMKIN

YING WANG, XUYI CAI, AND XIANDONG ZHAO

HAN CAI AND SONG HAN

SOONHOI HA, EUNJIN JEONG, DUSEOK KANG, JANGRYUL KIM, AND DONGHYUN KANG

MARK SANDLER AND ANDREW HOWARD

AMIR GHOLAMI, SEHOON KIM, ZHEN DONG, ZHEWEI YAO, MICHAEL W.
MAHONEY, AND KURT KEUTZER

Foreword

Whereas electronics and computing have provided our society with unprecedented means of advancing services in this millennium, the environmental cost of using electronic technology is becoming significant. For this reason, low-energy and low-power computing has become an important area of research and development. Moreover, the miniaturization of devices, for example phones and drones, requires small energy reservoirs (i.e., low-volume, low-weight batteries). The pioneering work on digital watches of the eighties has grown up by now to a full array of hardware and software design technologies to mitigate the energy consumption of processing and storage elements in many areas.

From an application perspective, the ability of recognizing situations and actors, possibly within a complex environment, has become the key element in creating advanced systems in many domains, such as security, automated driving, and surveying. There has been a tremendous growth in the capabilities of image recognition systems in both hardware and software, and the presence of such systems is now almost ubiquitous. Nevertheless, the complexity of recognition requires a corresponding energy cost. As in the case of other electronic systems, the energy consumption may be significantly high and be an impediment to a wide use of image recognition in some domains.

As a result of the aforementioned considerations, the search for low-power computer vision systems is a key problem in both the research and development fields. There is a wide gap between the ideal minimum energy cost solutions and the current realizations. This gap is hard to quantify, as many factors come into play, ranging from the non-ideality of electronic devices (e.g., leakage current) to the choice of heuristic algorithms that approximate solutions because of the inherent computational complexity. On the bright side, this wide gap enables a continuous search for improvements within the entire design space spectrum, from circuits to algorithms, from hardware architectures to software programs.

The search for bettering energy efficiency would not be possible without realistic drivers and a world-wide participation of researchers. This is why the low-power computer vision challenge has been, and currently is, an important instrument for advancing the state of the art. The challenge was taken by some of the best groups in the world, and their effort has tackled the problem with different means and perspectives. Overall, this challenge has brought us very important results, that are fully documented in this book, and that will provide a strong impact on industry and academia.

Lausanne, March 2021
Giovanni De Micheli

Rebooting Computing and Low-Power Computer Vision

Since its start in 2013 as an initiative of IEEE Future Directions Committee, "Rebooting Computing" has provided an international, interdisciplinary environment where experts from a wide variety of computer-related fields can come together to explore novel approaches to future computing. The need for Rebooting Computing follows from the recognition that the exponential improvement in computing performance in previous decades was due primarily to transistor scaling in Moore's Law, but this is coming to an end. Radical alternative approaches are needed over the entire technology landscape, from basic devices and circuits to architectures to software, with applications from supercomputers to smartphones. Some possible newer approaches that are being explored include neuromorphic computing, approximate and stochastic computing, quantum and cryogenic computing, low-power reversible and adiabatic computing, and computing based on non-volatile memories, analog and optical systems. The initiative has now evolved to become a Task Force within the Computer Society of IEEE and continues its mission unabated.

"Rebooting Computing" spawned many innovations, including the Low-Power Image Recognition Challenge (LPIRC) in 2015, the brainchild of Prof. Yung-Hsiang Lu. LPIRC ran for several years with ever-improving performance by the teams demonstrating subsystems for image recognition at the lowest possible power. Importantly, the competition involved a multitude of students, providing inspiration and motivation to students worldwide. LPIRC was renamed as the Low-Power Computer Vision Challenge (LPCVC) in 2020 when video was also included. These challenges evaluate both accuracy and energy consumption of systems that can recognize and understand images or videos. Over the six years since the inception of the Challenge, more than 100 teams have participated. The teams have sponsorship and participation from industry, including Facebook, Google, Xilinx, ELAN Microelectronics, Amazon, Qualcomm, and Bytedance.

This book contains the collection of the solutions of the winners of the Challenge. The authors compare different options, making computer vision more efficient and explaining important design decisions. The information provides deep insight for researchers and practitioners.

Elie K. Track, CEO of nVizix LLC,

Founding Co-Chair of the IEEE Rebooting Initiative

Editors

George K. Thiruvathukal is a professor of Computer Science at Loyola University Chicago, Illinois, USA. He is also a visiting faculty at Argonne National Laboratory. His research areas include high performance and distributed computing, software engineering, and programming languages.

Yung-Hsiang Lu is a professor of Electrical and Computer Engineering at Purdue University, Indiana, USA. He is the first director of Purdue's John Martinson Engineering Entrepreneurial Center. He is a fellow of the IEEE and distinguished scientist of the ACM. His research interests include computer vision, mobile systems, and cloud computing.

Jaeyoun Kim is a technical program manager at Google, California, USA. He leads AI research projects, including MobileNets and TensorFlow Model Garden, to build state-of-the-art machine learning models and modeling libraries for computer vision and natural language processing.

Yiran Chen is a professor of Electrical and Computer Engineering at Duke University, North Carolina, USA. He is a fellow of the ACM and the IEEE. His research areas include new memory and storage systems, machine learning and neuromorphic computing, and mobile computing systems.

Bo Chen is the Director of AutoML at DJI, Guangdong, China. Before joining DJI, he was a researcher at Google, California, USA. His research interests are the co-optimization of neural network software and hardware as well as landing AI technology in products with stringent resource constraints.

I

Introduction

CHAPTER 1

Book Introduction

Yung-Hsiang Lu

Purdue University

George K. Thiruvathukal

Loyola University Chicago

Jaeyoun Kim

Google California

Yiran Chen

Duke University

Bo Chen

Da-Jiang Innovations China

CONTENTS

DOI: 10.1201/9781003162810-1

1.1 ABOUT THE BOOK

The first IEEE Low-Power Image Recognition Challenge was held in 2015. Since then, winners have presented their solutions in conferences and published detailed studies in journals. After six years of competitions, there is a rich set of knowledge about how to make computer vision efficient running on embedded computers. The organizers decided to put together this book so that researchers, engineers, and practitioners can understand what methods worked well for winning the competitions.

The book is composed of three parts: Introduction, Winners' Solutions, and Invited Articles. The first part provides a brief history of the competitions and a survey of literature. The second part includes the articles from the winners. All winners were invited to contribute to this book; this part of the book includes the articles from the winners that accepted the invitations. The third part contains articles from leaders in low-power computer vision, including authors from industry and academia.

1.2 CHAPTER SUMMARIES

1.2.1 History of Low-Power Computer Vision Challenge

Yung-Hsiang Lu (Purdue University); Xiao Hu (Purdue University); Yiran Chen (Duke University); Joe Spisak (Facebook); Gaurav Aggarwal (Facebook); Mike Zheng Shou (Facebook Research), and George K. Thiruvathukal (Loyola University Chicago)

Abstract

This chapter describes the history of IEEE History of Low-Power Computer Vision Challenge 2015–2020.

Take-aways

- Describes the history of the IEEE Low-Power Computer Vision Challenge between 2015 and 2020.

- Explains the methods to select winners and lists the winners over these years.

1.2.2 Survey on Energy-Efficient Deep Neural Networks for Computer Vision

Abhinav Goel (Purdue University); Caleb Tung (Purdue University); Xiao Hu (Purdue University); Haobo Wang (Purdue University); George Thiruvathukal (Loyola University Chicago); Yung-Hsiang Lu (Purdue University)

Abstract

Deep Neural Networks (DNNs) are greatly successful in performing many different computer vision tasks. However, the state-of-the-art DNNs are too energy, computation, and memory-intensive to be deployed on most computing devices and embedded systems. DNNs usually require server-grade CPUs and GPUs. To make computer vision more ubiquitous, recent research has focused on making DNNs more efficient. These techniques make DNNs smaller and faster through various refinements and thus are enabling computer vision on battery-powered mobile devices. Through this article, we survey the recent progress in low-power deep learning to discuss and analyze the advantages, limitations, and potential improvements to the different techniques. We particularly focus on the software-based techniques for low-power DNN inference. This survey classifies the energy-efficient DNN techniques into six broad categories: (1)Quantization, (2)Pruning, (3)Layer and Filter Compression, (4)Matrix Decomposition, (5)Neural Architecture Search, and (6)Knowledge Distillation. The techniques in each category are discussed in greater detail in this chapter.

Take-aways

- Surveys the recent progress in low-power deep learning to analyze the advantages, limitations, and potential improvements to the different techniques.

- Focus on the software-based techniques for low-power DNN inference

1.2.3 Hardware Design and Software Practices for Efficient Neural Network Inference

Yu Wang (Tsinghua University); Xuefei Ning (Tsinghua University); Shulin Zeng (Tsinghua University); Changcheng Tang (Novauto); Yi Cai (Tsinghua University); Kaiyuan Guo (Tsinghua University); Shuang Liang (Novauto); Tianyi Lu (Novauto); Hanbo Sun (Tsinghua University); Tianchen Zhao (Beihang University)

Abstract

In this chapter, we introduce our efforts in accelerating neural network inference. From the hardware design aspect, we introduce the instructions-set-architecture deep learning accelerator to support all kinds of DNN models with customized ISA and optimized software compiler. And from the algorithm aspect, we introduce several practices we have used: sensitivity-based pruning without hardware model, quantization, iterative pruning with hardware model, and neural architecture search.

Take-aways

- Discusses hardware design: An instructions-set-architecture deep learning accelerator to support all kinds of DNN models with customized ISA and optimized software compile

- Discusses software practices: Sensitivity-based pruning without hardware model, quantization, iterative pruning with hardware model, neural architecture search.

1.2.4 Progressive Automatic Design of Search Space for One-Shot Neural Architecture

Xin Xia (Bytedance Inc); Xuefeng Xiao (ByteDance Inc); XING WANG (Bytedance AI Lab)

Abstract

Neural Architecture Search (NAS) has attracted growing interest. To reduce the search cost, recent work has explored weight sharing across models and made major progress in One-Shot NAS. However, it has been observed that a model with higher one-shot model accuracy does not necessarily perform better when stand-alone trained. To address this issue, in this paper, we propose Progressive Automatic Design of search space, named PAD-NAS. Unlike previous approaches where the same operation search space is shared by all the layers in the supernet, we formulate a progressive search strategy based on operation pruning and build a layer-wise operation search space. In this way, PAD-NAS can automatically design the operations for each layer. During the search, we also take the hardware platform constraints into consideration for efficient neural network model deployment. Extensive experiments on ImageNet show that our method can achieve state-of-the-art performance.

Take-aways

- Uses network architecture search methods to find better architectures for lower latencies and higher accuracy

- Formulates a search strategy to build a layer-wise operation search space through hierarchical operation pruning and mitigates weight coupling issue in One-Shot NAS.

- Compares the effects of different parameters on memory sizes, latency, and accuracy

1.2.5   Fast Adjustable Threshold for Uniform Neural Network Quantization

Alexander Goncharenko (Novosibirsk State University); Andrey Denisov (Expasoft); Sergey Alyamkin (Expasoft)

Abstract

The neural network quantization is highly desired procedure to perform before running neural networks on mobile devices. Quantization without fine-tuning leads to accuracy drop of the model, whereas commonly used training with quantization is done on the full set of the labeled data and therefore is both time- and resource-consuming. Real-life applications require simplification and acceleration of quantization procedure that will maintain the accuracy of full-precision neural network, especially for modern mobile neural network architectures like Mobilenet-v1, MobileNet-v2, and MNAS. Here we present two methods to significantly optimize the training with quantization procedure. The first one is introducing the trained scale factors for discretization thresholds that are separate for each filter. The second one is based on mutual rescaling of consequent depth-wise separable convolution and convolution layers. Using the proposed techniques, we quantize the modern mobile architectures of neural networks with the set of train data of only 10% of the total ImageNet 2012 sample. Such reduction of train dataset size and small number of trainable parameters allow to fine-tune the network for several hours while maintaining the high accuracy of quantized model (accuracy drop was less than 0.5%). Ready-for-use models and code are available at: https://github.com/agoncharenko1992/FAT-fast-adjustable-threshold.

Take-aways

- Describes ways how to get an 8-bit quantized network.

- The main idea is that simple min/max quantization with calibration works poor because of outliers which spoils thresholds of quantization.

- We can adjust this thresholds by using Straight-Through Estimators. Using some tips such as Batch Normalization folding and, channel equalization (more details you can found in the paper) we can get solution as good as training with quantization from scratch but with less data and way faster.

1.2.6 Power-efficient Neural Network Scheduling on Heterogeneous system on chips (SoCs)

Ying Wang (Institute of Computing Technology, Chinese Academy of Sciences); Xuyi Cai (Institute of Computing Technology, Chinese Academy of Sciences); Xiandong Zhao (Institute of Computing Technology, Chinese Academy of Sciences)

Abstract

The powerful deep neural networks (DNNs) have been propelling the development of efficient computer vision technologies for mobile systems such as phones and drones. To enable power-efficient image processing on resource-constrained devices, many studies have been dedicated to the field of low-power DNNs from different layers of the systems. Amongst the deep stack of low-power DNN systems, task scheduling also plays an essential role as the interfacing middleware between the algorithms and the underlying hardware. Especially when heterogeneous SoCs have been widely adopted in edge and mobile scenarios as the hardware solution, an efficient DNN task scheduler is needed to reduce the implementation overhead of DNN-based task and extract the most power from the SoC platform. This chapter will firstly exemplify DNN scheduling with the image recognition solution of LPIRC-2016 and introduce how to efficiently schedule a DNN-based visual processing task onto a typical heterogeneous SoC composed of general-purpose and specialized cores. After the elaborate task-level scheduling strategy, we will discuss the fine-grained DNN-wise scheduling policy on specialized DNN cores and show the effectiveness of memory-oriented DNN-layer scheduling. Last, since model quantization is an indispensable step to map a large-size neural network model onto the resource-thrifty mobile SoCs, we will discuss the implication of DNN quantization on the heterogeneous SoCs integrated with both integer and float-point cores, and then introduce the scheduler-friendly DNN quantizer for pure-integer hardware. Although most prior works on low-power DNNs focused their attention on efficient network and hardware architectures, it is shown that the scheduler-level optimization technology will also be critical to the energy-efficiency of the system, particularly when the algorithmic implementation is fixed and off-the-shelf hardware devices are adopted.

Take-aways

- Demonstrates the rank-1 solution of LPIRC2016 as a case study to introduce the basic coarse-grained scheduling techniques for DNN-based applications.

- Presents the memory-efficient fine-grained neural network scheduler on DNN processors.

- Introduces the scheduler-friendly quantization technique to reduce the overhead of neural network implementation on embedded SoCs.

1.2.7   Efficient Neural Architecture Search

Han Cai and Song Han (MIT)

Abstract

Designing efficient neural network architectures is a widely adopted approach to improve efficiency, besides compressing an existing deep neural network. A CNN (Convolutional Neural Network) model typically consists of convolution layers, pooling layers, and fully-connected layers, where most of the computation comes from convolution layers. For example in ResNet-50, more than 99% multiply-accumulate operations (MACs) are from convolution layers. Therefore, designing efficient convolution layers is the core of building efficient CNN architectures. This chapter first describes the standard convolution layer and then describes three efficient variants of the standard convolution layer. Next, we present three representative manually design efficient CNN architectures, including SqueezeNet, MobileNets, and ShuffleNets. Finally, we describe automated methods for designing efficient CNN architectures.

Take-aways

- Describes the standard convolution layer and then describes three efficient variants of the standard convolution layer.

- Presents three representative manually designed efficient CNN architectures, including SqueezeNet, MobileNets, and ShuffleNets.

- Describes automated methods for designing efficient CNN architectures.

1.2.8 Design Methodology for Low-Power Image Recognition Systems Design

Soonhoi Ha (Seoul National University); EunJin Jeong (Seoul National University); Duseok Kang (Seoul National University); Jangryul Kim (Seoul National University); Donghyun Kang (Seoul National University)

Abstract

In the development of an embedded image recognition system, there are many issues to consider, such as which hardware platform and algorithm to use, how to optimize the software with resource constraints and how to optimize multiple design objectives, and so on. This chapter presents a systematic design methodology that could be applied to the design of embedded systems with a concrete example of image recognition systems. Based on the proposed methodology, we could win the first prize in LPIRC (Low-Power Image Recognition Challenge) 2017. After selecting NVIDIA Jetson TX2 as the hardware platform and Tiny YOLO as the detection algorithm, we applied the well-known software optimization techniques in a systematic way, aiming to jointly optimize speed, accuracy, and energy. We have refined the methodology to choose a different algorithm on the same hardware platform and could build another winning solution in track 2 of LPIRC 2018. Recently new hardware platforms have been developed that contain CNN hardware accelerators as well as GPU (Graphics Processing Units), among which NVIDIA Jetson AGX Xavier is a representative example. Since it is a heterogeneous system that contains multiple hardware accelerators, how to exploit the computing power of those accelerators maximally becomes an important issue to consider in the proposed design methodology. We have developed a novel technique to maximally utilize multiple accelerators to achieve 21.7 times better score than our previous solution in LPIRC 2018, which is also presented in this chapter.

Take-aways

- First prize winning solution in LPIRC 2017 and in track2 of LPIRC 2018.

- Presents a systematic design methodology for the design of low-power image recognition systems.

- Demonstrates how to select the hardware platform and a neural network by considering the estimated performance.

- Demonstrates how to map the network onto the hardware platform aiming to maximize the throughput by pipelining.

- Shows how various software optimization techniques are then applied to each processing element.

1.2.9 Guided Design for Efficient On-device Object Detection Model

Tao Sheng and Yang Liu (Amazon)

The low-power computer vision (LPCV) challenge is an annual competition for the best technologies in image classification and object detection measured by both efficiency (execution time and energy consumption) and accuracy (precision/recall). Our Amazon team has won three awards from LPCV challenges: 1st prize for interactive object detection challenge in 2018 and 2019 and 2nd prize for interactive image classification challenge in 2018. This paper is to share our award-winning methods, which can be summarized as four major steps. First, 8-bit quantization friendly model is one of the key winning points to achieve the short execution time while maintaining the high accuracy on edge devices. Second, network architecture optimization is another winning keypoint. We optimized the network architecture to meet the 100ms latency requirement on Pixel2 phone. The third one is dataset filtering. We removed the images with small objects from the training dataset after deeply analyzing the training curves, which significantly improved the overall accuracy. And the forth one is non-maximum suppression optimization. By combining all the above steps together with the other training techniques, for example, cosine learning function and transfer learning, our final solutions were able to win the top prizes out of large number of submitted solutions across worldwide.

Take-aways:

- Discusses the methods involved in the winning solutions over the years.

- Explains the impacts of each method (quantization, architecture search, hyperparameter tuning)

- Reduces the resolutions to improve performance

1.2.10 Quantizing Neural Networks for Low-Power Computer Vision

Markus Nagel (Qualcomm); Marios Fournarakis (Qualcomm); Rana Ali Amjad (Qualcomm); Yelysei Bondarenko (Qualcomm); Mart van Baalen (Qualcomm); Tijmen Blankevoort (Qualcomm)

Abstract

Over the last years, Neural Networks (NNs) have been widely adapted in Computer Vision (CV) applications. While for many tasks they outperform traditional CV algorithms they often come at a high compute cost. Even mobile friendly architectures such as MobileNet still require hundreds of million floating point operations. To further reduce the energy efficiency and latency of NNs, quantization can be used to replace the original floating-point operations with low bit fixed-point operations. In this chapter we introduce NN quantization for low-power computer vision. Afterward we highlight recent advances in post-training quantization, a class of algorithms that can be applied to pretrained NNs and do not require any expert knowledge. In the last part we will focus on quantization-aware training, a technique that trains NNs with simulated quantization operations.

Take-aways

- Introduces neural network quantization

- Serves as a practical guide to quantization simulation with HW considerations

- Introduces state-of-the-art post-training quantization (PTQ) techniques that are easy to use.

- Introduces state-of-the-art quantization-aware training (QAT) approaches that result in best performance.

- Defines standard PTQ and QAT pipeline and evaluates them on several computer vision models and tasks.

1.2.11 A Practical Guide to Designing Efficient Mobile Architectures

Mark Sandler and Andrew Howard (Google)

Abstract

In this chapter we overview a set of basic techniques that can be applied when designing and fine-tuning efficient architectures. We establish basic principles that practitioners can use when adapting existing architectures to particular applications. While a lot of modern research has been dedicated to network architecture search, the basic design principles are often poorly understood. Our goal here is to build a solid foundation and demystify the reasoning about image neural networks from a practical perspective. From our experience, such a foundation is indispensable for both designing new architecture search spaces, as well as for practical tuning of existing architectures to new hardware and/or problems, without relying on opaque Network Architecture Search (NAS) techniques.

Take-aways

- Introduces a set of basic techniques for adapting and fine-tuning existing model architectures to different hardware and problems.

- Provides an in-depth overview of several types of multipliers that enable a user to independently adjust resource consumption such as model size, memory requirements, and energy consumption.

- Demonstrates more specialized ways to fine-tune individual layers.

- Demonstrates ways to phase in custom nonlinearities that have limited support on existing hardware.

1.2.12   A Survey of Quantization Methods for Efficient Neural Network Inference

Amir Gholami (UC Berkeley); Sehoon Kim (University of California, Berkeley); Zhen Dong (UC Berkeley); Zhewei Yao (University of California, Berkeley); Michael Mahoney (University of California, Berkeley); Kurt Keutzer (EECS, UC Berkeley)

Abstract

As soon as abstract mathematical computations were adapted to computation on digital computers, the problem of efficient representation, manipulation, and communication of the numerical values in those computations arose. Strongly related to the problem of numerical representation is the problem of quantization: in what manner should a set of continuous real-valued numbers be distributed over a fixed discrete set of numbers to minimize the number of bits required and also to maximize the accuracy of the attendant computations? This perennial problem of quantization is particularly relevant whenever memory and/or computational resources are severely restricted, and it has come to the forefront in recent years due to the remarkable performance of Neural Network models in computer vision, natural language processing, and related areas. Moving from floating-point representations to low-precision fixed integer values represented in four bits or less holds the potential to reduce the memory footprint and latency by a factor of 16x; and, in fact, reductions of 4x to 8x are often realized in practice in these applications. Thus, it is not surprising that quantization has emerged recently as an important and very active sub-area of research in the efficient implementation of computations associated with Neural Networks. In this article, we survey approaches to the problem of quantizing the numerical values in deep Neural Network computations, covering the advantages/disadvantages of current methods. With this survey and its organization, we hope to have presented a useful snapshot of the current research in quantization for Neural Networks and to have given an intelligent organization to ease the evaluation of future research in this area.

Take-aways

- As soon as abstract mathematical computations were adapted to computation on digital computers, the problem of efficient representation, manipulation, and communication of the numerical values in those computations arose.

- Strongly related to the problem of numerical representation is the problem of quantization, which is the main focus of this chapter.

- We will first introduce the basic concepts of quantization, and then discuss the advanced methods, as well as open problems in this area.

CHAPTER 2

History of Low-Power Computer Vision Challenge

Yung-Hsiang Lu and Xiao Hu

Purdue University

Yiran Chen

Duke University

Joe Spisak, Gaurav Aggarwal, and Mike Zheng Shou

Facebook

George K. Thiruvathukal

Loyola University Chicago

CONTENTS

2.1 REBOOTING COMPUTING

The "Moore's Law" has been one of the most profound driving forces of modern technologies. The law is an observation (not a physics law) that semiconductor technologies can double densities approximately every one to two years. This technology advancement is the foundation shrinking computers from mainframes to desktops to mobile phones. There have been numerous discussions about the "post Moore's Law" era—when the steady improvements in semiconductor density hit the physics limits and cannot grow at the same rate any more. The IEEE Future Directions started a new initiative called "Rebooting Computing" to think about future computer designs when the Moore's Law eventually ends. The first Rebooting Computing Summit was held in Washington DC in 2013. In this inaugural summit of Rebooting Computing, co-chair Elie K. Track asked each attendee to volunteer for one task after the meeting. David Kirk from Nvidia and Yung-Hsiang Lu from Purdue University suggested the idea of organizing a competition that would require solutions beyond available technologies. This competition would demonstrate artificial intelligence using only ambient energy such as sunlight, vibration, or wind. When this chapter is written in 2021, solar-powered cameras are already available. However, to the authors' knowledge, these cameras are unable to perform sophisticated vision tasks.

Readers are encouraged to visit the website `https://lpcv.ai/` about the most recent competitions and speeches on low-power computer vision.

2.2 LOW-POWER IMAGE RECOGNITION CHALLENGE (LPIRC): 2015–2019

The first challenge was to define "intelligence". In 2012, a new era of computer vision had started: deep neural networks achieved impressive progress in the ImageNet competition. Alexander C. Berg was one of the organizers of ImageNet. Lu and Berg discussed how to design a new competition. This new competition would build upon ImageNet's success using image recognition to judge intelligence, adding low energy consumption as a second factor. The competition used energy (Joules) as a parameter to recognize the importance of time to complete a task. Each solution had ten minutes to process all data. Using power (Watts) alone can be misleading because a solution can be low-power but run for a very long time thereby still consuming more energy, e.g., larger battery drain. The score is computed as the recognition accuracy divided by energy consumption:

$$\text{score} = \frac{\text{accuracy}}{\text{energy consumption}}. \tag{2.1}$$

Figure 2.1 (a) shows an example of an input image and two objects to be recognized: a bird and a frog. ImageNet has 1,000 categories of objects, including human, dog, cat, airplane, etc. An object is considered

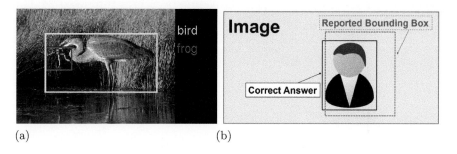

(a)   (b)

Figure 2.1 (a) A sample from ImageNet. Object detection needs to recognize the objects correctly and mark their locations. (b) The location of a recognized object is evaluated by using intersection over union (IoU) of the two bounding boxes.

successfully recognized if two conditions are met: (1) The category is correct. Recognizing a dog as an airplane is considered a failure. (2) The location of the object is marked correctly. Each recognized object is marked by a *bounding box*: a rectangle that encloses the object. Figure 2.1 (b) shows two bounding boxes: one reported by a computer program and the other marked by a human (treated as the correct answer). The bounding box by the computer program is accepted if the *intersection over union* exceeds 50%, as defined below:

$$\frac{\text{correct answer} \cap \text{reported}}{\text{correct answer} \cup \text{reported}} \geq 0.5. \tag{2.2}$$

Table 2.1 defines four possible outcomes of object detection. If an object is present in an image and it is detected, this is called *true positive*. If a detected object is not actually in an image, this is called *false positive*. Precision is defined as

$$\text{precision} = \frac{\text{true positive}}{\text{true positive} + \text{false positive}}. \tag{2.3}$$

TABLE 2.1 Four Possible Outcomes of Object Detection

		Detection	
		Yes	No
Truth	Yes	True Positive	False Negative
	No	False Positive	True Negative

TABLE 2.2 From 2015 to 2018, the Winners' Scores Improved 24 Times

Year	mAP	Energy (WH)	Score	Ratio
2015	0.02971	1.634	0.0182	1.00
2016	0.03469	0.789	0.0440	2.42
2017	0.24838	2.082	0.1193	6.56
2018	0.18318	0.412	0.4462	24.51

The test images contain objects of different categories (such as humans and cars). The average precision is calculated for each categories. Then, the mean of all categories is calculated; this is called the *mean average precision (mAP)*. The accuracy of the competition is measured by mAP across all test images.

In 2015, 5,000 images were used and no team finished all 5,000 images within 10 minutes. The champion obtained 0.02971 mAP with 1.634 Watt-Hour energy consumption; their score was 0.0182. Since 2016, 20,000 images were used. The 2018 champion finished 20,000 images within 10 minutes and obtain 0.1832 mAP with 0.412 Watt-Hour energy consumption. The score was 0.4446, improvement of 24 times since 2015. Table 2.2 compares the winners' scores. As a reference, the winner of the 2017 ImageNet competition (without time or energy restrictions) had an accuracy of 0.731 mAP. In 2015–2018, LPIRC had no restrictions on hardware or software and it was an onsite competition. Contestants brought a wide range of systems: laptops, phones, desktop with GPUs, embedded computers, reconfigurable systems, tablets, etc. To encourage more participation without the need of travel, starting from 2018, contestants could use pre-selected hardware and submit their solutions online. Online submissions have another advantage: contestants could submit multiple solutions and the best scores are used. Starting from 2019, the competition is online only with pre-selected hardware.

2.3 LOW-POWER COMPUTER VISION CHALLENGE (LPCVC): 2020

In 2020, LPIRC was renamed to Low-Power Computer Vision Challenge (LPCVC) by adding video taken by an unmanned aerial vehicle (UAV, also called unmanned aerial systems UAS, or drone). Figure 2.2 shows two sample frames of a video taken by a UAV inside a building. The challenge needs to process the video and provide question-answer pairs. Each answer is one or multiple English letters or numbers. The corresponding

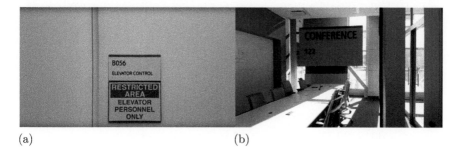

(a)   (b)

Figure 2.2 Two sample frames of a video taken by a UAV.

answer is the other English letters or the numbers in the same frame. For example, if the question is "B056", the correct answer is "ELEVATOR CONTROL RESTRICTED AREA ELEVATOR PERSONNEL ONLY". For the question "CONFERENCE", the answer is "122". The answers are case-insensitive. The *Levenshtein distance* is used to measure the correctness of the answers. The videos were pre-recorded and the same clip was made available to all participants for offline processing on a pre-determined hardware platform: Raspberry Pi 3B+. The software framework is PyTorch.

2.4   WINNERS

Since 2015, 108 teams submitted more than 500 solutions. The champions' solutions often outperformed the best solutions available in literature. Table 2.3 shows the workshops and special sessions of the competitions for the competitions and the winners. Staring from 2018, the challenge includes multiple tracks. These tracks were simultaneous, and teams could participate in more than one track. Each track had a unique challenge and potentially used different hardware and software platforms. The challenge has been open to the public and there is no particular limit on the team size or affiliation to any academic or professional institute. Teams came from academia, industry, as well as collaboration of both. The challenge was held twice in 2019. The following are the champions of each year. Winners' solutions have been published in [1, 2, 3, 4, 5, 6, 7]. The followings are the lists of winners:

- 2015: Tsinghua University and Huawei (China), Institute of Automation of Chinese Academy of Sciences and Huawei (China), Tsinghua University and Huawei (China).

TABLE 2.3 The Challenges Are Co-located with Major Conferences

Conference	Event
2015/06 DAC	Onsite competition
2015/11 ICCAD	2015 winners presented solutions
2016/06 DAC	Onsite competition
2017/01 ASPDAC	2016 winners presented solutions
2017/07 CVPR	Onsite competition
2018/03 DATE	2017 winners presented solutions
2018/06 CVPR	Onsite competition
2018/12 NeurIPS	2018 winners presented solutions
2019/03 AICAS	2018 winners presented solutions
2019/06 CVPR	2018 winners presented solutions
2019/10 ICCV	2019 winners presented solutions
2020/06 CVPR	2019 winners presented solutions
2021/05 AICAS	2020 winners presented solutions

- 2016: Chinese Academy of Science (China), Tsinghua University (China).

- 2017: Seoul National University (S Korea), Korea Platform Service Technology and Electronics and Telecommunications Research Institute (S Korea), Watrix (China).

- 2018 (three tracks):

  - Real-Time Image Classification with Low latency: Expasoft (Russia), Qualcomm (Canada).

  - Interactive Image Classification: Expasoft (Russia), Amazon (USA).

  - Interactive Object Detection: Amazon (USA), Qualcomm (Canada).

- 2019-1 (four tracks):

  - Image Classification using Field Programmable Gate Arrays: IPIU (China)

  - Image Classification on Mobile Accelerator: MIT HAN Lab (USA)

  - Object Detection: Amazon (USA), Tsinghua University (China), Qualcomm (Canada)

- Image Classification: Alibaba (China), Expasoft (Russia), MIT HAN Lab (USA).

- 2019-2 (two tracks):

    - Object Detection: MIT HAN Lab & Dawnlight (USA and China)

    - Image Classification: MIT HAN Lab & Dawnlight (USA and China)

- 2020 (five tracks):

    - Video using PyTorch: ByteDance (China), Texas A & M University and Kwai (USA), Stony Brook University (SUNY Korea).

    - Image Classification using Field Programmable Gate Arrays: MIT HAN Lab (USA), National Chiao Tung University (Taiwan), CAS Institute of Computing Technology (China).

    - TFLite Interactive Object Detection using CPU: MIT HAN Lab (USA), Shanghai Jiao Tong University (China), Novauto (China).

    - TFLite Real-Time Image Classification using CPU: Baidu and Tsinghua University (China), Beihang University (China), ByteDance (China).

    - TFLite Real-Time Image Classification using DSP: ByteDance (China), Beihang University (China), MIT HAN Lab (USA), National Chiao Tung University and National Tsing Hua University (Taiwan).

2.5 ACKNOWLEDGMENTS

The editors wish to express deep appreciation to the leaders of IEEE Rebooting Computing, including Elie K. Track, Terence Martinez, Alan Kadin, Tom Conte, Bichlien Hoang, William Tonti, Erik P. DeBenedictis, Bruce Kraemer, Scott Holmes, David Mountain, and Paolo Gargini. LPIR-C/LPCVC sponsors include IEEE Rebooting Computing, IEEE Council on Electronic Design Automation, IEEE Council on Superconductivity, IEEE Circuits and Systems Society, IEEE GreenICT, ACM Special Interest Group on Design Automation, Google, Facebook-PyTorch, Nvidia,

Xilinx, ELAN Microelectronics, Mediatek, and National Science Foundation. Any opinions, findings, and conclusions or recommendations expressed in this article are those of the authors and do not necessarily reflect the views of the sponsors.

CHAPTER 3

Survey on Energy-Efficient Deep Neural Networks for Computer Vision

Abhinav Goel, Caleb Tung, Xiao Hu, Haobo Wang, and Yung-Hsiang Lu

Purdue University

George K. Thiruvathukal

Loyola University Chicago

CONTENTS

DOI: 10.1201/9781003162810-3

Deep neural networks (DNNs) are greatly successful in performing many different computer vision tasks. However, state-of-the-art DNNs are too energy, computation, and memory-intensive to be deployed on most mobile and embedded devices. DNNs usually require server-grade CPUs and GPUs. To make computer vision more ubiquitous, recent research has focused on making DNNs more efficient. These techniques make DNNs smaller and faster through various refinements and thus are enabling computer vision on battery-powered mobile devices. Through this chapter, we survey the recent progress in low-power deep learning to analyze the advantages, limitations, and potential improvements to the different techniques. We particularly focus on the software-based techniques for low-power DNN inference. This survey classifies the energy-efficient DNN techniques into six broad categories: (1) Quantization, (2) Pruning, (3) Layer and Filter Compression, (4) Matrix Decomposition, (5) Neural Architecture Search, and (6) Knowledge Distillation. The techniques in each category are discussed in greater detail in this chapter.

3.1 INTRODUCTION

Deep Neural Networks (DNNs) are a class of machine learning algorithms that have led to significant breakthroughs in many computer vision tasks [8, 9, 10, 11]. The state-of-the-art performance of DNNs is mainly attributed to their ability to train billions of parameters for fitting complex functions [12]. However, the tremendous accuracy gains come with high costs in terms of computation and energy [13]. Visual Geometry Group (VGG-16) is a popular DNN used in computer vision and requires 15 billion operations to process a single image [8]. This amount of computation is beyond the scope of most embedded devices and is often limited to servers with high-performance GPUs or accelerators. In one experiment, we observe that one of the fastest object detectors, YOLOv3 [11], requires up to 4.9 seconds to process a single image on an Intel "Kaby Lake" Core i7 processor without a GPU. This experiment illustrates the tremendous computing requirements of DNNs and the difficulty to deploy them on resource-constrained devices.

Billions of mobile and embedded devices are now equipped with high definition cameras [14]. These devices are not usually equipped with GPUs. To process the images or videos captured by these cameras, computer vision practitioners often offload computing to the cloud [15]. However, many applications cannot be offloaded, e.g., computer vision deployed on drones flying in areas without reliable network coverage, or in satellites where offloading is too expensive [16]. Privacy concerns also limit the applicability of cloud-based solutions [17]. Lowering the computation requirements of computer vision would enable more embedded devices to process data without offloading [10]. Research has shown that a considerable amount of a DNN's computation is redundant when performing computer vision. These redundant operations can be removed from a DNN to increase efficiency without notable losses in accuracy [10, 18].

This chapter describes the influential research on energy-efficient DNNs. This chapter uses results reported in the existing literature to analyze and compare the different techniques. We focus specifically on software-based techniques for low-power DNN inference. Hardware-based techniques and low-power DNN training are beyond the scope of this chapter. This survey extends our previous paper [19]. In this chapter, we discuss the various low-power DNN techniques in greater detail. In doing so, we classify the techniques into finer categories and more accurately describe the intuition supporting the different design decisions.

This chapter classifies the methods for low-power computer vision into six broad categories. Each category uses a specific method to improve DNNs for efficient computer vision. We provide a short description of the techniques in the following list. Table 3.1 provides a high-level comparison of the techniques.

1. **Parameter Quantization:** Reduces the number of bits used to store the DNN parameters and activations. Quantized DNNs require less memory and have simplified computation. Unquantized DNNs usually use 32-bit single-precision floating point (float32) parameters, activations, and operations. Figure 3.1a shows representations of different quantization formats.

2. **Deep Neural Network Pruning:** Identifies and removes unnecessary connections from a DNN to improve efficiency. Pruning

TABLE 3.1 Summary of the advantages and disadvantages of the low-power computer vision techniques.

Technique	Advantages	Disadvantages
Quantization	High accuracy with small model size. Efficient arithmetic operations.	Only few quantization formats are available on most hardware.
Pruning	No accuracy loss for up to 50% pruning.	Sparse DNNs have negligible speedups on GPUs and CPUs.
Layer Compression	High accuracy. Orthogonal to other low-power techniques.	Compact convolutions are memory-inefficient. Many activations are stored at a time.
Decomposition	Decomposed matrices require significantly fewer memory accesses.	Decomposition does not always exist. Approximate decompositions may have poor accuracy.
Architecture Search	State-of-the-art accuracy.	Prohibitively high training costs.
Knowledge Distillation	Low computation cost with few DNN parameters.	Restrictions on DNN structure may lead to inefficient DNNs.

also makes DNN operations more sparse. Figure 3.1(b) shows an example of DNN pruning.

3. **Deep Neural Network Layer and Filter Compression:** Uses compact DNN layers that reduce the number of operations and memory without any accuracy losses. Figure 3.1(c) illustrates depthwise separable convolutions (a commonly used compact convolution layer).

4. **Parameter Matrix Decomposition:** Decomposes large DNN layers into smaller layers to increase inference speed. Only informative neurons are retained in the smaller layers. Figure 3.1(d) depicts an example of parameter matrix decomposition. Here, $m >> k$ and $n >> k$.

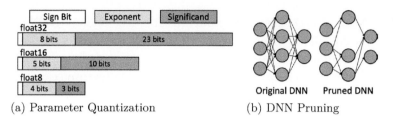

(a) Parameter Quantization

(b) DNN Pruning

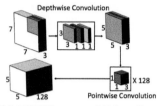

(c) DNN Layer and Filter Compression

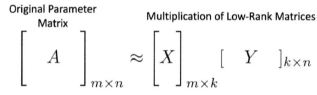

(d) Parameter Matrix Decomposition

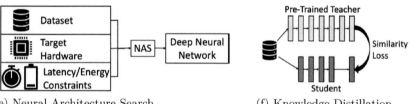

(e) Neural Architecture Search

(f) Knowledge Distillation

Figure 3.1 Illustrations of low-power computer vision techniques. (a) Quantization: More quantization (smaller bit-widths) reduce memory requirements. (b) Pruning: Redundant parameters and connections are removed from DNNs. (c) Layer Compression: Depthwise separable convolution is a compact convolution layer that replaces conventional convolution layers. (d) Matrix Decomposition: A large parameter matrix: A, is decomposed into multiple smaller matrices: X and Y, to reduce memory. (e) Architecture Search: DNN models that satisfy required constraints, given a specific dataset and target device are built automatically. (f) Knowledge Distillation: A small student DNN is trained using knowledge from a large pre-trained DNN.

5. **Neural Architecture Search:** Constructs many DNNs with different combinations of layers till a DNN architecture that meets the required energy and accuracy constraints is found. Figure 3.1(e) shows the general workflow of neural architecture search.

6. **Knowledge Distillation:** Teaches a small, compact, and efficient DNN (student) to mimic the outputs of a more computation-heavy larger DNN (teacher). Figure 3.1(f) shows an example of knowledge distillation from a teacher to a student.

The rest of the chapter is organized as follows: Section 3.2 provides a background on DNNs, the associated computation costs, and the need for low-power solutions. Sections 3.3–3.8 describe the six broad low-power DNN techniques. Section 3.9 compares the accuracy and efficiency tradeoff with the different low-power DNN techniques. Section 3.10 discusses general guidelines for developing low-power computer vision solutions for different applications. Evaluation metrics to measure the performance of low-power techniques are described in Section 3.11. This article is summarized and concluded in Section 3.12.

3.2 BACKGROUND

3.2.1 Computation Intensity of Deep Neural Networks

Deep Neural Networks (DNNs) are a class of machine learning algorithms [12]. In computer vision, DNNs are used for everything from recognizing images and detecting objects in video to cancer tumor research and autonomous driving [8, 20, 11]. There are many ways to train machine learning models. LeCun et al. [21] found that DNNs can achieve very high accuracy when trained using the backpropagation algorithm.

Although DNNs can be highly accurate, they carry a disadvantage: DNNs are compute-intensive. During inference, passing a single input image through a DNN requires many matrix operations, often on floating-point values [22]. Although there exist alternative computer vision algorithms with lower computational requirements, they are not desirable because they are less accurate than DNNs. For image classification, Table 3.2 shows that the accuracy obtained by the less compute-intensive algorithms is much poorer than that of DNNs. Therefore, computer vision applications rely heavily on DNNs.

The high power consumption of DNNs is due to the large number of parameters and arithmetic operations. In every layer, the parameters and

TABLE 3.2 Comparison of different techniques on the CIFAR-10 dataset. The DNNs (DenseNet and ResNet50) outperform the other machine learning techniques significantly. **NBC**: Naive Bayes' Classifier, **KNN**: K-Nearest Neighbors. Data source: Goel et al. [23]

Method	Decision Tree	NBC	KNN	Random Forests	DenseNet	ResNet50
Accuracy	0.27	0.29	0.41	0.49	0.94	0.93

feature maps are loaded from memory to perform the required operations. Billions of parameters and feature maps cannot be stored in cache at the same time. Thus, when the required parameters are not available in cache (cache-miss), the parameters are fetched from DRAM. A DRAM access is an expensive operation in terms of energy consumption and latency [24]. Table 3.3 shows the energy costs and latency associated with different arithmetic and memory operations on 45 nm technology. A DRAM access consumes $\sim$ 2,600 pJ. This is significantly greater than the 20 pJ consumed during an L1 cache access. Due to the large number of arithmetic operations performed in most DNNs (e.g. VGG-19 performs 2.0×10^{10} arithmetic operations per image of size 224×224 pixels), the arithmetic operations also add up to contribute to the overall energy consumption of DNNs.

3.2.2   Low-Power Deep Neural Networks

High power consumption becomes problematic when a DNN must be deployed on an embedded device with limited computing resources. Many applications, such as autonomous drone navigation [25] and surveillance [26], require on-device computer vision. Offloading computing to the cloud is not a desirable option in these applications because of latency constraints and the requirement of extensive network infrastructure.

This survey highlights the state-of-the-art methods that improve the energy efficiency of computer vision. Increasing the energy efficiency of

TABLE 3.3 Comparison of DNN arithmetic and memory operations on 45 nm technology. Data source: Nazemi et al. [24]

Type of Operation	Energy Consumption (pJ)	Latency (clock cycles)
float32 Add	0.9	1.0
float32 Multiply	3.7	1.0
L1 Data Cache Access	20.0	5.0
DRAM Access	2,600.0	422.0

DNNs includes reducing the number of memory accesses and arithmetic operations. The number of memory accesses is lowered by reducing the DNN memory requirement (model size and number of bits used to represent each parameter). The arithmetic operations are reduced by compressing feature maps, performing kernel transformations, or using smaller filters. Varying a DNN's memory requirement or number of operations usually leads to a tradeoff between energy efficiency and accuracy. It is important to limit the accuracy loss when designing low-power DNNs for computer vision applications. Techniques discussed in this survey propose methods to reduce the memory requirement and/or number of operations with marginal accuracy losses.

The Low-Power Computer Vision Challenge (formerly known as the Low-Power Image Recognition Challenge) [16, 27, 28, 29, 30, 31, 32] was organized to help create low-power computer vision solutions beyond the available technologies. The challenge first included image classification and object detection problems, and since then has expanded to include videos captured from drones.

3.3 PARAMETER QUANTIZATION

DNNs often use single-precision 32-bit floating point (float32) parameters and activations. However, the high precision of the float32 format is not always required to maintain DNN accuracy. One simple method to reduce the memory requirement of a Deep Neural Network (DNN) is quantization. During quantization, float32 parameters and activations are mapped to lower precision values (e.g. float16, int8, etc.). This technique is useful because it can reduce the memory requirement of any existing DNN without having to build the model from scratch [33]. Quantization usually reduces the inference time by 50% – 60% because of the fewer memory accesses and the reduced precision for computations. There are two major methods to perform quantization: (1) Quantization Aware Training, and (2) Post-Training Quantization. Quantization Aware Training (QAT) techniques include quantization into the DNN training process. Training with reduced bit-widths ensures that the accuracy loss in the quantized DNN is only marginal. DNNs employing QAT techniques need to be retrained multiple times, making the training process very expensive [10]. In Post-Training Quantization (PTQ), the parameters and activations of a trained DNN are quantized without any retraining. PTQ techniques use small amounts of unlabeled data to quantize DNNs quickly. However, PTQ techniques usually suffer from significant accuracy losses [34, 35].

TABLE 3.4 Comparison of DNN accuracy before and after 4-bit quantization. Quantization Aware Training (QAT) results in smaller accuracy losses than Post Training Quantization (PTQ). Data source: Krishnamoorthi et al. [22]

Deep Neural Network	Baseline Accuracy	PTQ Accuracy	QAT Accuracy
MobileNetV2 [36]	71.9%	0.1%	62.0%
ResNet50 [37]	75.2%	54.0%	73.2%
MNasNet [38]	74.0%	36.0%	70.0%
InpectionV3 [39]	78.0%	71.0%	76.0%
ResNet152 [37]	77.8%	74.0%	74.0%

The difference in accuracy between QAT and PQT is seen in Table 3.4. For 4-bit quantization (reducing precision from float32 to a fixed-point 4-bit format), QAT outperforms PQT consistently [22].

Quantization Aware Training: QAT techniques include DNN quantization into the training process [40, 41]. Han et al. [10] perform QAT through multiple training-quantization cycles, i.e., the DNNs are trained, quantized, and are trained again. Although the training-quantization cycles help obtain high accuracy, the increase in training time is substantial. Courbariaux et al. [40] use QAT to evaluate the relationship between a DNN's energy consumption, accuracy, and quantization level. The authors show that as the quantization level increases (fewer bits), the energy consumption and the DNN accuracy decrease. CompactNet [42] and FLightNN [43] use these relationships to design cost functions that determine the optimal quantization format for each DNN parameter, given an accuracy constraint. These techniques quantize DNNs while minimizing the memory requirement for a target accuracy. An extreme case of quantization is seen in Binarized DNNs [44, 45]. Here, a single bit is used to represent DNN parameters and activations to significantly improve efficiency. However, Binarized DNNs are inaccurate. Relaxed Quantization [46] is another form of QAT that learns a quantized DNN through gradient descent. This technique does not require training-quantization cycles to obtain high accuracy. The reduced training time of Relaxed Quantization comes with a tradeoff in flexibility, as the quantization bit-width is determined before training and all parameters are quantized to the same bit-width.

Post Training Quantization: A major drawback of QAT techniques is their reliance on labeled training data and expensive DNN training-quantization cycles. This is a significant limitation because many applications (e.g., detecting injured workers) may not have labeled data. Furthermore, users of the applications may not have the required computing resources to train the quantized DNNs [47]. PTQ techniques perform quantization without labeled training data or retraining. PTQ can be performed by rounding float32 values to the nearest float16 or int8 values [22, 34]. However, as seen in Table 3.4, this simple rounding scheme results in significant accuracy losses after quantization. AdaRound [34] performs PTQ through a novel rounding technique, rather than using the default round-to-nearest operation. Although round-to-nearest minimizes the difference per-parameter in the parameter matrix, the authors prove that it is sub-optimal for PTQ. The AdaRound scheme increases the accuracy of PTQ significantly. Nagel et al. [35] show that most PTQ techniques fail because of the difference in the parameter values across channels in a layer. For example, if the parameters in channel $X \in (-0.5, 0.5)$, and channel $Y \in (-128, 128)$, then quantization will round most of channel X's parameters to 0. The authors propose Data-Free Quantization (DFQ), a technique that removes outliers and uses a bias-correction term to normalize the parameter distributions before performing PTQ for high accuracy. ZeroQ [47] generates synthetic data based on the batch-normalization layer's statistics to assist in PTQ. The synthetic data captures the impact of quantization on the DNN layers, and is used to automatically select the optimal bit-precision for each layer. ZeroQ is more expensive to perform than AdaRound and DFQ, but obtains higher accuracy with lower bit-widths.

Potential Future Work: (1) DNNs employing QAT techniques need to be retrained and quantized multiple times, making the training process very expensive [48]. The training cost must be reduced to make these techniques more practical. PTQ techniques lower the training costs but fall short in terms of accuracy. Improving the accuracy of PTQ techniques is critical for their wide deployment. (2) Different layers in DNNs are sensitive to different features. A constant bit-width for all layers can lead to accuracy drops [43, 47]. In order to select a different parameter precision for each connection of the DNN, the parameter precision can then be represented in a differentiable manner and be included in the training process. Thus, during training, each connection will learn its parameter value and the parameter precision. (3) General-purpose CPUs and GPUs only support a few standard bit-formats and bit-widths. Thus, most quantization techniques are only applicable on custom AI accelerator platforms. The deployability of quantization techniques will improve with the availability of standardized AI accelerators.

3.4 DEEP NEURAL NETWORK PRUNING

DNNs have billions of parameters and connections. However, not all parameters are important to the prediction accuracy of a DNN. The DNN model size and complexity can be reduced by removing the unnecessary parameters and connections [49]. Techniques employing DNN pruning use various methods to determine which DNN parameters can be removed without accuracy losses. An intuitive method to perform pruning would be to remove all parameters close to zero. Early works by LeCunn et al. [49] and Hassibi et al. [50] show that removing DNN connections based on importance performs better than removing parameters based on value. The importance of a parameter is quantified by the DNN accuracy drop when the parameter is removed. If an accuracy drop is observed when a parameter's outputs are masked, then the parameter is marked as important [51]. Quantifying the importance of every parameter is computationally expensive. These techniques are not used to prune large convolutional DNNs.

Pruning can also be included in the DNN training process. Iterative pruning techniques [52] prune DNNs in every training epoch to simplify the DNN. Iterative pruning is depicted in Figure 3.2. Han et al. [52] obtain pruned DNNs using a two step process. First iterative pruning is used with a new loss function to identify and remove the unimportant connections and parameters. This step is followed by the training of the sparse DNN with the cross-entropy loss function to regain the lost accuracy. The authors show that up to 50% of AlexNet's [53] parameters can be removed without any accuracy losses. However, more pruning leads to significantly degraded performance. Li et al. [18] use iterative pruning to prune convolution filter channels, instead of pruning individual

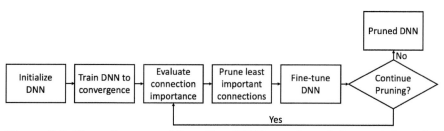

Figure 3.2 Flow diagram of iterative DNN pruning. After training, the importance of the connections is evaluated. The least important connections are removed and the DNN is fine-tuned to regain accuracy. This process is repeated multiple times.

connections or parameters. This pruning technique is useful because it does not require specialized hardware or software to handle sparse matrix operations. However, the drawback with channel-level pruning is that the accuracy losses are significant when compared with connection-level pruning.

Iterative pruning requires significant amounts of computing resources for training. Easy-to-use alternatives to iterative pruning have been proposed in Yu et al. [54], and Srinivas et al. [55]. Yu et al. [54] use an algorithm that propagates importance scores to measure the importance of each parameter with respect to the final output during training. Without any fine-tuning, this technique prunes 50% of the DNN with a marginal accuracy loss. Analogous to Post Training Quantization, Srinivas et al. [55] propose a data-free pruning method. Without any training data, the authors propose a method that combines neurons with similar inputs and outputs. After the neurons and the associated parameters are combined, the parameter weights are normalized to ensure correctness of the combined operation. This technique reduces the training time, but suffers from notable accuracy losses.

A dynamic pruning method is seen in tree-based hierarchical DNNs [56, 57]. Here, multiple small DNNs in the form of a tree work together to perform a computer vision task. During inference, intermediate classifications are performed at every level of the hierarchy. Each intermediate classification selects one child DNN to further process the image, thus dynamically pruning off the other children for increased efficiency. Tree-CNN [58] demonstrates how hierarchical DNNs can be constructed and trained for datasets of different sizes. Panda et al. [56, 57] use various similarity metrics to construct hierarchies of small DNNs for image classification. Goel et al. [23, 59] propose a method to quantify the visual similarity between categories to build a hierarchy. The authors show that hierarchies based on visual similarities outperform the hierarchies based semantic similarities for computer vision tasks. Tree-based DNNs lower the memory requirement, the energy consumption, and the number of operations. The accuracy losses with these DNN architectures are a significant limitation yet to be overcome.

Potential Future Work: (1) Although pruning techniques identify and remove the unimportant connections, they create sparsity in DNNs. Sparse matrices require special data structures and are difficult to map onto general purpose GPUs [60, 61]. Most deep learning libraries perform pruning by zeroing out parameters with a mask [62]. Such pruning implementations do not reduce the memory or computation requirements of DNNs. Channel-level pruning [18] does not suffer from this drawback because it can be performed without any special data structures. Data-free and accurate channel pruning techniques are essential for future progress. (2) Many pruning techniques require a manually set pruning percentage or number of pruning iterations. This makes it challenging to obtain consistent results across datasets and applications. Future pruning techniques should treat the pruning percentage as a hyper-parameter that can be learned and optimized for different datasets and different applications [63].

3.5 DEEP NEURAL NETWORK LAYER AND FILTER COMPRES-SION

Standard DNN convolutions perform arithmetic operations across two spatial dimensions and the one depth dimension in every layer. The convolution operation is implemented as a tensor multiplication between the parameter tensor and the input tensor. For an input tensor of shape $1 \times 12 \times 12 \times 3$ and a parameter tensor of shape $256 \times 3 \times 3 \times 3$, the standard convolution operation performs $995, 328$ ($= 1 \times 12 \times 12 \times 3 \times 256 \times 3 \times 3$) multiplication operations, with padding $= 1$. Smaller convolution filters perform considerably fewer operations than larger filters. For example, a 1×1 filter performs $\sim 90\%$ fewer operations than a 3×3 filter. However, removing all large convolution layers affects the translation invariance property of a DNN and lowers its accuracy [64]. SqueezeNet [13] is a DNN that successfully replaces most 3×3 filters with 1×1 filters. This architecture performs pooling and downsampling late in the DNN. By doing so, the 1×1 filters operate on large informative activations, and thus reduces the accuracy loss.

The MobileNet family [65, 9, 36] of DNN models use another architecture that employs compact DNN filters. These DNNs use depthwise separable convolutions (instead of standard 2D convolutions) to achieve state-of-the-art performance for low-power computer vision. These convolutions are performed in two steps: (1) depthwise convolution: applies a single convolutional filter on each input channel, and (2) pointwise convolution: performs 1×1 convolutions to create a linear combination of the depthwise convolutions. For the same example as above, with an input tensor of shape $1 \times 12 \times 12 \times 3$, first a $1 \times 3 \times 3 \times 3$ depthwise convolution is performed. This is followed by a $256 \times 1 \times 1 \times 3$ pointwise convolution. Here, a total of $114, 480$ multiplications are performed. Thus, the same convolution is performed with far fewer operations.

TABLE 3.5 Techniques employing DNN layer compression (MobileNets and SqueezeNets) require significantly fewer resources with small accuracy losses.

Method	FLOPs	# Parameters	ImageNet Accuracy
GoogleNet [68]	1,550 M	6.08 M	69.8%
VGG-16 [8]	15,300 M	138.00 M	71.5%
AlexNet [53]	720 M	60.00 M	57.2%
MobileNetV1 [9]	569 M	4.20 M	70.6%
MobileNetV2 [36]	300 M	3.40 M	72.0%
SqueezeNet-1.0 [13]	837 M	1.25 M	57.5%
SqueezeNet-1.1 [13]	360 M	1.25 M	58.0%

Bottleneck layers [37] are another technique commonly used to compress DNN layers. The bottleneck is a layer with fewer neurons/channels than the layers preceding and succeeding it. Bottlenecks compress activation maps to force DNNs to only learn the highly informative features. This reduces the memory requirement without accuracy losses [39]. CondenseNet [66] and ShuffleNets [67] both use the group convolution operation to reduce the number of operations by $\sim 5\times$. In group convolution, the convolutional filters are separated into multiple groups. Each group is responsible for a performing convolution on specific channels of the input. The group convolution is efficient because the number of operations decreases by a factor of the number of groups. ShuffleNets improve the accuracy of group convolutions by shuffling the filters between groups during training. The shuffle operation has regularization effect. A major advantage of these filter and layer compression techniques is that they are orthogonal to the previously described pruning and quantization techniques. The techniques can be used together to further reduce energy consumption.

Potential Future Work: (1) Depthwise separable convolutions are less efficient on general-purpose hardware because their arithmetic intensity is low [69]. The arithmetic intensity is the ratio of the number of operations to the number of memory accesses in a DNN. When the arithmetic intensity is low, the memory operations become a bottleneck for performance. The arithmetic intensity of depthwise separable convolutions can be increased by managing memory more effectively. By performing operations in an order that accounts for the cache locality, the number of memory accesses can be reduced. (2) To compensate for the smaller filter sizes used in the depthwise convolution operation, usually a large number of channels are required. This makes the layers highly parameter-efficient, but also leads to the formation of a large number of activation maps [70]. Furthermore, bottleneck layers perform two projection operations and thus further increase the number of activations by a factor of two. Storing a large number of activations requires a significant amount of memory. Group convolutions provide a good tradeoff between accuracy and memory without the drawbacks of the depthwise separable convolutions.

3.6 PARAMETER MATRIX DECOMPOSITION TECHNIQUES

Research has shown that it is possible to decompose a pre-trained DNN's parameter matrix into the multiplication of multiple low-rank matrices [71, 72]. Low-rank matrices require less memory and also require significantly fewer operations to perform the same task. There are multiple matrix decomposition techniques. Kolda et al. [73] show that most decomposition techniques can factorize DNN matrices. However, subsequent research has shown that the post-decomposition DNN accuracy varies considerably with different techniques. The low-rank parameter matrices obtained after Singular Value Decomposition (SVD) have significantly lower accuracy [74] than Canonical Polyadic Decomposition (CPD) [75]. The high accuracy and reduced resource requirements with CPD are promising, but the CPD does not always exist and may need to be approximated for large DNNs. The approximate matrix decompositions often have significant accuracy losses, even after fine-tuning with training data [76]. Moreover, CPD significantly increases the number of layers in a DNN (each parameter matrix is broken into at least four low-rank matrices). The DNN fine-tuning is challenging because of exploding or vanishing gradients. Overall, techniques that decompose pre-trained DNNs are not commonly used in practice because: (1) Compared with the original full rank DNNs, the capacities of low-rank DNNs are limited, thus causing difficulties in fitting large datasets through fine-tuning [77]. (2) The rank of each parameter matrix decomposition must be selected manually, thus making it difficult to find the optimal rank [78]. Thus, the research focus has shifted toward training low-rank DNNs from scratch instead of decomposing pre-trained DNNs.

Techniques that train compact DNNs with low-rank matrices can reduce the accuracy losses significantly. Alvarez et al. [78] and Xiong et al. [77] use nuclear norm regularizers (the sum of all singular values) in the training objective to reduce the rank in each step of the training process. The regularization term gradually pushes the parameter distribution into the low-rank form [77]. However, computing the regularization term includes the computation of the SVD. The SVD computation is computationally expensive, making the training of these DNNs very difficult. To apply these methods in large DNNs, Wen et al. [79] and Ding et al. [80] train low-rank DNNs in two steps. First, they use a regularizer to increase the correlation between filters during training. This is followed by a post-processing step: the highly correlated filters are removed to factorize DNNs with low accuracy loss. These methods do not require SVD computation during training.

Potential Future Work: (1) Low-rank factorization techniques require significant training resources for achieving high accuracy. Matrix decomposition of pre-trained DNNs does not work for all DNNs and all datasets. Understanding why some decompositions perform better than others will allow us to obtain high accuracy with pre-trained DNN decomposition. (2) Approximating the SVD operation during the training of compact DNNs will significantly make matrix factorizations more practical. However, the approximations should not impact the DNN accuracy.

3.7 NEURAL ARCHITECTURE SEARCH

With so many DNN architectures and optimization possibilities to choose from, it can be difficult for a human to manually select the best DNN for a given computer vision task. Neural Architecture Search (NAS) attempts to solve this problem by automating DNN architecture selection. Given accuracy requirements for a computer vision task, NAS automatically tweaks the hyperparameters of a DNN until it finds one that fits the requirements.

A key observation behind NAS is that a DNN's architecture can be represented by a string of text [81]. Therefore, a Recurrent Neural Network (RNN) can be used as a controller to automatically generate such string representations of DNNs. In pioneer NAS research, reinforced learning optimizes the RNN controller so that it ultimately generates a DNN architecture that satisfies the requirements of the computer vision task. Each generated DNN is trained and tested. The DNN's test performance is then used to calculate the RNN controller's reward function so that it can generate a more suitable DNN the next time. Early NAS designs like NAS-RL [82] and MetaQNN [83] show that using such automated network architecture design can be just as good as human design; they achieve image classification accuracy on par with manually designed DNNs.

TABLE 3.6 Comparison of NAS methods for selecting efficient computer vision models

Method	Search Time	ImageNet Accuracy	# Parameters	FLOPs
MobileNetV2 [36]	–	72.0%	4.2 M	300 M
AmoebaNet-A [84]	3,150.0 days	74.5%	5.1 M	555 M
NASNet-A [38]	2,000.0 days	74.0%	5.3 M	564 M
MNasNet-A1 [85]	1,666.0 days	74.0%	3.9 M	312 M
FBNet-B [86]	9.0 days	74.1%	4.5 M	375 M
ProxylessNAS [87]	8.3 days	75.1%	5.7 M	465 M
RENASNet [88]	6.0 days	75.7%	5.4 M	580 M
DARTS [89]	4.0 days	73.1%	4.9 M	595 M
PC-DARTS [90]	3.8 days	75.8%	5.3 M	1200 M
BNAS [91]	2.6 days	71.3%	4.3 M	–
AutoNL [92]	1.3 days	76.5%	4.4 M	267 M

State-of-the-art NAS image classification solutions include AmoebaNet-A [84] and NASNet-A [38], with CIFAR-10 test error hovering in the 3% neighborhood.

NAS can be used to automatically select efficient, low-power architectures for computer vision. Metrics such as parameter count and latency can be added to the reward function so that the NAS controller factors them in during search. By adding desired parameter count and latency as new hyperparameters, NAS algorithms select architectures that achieve a balance between accuracy and efficiency. MNasNet [85] builds on NASNet with a multi-objective (both accuracy and latency) reward function in the controller to find efficient DNN architectures that run 80% faster than MobileNetV2, a human-designed architecture. Loni et al. [93] focus their optimization strategy on the number of parameters, allowing their NeuroPower NAS to generate CIFAR-10 models with fewer than 1 million parameters. Chen et al. take a different approach: they restrict the search space to exclusively contain only binarized convolutions [91]. The accuracy of these automated solutions either rival or exceed those of human-designed architectures.

Unfortunately, achieving good results with NAS can be computationally expensive. NAS must train and evaluate many DNNs as it searches for a suitable DNN, a very time-consuming task. The human-free, automated success of MNasNet comes at the cost of 40,000 GPU hours, or 4.6 GPU-years in search time [85]. Thus, most emerging work focuses on accelerating the NAS search time. Early efforts reduced computation by employing DNN weight sharing from one search iteration to the next [94]. More recent optimization frameworks like DARTS [89] and PC-DARTS [90] treat the search space as differentiable. That allows NAS to use efficient gradient descent algorithms to traverse the search space faster. Techniques like RENASNet [88] and AutoNL [92] search neural networks block-by-block, making small mutations to each block over time. For low-power DNNs in particular, search time can also be improved by using easier-to-optimize proxy tasks. FBNet [86] optimizes over a smaller dataset as a proxy task for efficient architectures, completing its search ~ 95% faster than MNasNet. However, proxy task-optimized DNNs do not always guarantee suitable accuracy [87]. Proxyless-NAS [87] uses path-level pruning to cut down on candidate architectures while handling objectives with gradient descent algorithms. Single-Path NAS [95] builds on this idea, reducing the architecture search time from 300 GPU-hours to 4 GPU-hours. A comprehensive survey about NAS and its applications is available in Ren et al. [81].

Potential Future Work: (1) Over time, NAS has become less hands-free. The early NAS algorithms did not need manual participation once search began, but modern algorithms require more human interaction to help shrink the search space for faster search. Attempting to balance the need for fast search of efficient DNN models and hands-free functionality is suitable for future research. (2) NAS research is currently heavily focused on producing image classifiers. However, low-power computer vision models must be capable of achieving other tasks, like object detection and semantic segmentation. Therefore, designing NAS to search for Region Proposal Networks and single-shot detectors should be a priority.

3.8 KNOWLEDGE DISTILLATION

Large DNNs are more accurate than small DNNs because the greater number of parameters allows them to learn complex functions [96, 97]. Smaller DNNs usually do not have enough parameters to learn these complex functions [98]. Some methods train small DNNs to learn complex functions by making the small DNNs mimic larger pre-trained DNNs [99]. These techniques transfer the "knowledge" of a large DNN to a small DNN through a process called Knowledge Distillation [100, 101, 102].

There are three major methods through which knowledge distillation can be performed. (1) The small student DNN is trained on data labeled by a large DNN instead of the labels available in the training dataset [103, 104]. The data labeled by the large DNN (softmax output) contains information that is useful for training the small DNN. For example, in an

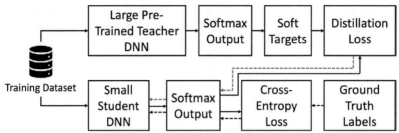

Figure 3.3 Knowledge distillation techniques use a teacher DNN to assist in the training of a smaller student DNN. The soft targets generated by the pre-trained teacher DNN are distilled to the student DNN during training. Solid lines indicate forward propagation path. Dotted lines mark the backward propagation path.

image classification task, the teacher may have high softmax outputs for two or more categories for a single input image, thus implying that those categories share some visual features. (2) The student DNN mimics the activation maps of specific layers from the large teacher DNN [105, 106]. The techniques relying on activation-based knowledge distillation achieve high accuracy, but they require strict assumptions on the structure of the student. The activation maps of the student and the teacher need to be of the same size, and thus leads to the formation of inefficient DNNs. (3) The highly correlated filters and layers in the teacher DNN are combined in the small student DNN [99, 107]. The correlation between DNN filters are usually computed using inner products. Singular Value Decomposition (SVD) is also used to obtain more accurate correlations at the expense of increased training costs.

The model size discrepancy between the large teacher DNN and a small student DNN will always cause an accuracy drop [108]. To preserve accuracy while performing knowledge distillation, some techniques deviate from the traditional teacher-student knowledge distillation. Mirzadeh et al. [108] use a teacher-assistant to act as a buffer between the small student DNN and the large teacher DNN. Nowak et al. [109] propose a method that transfers the knowledge learned by multiple teacher layers to a single student layer. These methods improve the student DNN's accuracy, but require more complex training procedures and more resources for training.

Potential Future Work: (1) Most knowledge distillation methods focus either on new ways to represent the knowledge of the teacher DNN, or on new distillation techniques to help the student learn more effectively.

More work on the efficient design of the teacher-student architectures is required [99]. Theoretical analysis has shown that small student DNNs are capable of learning the same functions as large teacher DNNs [100]. However, most practical techniques show significant accuracy drops when using small student DNNs [108]. Knowledge distillation will become very useful for low-power computer vision when this accuracy drop can be overcome. (2) To build low-power DNNs, knowledge distillation could be used along with other techniques like decomposition, pruning, or compression. However, the combination of these techniques with knowledge distillation has not yet been explored thoroughly.

3.9 ENERGY CONSUMPTION—ACCURACY TRADEOFF WITH DEEP NEURAL NETWORKS

Since large and densely connected DNNs can fit datasets better than smaller DNNs [77], achieving high accuracy often demands large model sizes and thus high energy consumption. Across the energy-efficient computer vision techniques discussed so far, researchers try to balance the trade-off between accuracy and energy efficiency.

Figure 3.4 approximates the impact of the various low-power computer vision techniques on the accuracy (test error) and energy consumption of DNNs. The *Baseline* (Cluster B) techniques in Figure 3.4 refer to the popular DNN architectures, such as VGG [8] and ResNet [37] deployed without any improvements. *Neural architecture search* (NAS) can be used to build large DNNs [84, 38] to obtain state-of-the-art accuracy (cluster A). Recent NAS techniques [111, 85] include both efficiency and accuracy as requirements in the search process. These techniques (cluster D) can achieve high accuracy with low energy consumption. *DNN compression* techniques have become more accurate over the years without increases in energy consumption. SqueezeNet [13] in 2016 (Cluster H), MobileNetV2 [36] and ShuffleNet [67] in 2018 (Cluster F), and MobileNetV3 [65] and EfficientNet [110] in 2020 (Cluster C) show this trend. Early *1-bit quantization* techniques [44, 45] that have high efficiency but low accuracy are represented in Cluster L, while more recent quantization techniques [118] (Cluster I) are still efficient and also achieve higher accuracy.

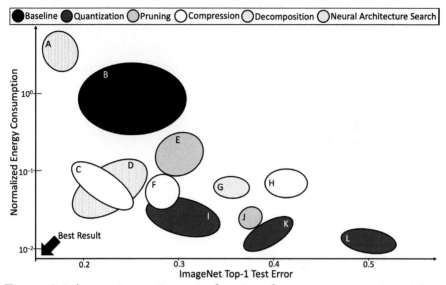

Figure 3.4 Approximate impact of various low-power computer vision techniques on accuracy and energy consumption of DNNs. Each cluster corresponds to a group of techniques published at similar times with comparable performance. Techniques in the bottom-left of the plot achieve the best performance: high accuracy (low error) and low energy consumption. Cluster A: Neural architecture search techniques like NASNetA-Large [38] and AmoebaNetA-Large [84]. These techniques construct large DNN models to achieve state-of-the-art accuracy. Cluster B: Popular DNN architectures like VGG [8] and ResNet [37] without any optimizations. Cluster C: DNN compression techniques published in 2019 can reduce energy consumption without significant accuracy losses. These techniques include MobileNetV3 [65] and EfficientNet [110]. Cluster D: Low-power neural architecture search techniques that include efficiency and accuracy as requirements in the search process, e.g., Once-for-all [111] and MNas-Net [85]. Cluster E: DNN Pruning techniques like Deep Compression [10] and ThinNet [112] increase DNN efficiency at the expense of accuracy. Cluster F: Less recent DNN compression techniques like MobileNetV2 [36] and ShuffleNet [67]. Cluster G: Matrix decomposition techniques seen in Trained Rank Pruning [113]. Cluster H: Pioneer DNN compression techniques like SqueezeNet [13]. Cluster I: Recent parameter quantization techniques can reduce energy consumption significantly. These techniques include Hardware Aware Quantization [114]. Cluster J: Early iterative pruning of baseline DNNs [115]. These techniques achieve high efficiency, but incur significant accuracy losses. Cluster K: Early quantization aware training on baseline DNNs [116]. Cluster L: 1-bit quantization techniques like BinaryConnect [117] and XNOR-Net [45]. These techniques are the most efficient, but also have considerable accuracy losses.

3.10   GUIDELINES FOR LOW-POWER COMPUTER VISION

Techniques to build low-power computer vision solutions are often complementary and can be used with each other. In this section, we discuss some methods to increase the efficiency of computer vision solutions.

3.10.1   Relationship between Low-Power Computer Vision Techniques

Many low-power computer vision techniques are often used together. Techniques are used together because their effects are orthogonal and the DNN efficiency can be increased. Often, training procedures for new low-power computer vision techniques are inspired from existing techniques. In the section, we provide some examples of the relationship between different techniques.

Quantization and Pruning are often used together in many applications [119, 10, 120]. This is because the iterative training scheme is suitable for both techniques. Han et al. [10] show that pruning alone reduces the DNN model size by $\sim 88\%$. The authors also show that pruning and quantization are orthogonal, and when used together, they reduce the model size by $\sim 96\%$. However, not all quantization schemes are compatible with all pruning techniques. Some quantization approaches retain accuracy by increasing the number of channels to compensate for the reduced precision [121]. Thus, using channel pruning [18] along with quantization may lead to significant accuracy losses. Research is still required to understand if quantized models with many channels perform better than channel-pruned models with higher precision.

Knowledge distillation builds small student models that learn from large pre-trained teacher models. Mishra and Marr et al. [122] and Kim et al. [123] show that knowledge distillation can also be used to transfer knowledge from a high-precision DNN to a low-precision student DNN. This is possible when the student and the teacher have the same DNN architecture. Using knowledge distillation for quantization is particularly useful because it does not require the expensive training-finetuning cycles. Results show that the quantized student DNN can reach the same accuracy levels as that of the high-precision teacher DNN.

Neural Architecture Search (NAS) builds DNN models based on accuracy and performance constraints for a target hardware device. NAS is usually used as a stand-alone operation because no subsequent optimizations are required. Because the time and resources required by NAS to search for and train a DNN is prohibitively exorbitant, many techniques have been developed to reduce the search time. ProxylessNas [87] is one such technique that uses DNN pruning to reduce the number of candidate DNNs for evaluation and speed up search. Moreover, the methods used in NAS have been adapted to perform DNN quantization. Hardware Aware Quantization [114] leverages reinforcement learning (commonly used in NAS) to automatically determine the optimal mixed-precision quantization scheme for pre-trained DNNs running on specific devices.

DNN matrix decomposition can be combined with pruning to introduce structured sparsity into the low-rank approximations. The sparsity constraints remove the less-informative neurons and connections to enable more efficient inference [124].

The filter compression techniques used in MobileNets [9, 36, 65] serve as the backbone for low-power computer vision. These techniques obtain state-of-the-art accuracy and are often used with both pruning and quantization for higher efficiency. Table 3.7 shows the performance of MobileNetV2 with different levels of pruning. Approximately 50% of MobileNet's parameters can be pruned with a marginal accuracy loss.

3.10.2 Deep Neural Network and Resolution Scaling

DNN scaling allows us to change the accuracy, computation, and energy costs of a DNN. DNN scaling has three dimensions: (1) changing the width (number of channels), (2) changing the depth (number of layers), and (3) changing the resolution (size of input). Wide DNNs can capture fine-grained features (e.g. texture of person's hair), but often have difficulties in learning high-level features (e.g. color of person's clothes). Increasing the DNN depth increases the accuracy at first. However, the accuracy gain quickly hits a point of diminishing return and deeper models become redundant. Higher input resolutions considerably increase accuracy, but the activation map size and number of operations grow quadratically.

TABLE 3.7 Performance of MobileNetV2 [36] with different levels of pruning. Data source: Zhu et al. [125]

Pruning%	# Parameters	ImageNet Accuracy
0%	4.21 M	70.6%
50%	2.13 M	69.5%
75%	1.09 M	67.7%
90%	0.46 M	61.8%
95%	0.25 M	53.6%

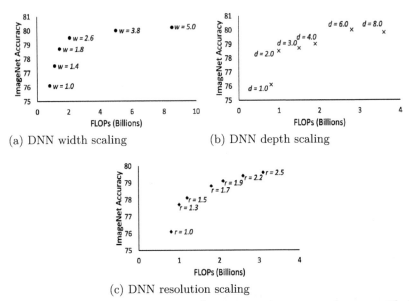

(a) DNN width scaling

(b) DNN depth scaling

(c) DNN resolution scaling

Figure 3.5 The DNN accuracy and computation costs change with the DNN width, depth, and input resolution. The data-point $d = 2.0$ corresponds to a DNN with twice the number of layers as the baseline DNN. The data-points $w = 1.0$, $d = 1.0$, and $r = 1.0$ correspond to the same baseline DNN. Data source: Tan et al. [110].

When designing DNNs for low-power devices, DNN scaling can reduce the computation costs. Tan et al. [110, 126] show that scaling a DNN only along one dimension at a time (either width, depth, or resolution) leads to significant accuracy losses. As seen in Figure 3.5, upon reducing the width from $r = 2.5$ to $r = 1.7$ the accuracy drop is considerable. Balancing the scaling along all dimensions is required to reduce the DNN size without accuracy losses.

3.11 EVALUATION METRICS

Low-power DNNs for computer vision need to be evaluated on multiple aspects. We list some of the major metrics that should be considered.

3.11.1 Accuracy Measurements on Popular Datasets

CIFAR [127], SVHN [128], Extended MNIST (EMNIST) [129], Caltech-256 [130], and ImageNet 2012 [131] datasets are commonly used for image classification. CIFAR, SVHN, and EMNIST contain centered and fixed-size images. Images in datasets like ImageNet 2012 [131] and Caltech-256 [130] are of different sizes and represent real-life images more closely. For image classification, accuracy is reported in terms of classification error on the testing/validation/holdout dataset. For object detection and object counting tasks, results are generally reported on the COCO [132], Pascal VOC [133], and ImageNet detection [131] datasets. These datasets contain images with multiple objects. Mean average precision (mAP) is used to measure the accuracy of object detection solutions. Root mean squared error is used to compare object counting techniques. Semantic segmentation tasks are evaluated on the COCO [132] and CamVID [134] datasets. There exist other datasets for specific computer vision tasks. For example, Market-1501 [135], MARS [136], and VRAI [137] are used for object re-identification. The KITTI [138] and Cityscapes [139] datasets are used for tasks related to autonomous driving.

3.11.2   Memory Requirement and Number of Operations

The number of parameters is generally associated with the memory requirement of the DNN. It is also important to consider the memory required to save the activations of the layers. Some DNNs have fewer parameters, but require more memory for activations. For example, VGG-16 uses 528 MB to store its parameters and requires 58 MB for storing the activation maps (when batch size $= 1$). DenseNet requires only 77 MB to store the DNN parameters, but uses 196 MB to store the layer activations during inference of a single image. This trend is seen in DNNs that use skip connections because multiple activation maps are stored in memory concurrently. When using quantization, the memory requirement is proportional to the number of bits used for storing parameters and activations. To report the size of a model agnostic to the quantization level, one should report the number of parameters in the DNN model.

The number of operations of a DNN should be evaluated to find the computation costs. Different DNN designs can have a very different number of operations while requiring similar memory and achieving similar accuracy. SqueezeNet-1.0 and SqueezeNet-1.1 both require 5 MB to store parameters [13]. However, SqueezeNet-1.0 performs 837 million operations, SqueezeNet-1.1 performs 360 million operations. Convolution layers perform more operations than FC layers, while FC layers have more parameters. Thus, the number of operations can be reduced by decreasing the size of convolution kernels or the number of channels.

TABLE 3.8 Energy consumption and inference time comparison of MobileNetV1 on Google Pixel-2 with different optimizations. Data source: Alyamkin et al. [16]

Image Resolution	Data Type	Accuracy	Inference Time (ms)
224 × 224	float32	70.2	81.5
224 × 224	int8	65.5	68.0
128 × 128	int8	64.1	28.0

3.11.3 On-device Energy Consumption and Latency

The number of parameters and operations are not always proportional to the energy consumption and the latency of a DNN [16]. DNNs are sometimes better suited for different devices. Some DNNs are faster on GPUs because they benefit from the high degree of parallelism, other DNNs are better suited for CPUs because of their memory access patterns. Furthermore, different optimizations have a different impact on the on-device energy consumption and latency. As seen in Table 3.8, changing from float32 to int8 decreases the latency of MobileNetV1 [9] from 81.5 ms/image to 68 ms/image on the Google Pixel 2. Reducing the input resolution from 224 × 224 to 128 × 128 further reduces the latency to just 28 ms/image. To measure the energy consumption and the latency, DNNs should be deployed on the target device. Power meters should be used to measure the device's power and energy consumption.

3.12 SUMMARY AND CONCLUSIONS

Deep Neural Networks (DNNs) are the most powerful tools for computer vision. However, the state-of-the-art DNNs require enormous amounts of memory, energy, and computational resources to process images. This makes DNNs difficult to deploy on resource-constrained embedded devices. To make DNNs edge-friendly, researchers have built techniques that make DNNs smaller by removing redundant operations. Through this chapter, we survey the research on energy-efficient DNNs and classify the research into six categories of techniques: Parameter Quantization, Deep Neural Network Pruning, Deep Neural Network Layer and Filter Compression, Parameter Matrix Decomposition Techniques, Neural Architecture Search, and Knowledge Distillation. We describe the techniques in each category along with their strengths and limitations. We also point to potential future work that will help improve the field of low-power computer vision.

Through the analysis presented in this survey, it is clear that no single technique is best suited for increasing the efficiency of all applications. One must consider the availability of training data, the costs and resources required for training, the accuracy constraints, and the hardware compatibility before selecting a low-power DNN. Often combinations of techniques are used to better results. Continued research on improving the state-of-the-art low-power techniques will make computer vision more ubiquitous.

II

Competition Winners

CHAPTER 4

Hardware Design and Software Practices for Efficient Neural Network Inference

Yu Wang, Xuefei Ning, Shulin Zeng, Yi Cai, Kaiyuan Guo, and Hanbo Sun

Tsinghua University

Changcheng Tang, Tianyi Lu, and Shuang Liang

Novauto

Tianchen Zhao

Beihang University

CONTENTS

DOI: 10.1201/9781003162810-4

4.1 HARDWARE AND SOFTWARE DESIGN FRAMEWORK FOR EFFICIENT NEURAL NETWORK INFERENCE

4.1.1 Introduction

Deep learning and neural networks are the keywords in the current artificial intelligence (AI) field. Deep learning is showing dominant performance in applications like image classification [140] and speech recognition [141], which makes it the top candidate for real-world AI applications. However, as the computation power of devices is limited, the application of deep learning to resource-constrained scenarios is hindered by its large computation complexity. To address this issue, researchers and industries around the world have been working on customized hardware acceleration solutions [142]. And we has maintained a webpage at `http://nicsefc.ee.tsinghua.edu.cn/projects/neural-network-accelerator/` that gives a comprehensive summary chart of neural network accelerators.

We believe that, to build an efficient system for deep learning, we must consider software-hardware co-design since software and hardware are coupled in deep learning. Considering both optimizations in software and hardware, we propose a design flow, shown in Figure 4.1. In our understanding, there are three major factors that affect how efficient a deep learning system is: **workload**, **peak performance**, and **efficiency**, which corresponds to the upper part of Figure 4.1.

A smaller workload with the same precision is always welcomed, as depicted by the down arrow in the left part of Figure 4.1. However, the change of workload may affect the hardware design. For example,

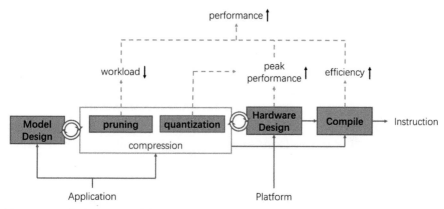

Figure 4.1 Software and hardware co-design optimization flow from model design to instruction generation. There are three major ways to imporve the overall performance of a deep learning system: reduce workload, increase peak performance, and improve computational efficiency.

replacing direct 2-d convolution with the fast algorithm in CNN like Winograd changes the ratio between multiplication and addition and also changes data access pattern. Besides, exploring the sparsity in neural networks even changes the data description format and entire computing system, i.e., from dense matrices to sparse matrices.

A higher peak performance is always wanted, as shown by the up arrow in the middle part of Figure 4.1. However, as peak performance is usually proportional to computation unit number and frequency of the system, a higher peak performance often results in higher cost and power. One way for increasing the peak performance while saving cost is to simplify the operation, such as using fewer bits to represent data and weight in neural networks: The robustness of deep learning algorithms makes it possible to use 16bit, 8bit, and even fewer-bit fixed-point operations to replace 32-bit floating-point operations while introducing negligible accuracy loss. This trade-off between peak performance and variable precision influence both algorithm design and hardware design.

Efficiency reflects how well we utilize the computation units, thus a higher efficiency would help improve the overall performance, as illustrated by the up arrow in the right part of Figure 4.1. An elegant memory system design to feed the computing units with enough data is the key to high efficiency. To achieve this, we need to tackle both on-chip memory and external memory system design. For the on-chip memory part, it is necessary to explore data locality and data reuse to make data stay in the cache as long as possible. For the external memory part, increasing the bandwidth definitely helps increase the efficiency but also leads to higher cost and power. With the same theoretical bandwidth, we need to increase the burst length to fully utilize it, i.e., organize data storage to match hardware requirements. The data simplification method also reduces the data bit-width and thus reduces bandwidth requirement.

Taking all three factors into account helps to design a highly efficient deep learning system. Besides, since deep learning is evolving rapidly, taping out a certain design may not be a good choice for a commercial product. In this case, general-purpose processors or specialized hardware accelerators with enough flexibility and scalability for reprogramming are favorable. Field-programmable gate array (FPGA), with inherent reconfigurability, which provides the chance to explore all three levels of design and the ability to incorporate state-of-the-art deep learning techniques into a product within a short design time, has a good potential to become a mainstream deep learning processing platform.

4.1.2 From Model to Instructions

For CNNs, early models first applied several convolution (Conv) layers sequentially to the input image to generate low-dimension features and then several fully connected layers as the classifier. Current networks such as ResNet [140] and the inception module in GoogLeNet [143] used different branches and parallel layers in the network to achieve multi-scale sampling and avoid vanishing gradients. The model size ranges from less than 10 layers to more than 100 layers for different tasks.

A system must be flexible enough to execute different neural network models. To achieve this, a flexible description is needed. Caffe, Tensor-Flow, and other deep learning frameworks provide an efficient interface on CPU and GPU platforms, since there are determined instruction-set-architecture (ISA) designs and programming languages. But for the specialized systems, we need a tool, and also an intermediate representation to bridge these frameworks and the hardware accelerators. We design the customized hardware accelerator with a coarse-grained ISA considering the patterns of neural network computation to achieve high efficiency while leaving the interface flexible. In this way, we can map different networks onto it. Meanwhile, algorithm researchers and hardware developers can work simultaneously, making the iteration of products fast and efficient.

We implement a coarse-grained instruction interface for our design. For CPUs or GPUs, the instructions are fine-grained, usually with a single operation on a scalar or a vector. Fine-grained instructions lead to the highest flexibility. But considering the specialty of neural networks, this may not be an efficient way. For example, the computation of a neural network is usually full of loops, we try to partition the loops into small blocks such that each block can be done by hardware. For CNN, each block may be a set of 2-d convolutions. Thus, a coarse-grained ISA greatly reduces the instruction size while remains the hardware efficiency. We also use instructions to describe data transfers between on-chip cache and off-chip memory. This enables the compiler to do static scheduling to achieve a balance between the computation and I/O.

Again we refer to Figure 4.1 to show our design flow in terms of the software and hardware co-optimization. First, the deep learning models are designed for the target application according to the specification of the system, either manually or by the newly developed neural architecture search methods. Then the model is optimized for hardware acceleration. The optimization step usually consists of model compression to reduce the workload and data quantization to increase the peak performance. Both these two steps are conducted by automatic tools but need developers to decide the best decision considering the accuracy loss and hardware performance gain. Next, the specialized dataflow and computation units are carefully designed according to the software compression techniques used. These three steps are done iteratively to make sure the performance requirements of the target application are met. After hardware design, we use a customized compiler to convert the neural network model into instructions that are executed at runtime. Further optimization on scheduling is automatically done in the compiler to increase computation efficiency.

The rest of this chapter is organized as follows. In Section 4.2, we describe our proposed coarse-grained ISA-based CNN accelerator design, with the runtime workflow and extension support for upsampling layers. Then, we present quantitative results of the proposed hardware accelerator, including our winning practices on the DAC-SDC 2018 competition. Then, in Section 4.3, we describe the practices of neural network model optimization, including quantization, sensitivity analysis-based pruning (without hardware model), and iterative pruning scheme with hardware model. These practices are employed in our solutions for DAC-SDC 2018 and LPCVC 2019/2020. Finally, we discuss the neural architecture search framework developed by us and a hardware-aware NAS case study using this framework.

4.2   ISA-BASED CNN ACCELERATOR: ANGEL-EYE

In many applications, like object detection [144], face recognition [145] and stereo vision [146], CNN has shown its power and beats traditional algorithms where handcrafted models are used. Implementing this kind of algorithms on mobile devices will do great help to the robot or smart camera manufacturers. But in some cases, more than one network is needed in the algorithm. In [145], a cascaded CNN structure is proposed for face detection. In this algorithm, the first CNN goes over the whole image to drop useless proposals. The second CNN is applied on the preserved proposals. More proposals are dropped in this step. In this case, more than one CNN model is needed in the algorithm. The results of CNN influence the control flow of the algorithm. Simply implementing a CNN accelerator for this kind of application is not enough.

In this case, using multiple accelerators is possible but not a scalable solution if more models are involved. **So the CNN accelerator should be configurable at run-time.** As the execution of CNN can be decided by run-time results, **a host controller is needed to handle the control flow.** In this section, we extend our previous work [147] to a complete design flow for mapping CNN onto embedded FPGA. Four parts are included in this flow:

- A data quantization strategy to compress the original network to a fixed-point form, so as to increase peak performance of the hardware design.

- A parameterized and run-time configurable hardware architecture to support various networks and fit into various platforms, together with specialized dataflow and computation units to improve peak performance.

- A compiler is proposed to map a CNN model onto the hardware architecture, with block partition and memory mapping optimization to increase computational efficiency.

- A runtime workflow and extension support of upsampling layer is proposed to support more general deep learning applications such as object detection.

The first part of data quantization strategy will be introduced in the Section 4.3. In the following parts of Section 4.2, we will first introduce our proposed coarse-grained ISA-based hardware architecture design. Then, the compiler design and optimization methods will be described and analyzed. Finally, the runtime workflow and extension support of upsampling layer will be illustrated.

4.2.1 Hardware Architecture

As discussed above, the CNN accelerator should be run-time configurable to support as many applications and neural network models as possible. Our previous work [147] is limited to VGG models. In this work, a flexible instruction interface is proposed. The calculation of CNN is described with three kinds of instructions: LOAD, SAVE and CALC, corresponding to the I/O with external memory and the convolution operation. Most of the variations of state-of-the-art CNN models are covered with this instruction set. Each instruction is 128-bit or 192-bit and contains the following fields:

- **Operation code** is used to distinguish different instructions.

- **Dependency code** sets the flags for inter-instruction dependency and helps to parallelize different kinds of instructions. This enables scheduling before instruction execution.

- **Parameter** contains specific fields for each kind of instruction. For LOAD and SAVE instructions, address and size description for the data block in external memory and on-chip memory is set. The address information of on-chip memory helps the software fully utilize the limited on-chip memory. For CALC instructions, data block address and size in on-chip memory are set. Other flags for pooling, bias, and padding are also set.

A hardware architecture is proposed as shown in Figure 4.2 to support this instruction interface. It can be divided into four parts: PE array, On-chip Buffer, External Memory, and Controller.

PE Array: The PE array implements the convolution operations in CNN. Three levels of parallelism are implemented by PE array:

- **Kernel level parallelism.** Each PE consists of several convolution engines. Each convolution engine computes the inner product of the convolution kernel and a window of the image in parallel.

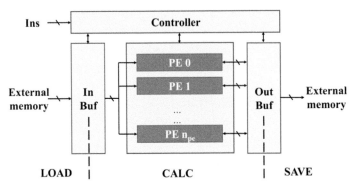

Figure 4.2 Overall hardware architecture of Angel-Eye. There are three parts of coarse-grained ISA-based hardware modules: LOAD, CALC, and SAVE, which corresponds to the input buffer, computation units, and output buffer.

- **Input channel parallelism.** Different convolution engines in each PE do convolution on different input channels in parallel. The results of different input channels are added together as CNN defines.

- **Output channel parallelism.** Different PEs share the same input channels, but not the convolution kernels, to compute different output channels in parallel.

A detailed structure of a single PE is shown in Figure 4.3. Within each PE, different convolvers calculate 2D convolution on different input

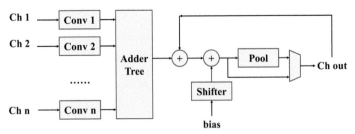

Figure 4.3 Hardware structure of a single processing element (PE), which consists of n 3×3 convolution engines (Conv) with n input channel parallelism. Each convolution engine achieves kernel level parallelism with the computation of the convolution kernel and a window of the image in parallel. Besides, different PEs compute different output channels in parallel.

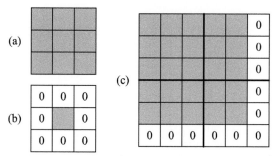

Figure 4.4 Using 3×3 convolver for general convolution: (a) 3×3 kernel, (b) 1×1 kernel by padding, and (c) 5×5 kernel by 4 3×3 kernels and padding.

channels in parallel. The most popular convolution kernel in state-of-the-art CNN models is of size 3×3. So we adopt the 3×3 convolution kernel in our hardware based on the line buffer design [148] This achieves the kernel level parallelism and makes good reuse of image data. Though the kernel is fixed, we are still available to support other kernel sizes as shown in Figure 4.4. For smaller kernels like 1×1 ones, the kernel is padded to 3×3 to be supported. For larger kernels like 5×5 ones, multiple 3×3 kernels are used to cover it. This means doing 3×3 convolution on the same image with slight deviation and add the result together.

With the help of data quantization, the multipliers and adders can be simplified to use fixed-point data with certain bit-width. To avoid data overflow, bit-width is extended for intermediate data. For our 8-bit design, 24-bit intermediate data is used. Shifters are used to align the bias with the accumulated data and cut the final result according to data quantization result for each layer.

On-chip Buffer This part separates PE Array with External Memory. This means data I/O and calculation can be done in parallel. Output buffer also offers intermediate result to PE Array if more than one round of calculation is needed for an output channel. As CNN is memory intensive, we need to efficiently utilize on-chip buffer. We introduce a 2-D description interface to manage the data, which is shown in Figure 4.5. Each of the image in the buffer is described with the following parameters: start address, width, height, and line step. This enables that software can fully utilize the on-chip buffer for different feature map sizes. With this interface, software can also implement the ping-pong strategy on these buffer by splitting the matrix with address. As for the dimension of input channel which is not shown in Figure 4.5, there are n identical

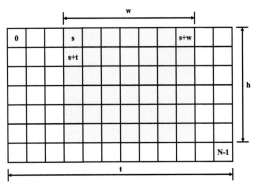

Figure 4.5 A 2-D data description example. An image of width w and height h is stored in a 1-D buffer of size N at start address s with line step t. The colored blocks denote the image.

input buffers to store n 2-d input feature maps seperately, and they are consistent with the input channel parallelism of the PE array.

External Memory For state-of-the-art CNN and the currently available embedded platforms, On-chip Buffer is usually insufficient to cache all the parameters and data. External memory is used to save all the parameters of the network and the result of each layer. In the proposed system, external memory is also used for the communication between the CNN kernel and the host CPU. Using a shared memory for data communication has the chance of reducing abundant data transportation.

Controller This part receives, decodes and issues instructions to the other three parts. Controller monitors the work state of each part and checks if the current instruction to this part can be issued. Thus the host can send the generated instructions to Controller through a simple FIFO interface and wait for the work to finish by checking the state registers in Controller. This reduces the scheduling overhead for the host at run-time. Other tasks can be done with the host CPU when CNN is running.

Figure 4.6 shows the structure of this part. Parallel execution of instructions may cause data hazard, since there are data dependency between different instructions. For example, to compute a tile of output feature map, a CALC instruction must wait for the completion of the corresponding LOAD instructions to load the input feature map and weights from off-chip DDR memory to on-chip memory. Then, the SAVE instructions needs to wait for the CALC instruction to finish before starting to transfer the results back to the off-chip DDR. In hardware, an instruction is executed if: (1) the corresponding hardware is free and

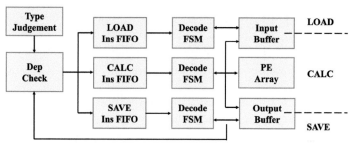

Figure 4.6 Structure of the controller to deal with the data hazard caused by the parallel execution of coarse-grained instructions among LOAD, CALC, and SAVE.

(2) the instructions it depends on have been finished. Condition 1 is maintained by LOAD Ins FIFO, CALC Ins FIFO, and SAVE Ins FIFO as shown in Figure 4.6. The instructions in the FIFOs are issued when the corresponding hardware is free. Condition 2 is maintained by checking the dependency code in Dep Check module.

4.2.2 Compiler

A compiler is proposed to map the network descriptor to the instructions. Optimization is done to deal with the high storage complexity of CNN. Some basic scheduling rules are followed in this compiler to fully utilize the data localization in CNN and reduce data I/O:

1 **Input channel first.** Sometimes, the input feature map needs to be cut into smaller blocks. We keep a set of loaded input feature map blocks in input buffer and generates as many output channels' intermediate results as possible. This means the convolution kernels are changing in this process. Usually, feature map is much larger than convolution kernels. So keeping the feature maps on-chip is better than keeping the convolution kernels.

2 **Output channel second.** When the feature maps are cut into blocks, we first calculate all the output blocks at the same position and then move on to the next position.

3 **No intermediate result out.** This means when the output buffer is full with intermediate results, we load a new set of input feature maps to input buffer and do accumulation on these output channels.

4 **Back and forth.** When a set of output buffer finishes the calculation, we have traversed all the input channels. The next round of
traverse is done in the opposite direction. This reduces a redundant
LOAD between two rounds of traverse.

Three steps are included in the compiling process:

Block partition. Since the on-chip memory is limited, especially for
embedded platforms, not all the feature maps and network parameters
for one layer can be cached on-chip. Thus we need to partition the
calculation of one layer to fit each block into the hardware. Different
partition strategies are analyzed, in order to achieve high efficiency, while
almost any kind of partition can be implemented with the instruction
set. The main problem of the partition is the bandwidth requirement.
Reducing I/O can reduce power consumption and saves the bandwidth for
other cooperative accelerators and the host in the system. To remain the
data I/O burst length, we require that the feature map is cut horizontally
for the row-major data format. Then the remained problem is to decide
how many rows are in a single block.

Suppose a layer has M input feature maps of size $f \times f$ and N output
feature maps of the same size. The convolution kernels are of size $K \times K$.
The buffer size for input, output, and convolution kernels are B_i, B_o and
B_w, respectively. r rows are in each feature map block. Since we do not
store intermediate result to DDR, the output amount is a constant to
a layer. We can generate the functions for the input amount of input
feature maps and convolution kernels as D_i in equation (4.2) and D_w in
equation (4.3).

$$R = \frac{B_i}{Mf} - K + 1 \tag{4.1}$$

$$D_i = \begin{cases} \frac{f}{r}(r + K - 1)fM & r \leq R \\ \frac{f}{r}\left\{[(r + K - 1)fM - B_i]\frac{Nrf}{Bo} + B_i\right\} & r > R \end{cases} \tag{4.2}$$

$$D_w = \begin{cases} MNK^2 & MNK^2 \leq B_w \\ \frac{f}{r}(MNK^2 - B_w) + B_w & MNK^2 > B_w \end{cases} \tag{4.3}$$

Equation (4.1) gives the boundary of the two branches for D_i. If
r rows are in a block, we get f/r blocks of a feature map. $r + K - 1$
rows are loaded for each block considering padding and overlap between
adjacent blocks.

If $r \leq R$, the blocks at the same position of all the input channels
can be buffered on-chip. Moving from one output channel to the next

will cost no extra data exchange with external memory. So each block is loaded only once and the total amount of input is according to the first branch of equation (4.2). If $r > R$, extra data exchange is needed. Consider the computation for one output block, all the input blocks at the same position are needed. If the previous output block is at the same position, the input blocks can be reused. The maximum reuse size is B_i. So data input amount for each output block is $(r + K - 1)fM - B_i$, except for the first output channel. To utilize output buffer, B_o/rf output channels are grouped together. This means each group can be totally buffered on-chip. So getting the blocks at the same position of all the output channels needs Nrf/B_o rounds of calculation. This corresponds to the second branch of equation (4.2).

For convolution kernels, if the total amount of data is larger than weight buffer, then extra data exchange is needed when moving from the blocks at one position to the next. Similar to the input feature maps, B_w data can be reused and we get the second branch of equation (4.3). This is the common case for our design.

The above functions do not consider the non-divisible situations, where equations (4.2) and (4.3) need to add rounding symbols, making the optimization process more complicated. Thus, in our compiler, a simulation is done to calculate all the input amount for each possible r. The r with the least input amount is selected. Since the design space of r is quite small, the time required for the traversal process is within an accepetable range. Three examples are shown in Figure 4.7.

As for case (a), only the first branch of D_i in equation (4.2) is satisfied. So the total input amount can be expressed as equation (4.4). r should be as large as possible in this case.

$$D_i + D_w = \frac{f}{r}\left[Mf(K - 1) + MNK^2 - B_w\right] + const. \qquad (4.4)$$

Case (b) is a typical layer in the middle of a CNN model where the number of channels is large and the feature maps are of middle size. The

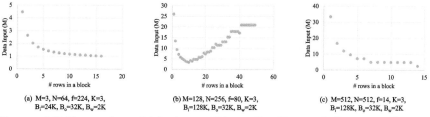

(a) M=3, N=64, f=224, K=3, B_i=24K, B_o=32K, B_w=2K

(b) M=128, N=256, f=80, K=3, B_i=128K, B_o=32K, B_w=2K

(c) M=512, N=512, f=14, K=3, B_i=128K, B_o=32K, B_w=2K

Figure 4.7 Examples of block partition. B_i, B_o, and B_w are effective values.

split condition R lies in the domain of r so both of the branches should be considered. For the second branch in equations (4.2) and (4.3), the total input amount can be expressed as equation (4.5). In this case, a local minimum solution can be found.

$$D_i + D_w = (B_i - B_w + MNK^2)\frac{f}{r} + \frac{MNf^3}{B_o}r + const. \qquad (4.5)$$

Case (c) is a typical layer at the end of a CNN model where the number of channels is large and the feature maps are small. Only the first branch in equation (4.2) is satisfied. So the solution is the same to case (a).

Note that B_i and B_o in case (a) are different from that in case (b) and case (c). Only three input channels are used in this layer while we have 16 input channels in hardware design. So B_i is only $3/16$ of the total input buffer size.

Memory Mapping. External memory space is allocated for the communication between host CPU and the CNN accelerator. First, input feature map memory space should be allocated. The feature maps should be in the row-major format with each channel stored continuously. Then, the memory space for the result of each layer should be allocated. The data format will be automatically handled by hardware. Only two blocks of memory are needed during the calculation of one layer, one for input and one for output. Thus the memory space for non-adjacent layer's result can be overlapped. The compiler supports the case if an intermediate layer's result is needed and preserves the space from rewritten by other layers. Then, memory space for convolution kernels and bias is allocated. This space is preserved during the whole process of CNN acceleration. Usually

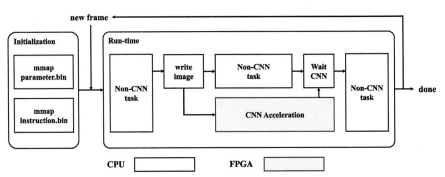

Figure 4.8 Run-time work flow of the proposed system on embedded FPGA.

this space is only initialized once before the first time for CNN acceleration. With the block partition result, the order of how the convolution kernels and bias are used is determined. A *parameter.bin* file filling the parameter memory space is generated according to this order.

On-chip memory is also allocated for input and output feature map blocks and also the convolution kernels according to the block partition result. After all the memory allocation, the corresponding address fields in the instruction sequence are filled.

Dependency Check. After memory mapping step, the instruction set can already finish the CNN calculation process. But data dependency check can find potential parallelism between calculation and data I/O. This step checks the data dependency among instructions and sets the flag bits in instructions to let the hardware explore the parallelism. The order of the instructions is also adjusted to make the most use of hardware parallelism.

4.2.3 Runtime Workflow

The run-time work flow of the proposed system is shown in Figure 4.8. In the initialization phase, the *parameter.bin* file generated by data quantization should be loaded into the memory according to the address given by compiler. Instructions should be prepared in the memory as well. At run-time, non-CNN tasks are run on the ARM core in the system. When CNN is to be called, the input image is first copied to the physical memory space allocated by the compiler, then the instructions are sent down to the accelerator. While the accelerator is working, other tasks can be executed with the host CPU. The host checks the state register of the accelerator to see if it is done. Then the algorithm goes on. Note that multiple CNN can be done within each frame while the graph is an example of one inference per frame.

4.2.4 Extension Support of Upsampling Layers

The upsampling layer makes copies for each element in the feature map, the straight-forward hardware implementation of the upsampling layer could result in a very expensive resource overhead due to the datamover and local cache memory. Actually, we don't have to design specific computation units for upsampling, but only need to load the feature map and save its elements to relevant memory addresses. In this way, only an address generator is required for upsampling support. Moreover, as

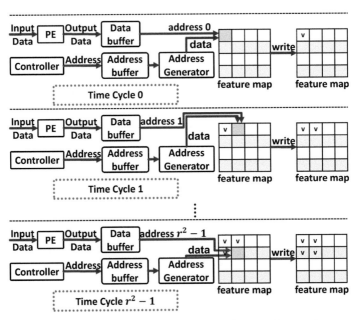

Figure 4.9 Fusion design for upsampling and convolution layers.

each element is copied to a regular square area, the address generator module can be made up of just a two-stage counter, which is with very low complexity.

Generally for CNN accelerators, different operators are implemented with independent hardware modules and invoked in different time slots, i.e., if an upsampling layer is inserted behind the convolution layer, it should execute the load-computing-save flow after the PE slot. However, as data accessing is a main source of power consumption and processing latency, such a concatenation processing flow will obviously reduce the efficiency performance. To guarantee the energy efficiency of our proposed CNN accelerator when running deep learning models for object detection such as Single Shot Multi-Box Detector (SSD), which is composed of several upsampling layers to increase detection accuracy, we make a fusion design for the convolutional layer and upsampling layer.

As shown in Figure 4.9, we make some modifications to the PE array: First, we add buffers for both output data and address to fit various upsampling configurations. In our design, not all layers are inserted with upsampling operations and the upsampling ratio varies according the applications. Second, we insert the upsampling address generator behind

PE array controller. Thus the output of PE array can be directly saved to relevant memory addresses to accomplish upsampling.

By modifying the conventional PE array structure, the additional data save-load cycle can be avoided. Moreover, as there are usually thousands of clock cycles in PE processing and only several cycles are required for multiple save process, the fusion design will not introduce any latency for PE array. Actually, compared with conventional SSD accelerator, the overhead of our architecture exists in data movement between on-chip and off-chip memories, as upsampling increases the quantity of feature map data. However, since only two convolutional layers are inserted with upsampling layers in our software design, the overall overhead can be negligible.

4.2.5 Evaluation

Two FPGA-based designs of the hardware architecture are carried out. A 16-bit version of the design is implemented on the Xilinx XC7Z045 chip which targets at high-performance applications. An 8-bit version is implemented on the Xilinx XC7Z020 chip which targets at low-power applications.

The hardware parameters and resource utilization of our design are shown in Table 4.1. All the results are generated by Vivado 2015.4 version after synthesis and implementation. By choosing the design parameters properly, we can fully utilize the on-chip resource. Note that we are not using all the resource on XC7Z020 because the design coexists with an HDMI display logic for our demo. Comparing the 8-bit and 16-bit version result on XC7Z045, we see that 8-bit version offers 50% more

TABLE 4.1 Hardware Parameter and Resource Utilization of Angel-Eye.

design	data	#PE	#Conv	FFs	LUTs	BRAMs	DSPs	Clock	Perf. (GOPS)
XC7Z045	16-bit	2	64	127653 (29%)	182616 (84%)	486 (89%)	780 (87%)	150MHz	187.8
XC7Z020	8-bit	2	16	35489 (33%)	29867 (56%)	85.5 (61%)	190 (86.4%)	214MHz	84.3
XC7Z030	8-bit	4	16	34097 (22%)	43118 (55%)	203 (77%)	400 (100%)	150MHz	105.2*
XC7Z045	8-bit	12	16	85172 (19%)	139385 (64%)	390.5 (72%)	900 (100%)	150MHz	292.0*

* The performance is estimated by simulation

TABLE 4.2 Performance Comparison of Angel-Eye on XC7Z045 and XC7Z020 with Other FPGA designs and GPU

	[149]	[150]	[151]	[152]	Ours		GPU	
Platform	Virtex 5 SX240t	Zynq XC7Z045	Virtex 7 VX485t	Virtex 7 VX690T	Zynq XC7Z045	Zynq XC7Z020	Nvidia Titan X	
Clock (MHz)	120	150	100	156	150	214	1000	
Bandwidth (GB/s)	-	4.2	12.8	29.9	4.2	4.2	336	
Data format	INT48	INT16	FP32	INT16	INT16	INT8	FP32[b]	FP32[c]
Power(W)	14	8	18.61	30.2	9.63	3.5	174	210
Performance (GOP/s)	16	23.18[a]	61.62	565.9	137	84.3 (CONV)	3544	5991
Energy Efficiency (GOP/s/W)	1.14	2.9	3.31	22.15	14.2	24.1	20.4	28.5

[a] The performance is of the face detector application in [150] where the 552M-op network is running at 42 frames per second.
[b] The batch size is 1.
[c] The batch size is 32.

parallelism while consuming less resource. This shows the importance of data quantization.

The VGG16 network is used to test the performance and energy efficiency of our design on XC7Z045 and XC7Z020 FPGAs. The result together with that of other FPGA designs for CNN and GPU is shown in Table 4.2. Some conclusions can be drawn from this comparison.

First, **precision greatly affects the energy efficiency.** Early designs [149][151] using 48-bit fixed-point data or 32-bit floating point data are with much lower energy efficiency. Comparing the estimated 8-bit design on XC7Z045 with the 16-bit version also gives this conclusion. These two designs utilize similar resource and run with the same clock frequency, thus should consume similar power. But the 8-bit design offers more than 50% performance improvement, which means the energy efficiency is better.

Second, **the utilization of the hardware is important.** The reported performance in [150] is 200GOPs when the network perfectly matches the 10×10 convolver design. But for the 5×5 and 7×7 kernels, the performance is down to 23 GOPs. As most of the computation in state-of-the-art neural networks is from 3×3 convolution. So the proposed design in this work should be better.

Third, **memory I/O affects the energy efficiency.** The energy cost of reading/writing data from/to memory is high. The design in [151] only implements channel parallelism to simplify the design of data path. But this strategy does not utilize the data locality in the convolution operations and leads to more data I/O. The design in [152] implements the whole AlexNet the large VX690T chip, where the intermediate result of each layer is not written back to memory. This further reduces data I/O and thus achieves higher energy efficiency compared with our 16-bit design. But this kind of design is hard to be scaled down to be deployed on embedded platforms with limited BRAM resources.

We also compared our design with desktop GPU using the VGG-16 network. Both batch 1 mode and batch 32 mode are tested. The batch 1 mode suffers about 41% performance loss compared with batch 32 mode. Our 8-bit design achieves even higher energy efficiency with the batch 1 mode on the large network. But the scale of GPU is too large for embedded platforms.

For the 8-bit version implementation on XC7Z020, two more tasks, YOLO, and face alignment are used for evaluation besides the VGG-16 network. We compare the performance of our design with the 28nm NVIDIA TK1 SoC and the latest NVIDIA TX1 SoC platforms. For YOLO and face alignment, CNN part is implemented on FPGA. The rest of the algorithms are handled by the integrated CPU in the SoC. Although a batched way of processing can fully utilize the parallelism of GPU on TK1 or TX1, it is not a good choice for real-time video processing because it increases latency. For some applications like tracking, the result of one frame is used for the computation on the next frame. This requires the frames to be processed one by one. So we do not use batch in our experiment. Performance comparison is shown in Table 4.3.

All the three platforms perform better on larger CNN models. But the proposed design offers a more stable performance. On YOLO and

TABLE 4.3 Performance Comparison of Angel-Eye on XC7Z020 with TK1 and TX1 on Different Tasks

	VGG		YOLO		Face Alignment	
	time/frame (ms)	performance (GOP/s)	time/frame (ms)	performance (GOP/s)	time/frame (ms)	performance (GOP/s)
7Z020	364	84.3	88.0	62.9	2.54	41.1
TK1(fp32)	347	88.4	150	36.9	14.3	7.3
TX1(fp32)	153	200	59.4	93.2	9.04	11.5
TX1(fp16)	96.5	318	42.4	131	2.18	47.9

face alignment tasks, Angel-Eye even offers better performance than TK1 and achieves similar performance as TX1. This is because the parallelism pattern of GPU does not fit into small network well. The running power of TK1 and TX1 are 10W while that of Angel-Eye on XC7Z020 is only 3.5W. So our design can achieve up to 16× better energy efficiency than TK1 and 10× better than TX1.

Performance of the 8-bit version on XC7Z030 and XC7Z045 is estimated with simulation. On XC7Z020, we measured the actual I/O bandwidth to be about 500MB/s. The estimation is based on this. XC7Z030 is with the same bandwidth and XC7Z045 doubles the bandwidth with an extra independent DDR port for FPGA. About 1.25x and 3.46x performance can be achieved by these two platforms compared with XC7Z020 with the help of more resource even with a conservative 150MHz estimated clock frequency.

4.2.6 Practice on DAC-SDC Low-Power Object Detection Challenge

We extend our first version of CNN accelerator design, i.e., Angel-Eye, to the second version called Aristotle with three computation parallelism dimensions support. The total computation parallelism of Aristotle can be calculated as below:

$$Parallelism = 2 \cdot PP \cdot ICP \cdot OCP \quad (OPs/cycle) \qquad (4.6)$$

where PP, ICP, and OCP represent the parallelism along the pixel, the input channel, and the output channel dimensions of the feature map, respectively. Corresponding to the convolutional computation module, PP equals to the number of PE P, meaning that each PE completes the computation of one pixel of feature map per cycle. Inside each PE, there are a total of OCP parallel computing channels, each corresponding to the computation of one output channel. For each computation channel, it can handle the multiply-accumulate (MAC) computation between the feature maps and weights of ICP input channels in each cycle, so the computation parallelism in equation 4.6 needs to be multiplied by two.

We demonstrated Aristotle CNN accelerator in the 55th IEEE/ACM Design Automation Conference System Design Contest (DAC-SDC) [153], a low-power object detection challenge for unmanned aerial vehicles (UAVs) applications. The goals of DAC-SDC are similar to LPCVC, that is, the software and hardware co-design optimization is needed for the target application. In addition, DAC-SDC provides an FPGA track and allows participants to customize the hardware architecture for the specific neural network architecture. Our Aristotle CNN accelerator design with pruning and quantization optimization techniques won the first prize award for the FPGA track of the contest. The hardware platform of the FPGA track is Xilinx PYNQ Z-1 FPGA SoC board, which is an embedded system of ZNYQ series platforms with Python support. It contains a Cortex-A9 processor, a 512MB DDR3 memory, and a ZYNQ XC7Z020-1CLG400C FPGA chip, which comprises 53K look-up-tables (LUTs), 220 DSPs, and 630KB BRAMs.

We implement one B1152 core, of which the computation parallelism is $2 \times 4 \times 12 \times 12$ operations per cycle, on the PYNQ Z-1 FPGA board running at 144 MHz, thus providing 166GOPs peak performance for CNN inference applications. It consumes 44627 (83.9%) LUTs, 57709 (54.3%) FFs, 110.5 (78.9%) BRAMs, and 220 (100%) DSPs of the total hardware resources. As for the connectivity between the Aristotle RTL IP and Cortex-A9 CPU, there is one 32-bit AXI-lite memory port for controlling and register I/O. Besides, there are two 64-bit AXI-4 memory ports used for feature maps and weights and one 32-bit AXI-4 memory port for instructions of DDR I/O. Our FPGA system design with SSD network delivers 0.624 Intersection over Union (IoU), 11.96 frames per second (FPS), and 4.2 W power, achieving 1.267 total score and winning the first prize in the FPGA track of in DAC-SDC'18. More practical results regarding the neural network model optimization will be discussed in Section 4.3.1.3.

4.3 NEURAL NETWORK MODEL OPTIMIZATION

In this section, we first describe the basic pratices of pruning and quantization in Section 4.3.1. These practices are employed in our solutions for DAC-SDC'18 and LPCVC'20. Then, Section 4.3.2 describes our hardware-aware pruning solution for LPCVC'19. Finally, we discuss on the architecture search framework that we have been recently developing, aw_nas. We also present a case study of running hardware-aware NAS for a Xilinx DPU.

4.3.1 Pruning and Quantization

Neural network computing is usually energy- and bandwidth-consuming, owing to the large number of weights involved in the computation. This makes it challenging to deploy neural networks on edge scenarios with limited power, memory, and bandwidth budget. Therefore, compression techniques are widely studied and applied since when researchers found that most neural networks can be heavily compressed without any loss of accuracy. Generally, the compression process includes two basic stages, and both of them will not interfere with each other in most cases. The first stage is network pruning, which removes the redundant connections, and keeps the informative parts to preserve the accuracy. The second stage is network quantization, which reduces the bit-width of the weights and neurons from 32-bit floating-point to 8-bit or even less fixed-point. By network pruning and quantization, the model size and computation amount can be highly compressed.

4.3.1.1 Network Pruning

A straightforward motivation of network pruning is that there exists a significant proportion of weights that are close to zero, which have no contribution to the computation [154]. The sparse nature of the neural networks provides great potentiality to prune the zero and redundant weights. We only consider structured pruning here because unstructured (irregular) pruning will introduce a lot of sparse matrix-matrix computation and be inefficient in most hardware. Structured pruning means that we completely remove a whole filter in convolutional layers or a whole weight vector connected to a neuron in fully-connected layers. For example, a convolutional layer with a $64\times64\times3\times3$ ($Cout\times Cin\times Kh\times Kw$) kernel can be pruned to a $32\times64\times3\times3$ kernel by removing 32 filters in the output channel. Correspondingly, its next layers will also remove the filters in the input channel with the same index, e.g., from $128\times64\times3\times3$ to $128\times32\times3\times3$. Previous work has discussed methods to learn structured sparsity in neural networks, e.g., by group lasso [155]. Moreover, an essential part of filter-level pruning is to measure the importance of the weights to determine the redundant weights that can be pruned. Several possible methods can measure the importance based on different criterions, e.g., based on the magnitude of weights [156], based on the sparsity of the activations after the ReLU function [157], or based on the reconstruction error of the next layer outputs [158], etc. Our pruning framework is shown in Figure 4.10, consisting of sensitivity analysis, weight pruning, and network finetuning. In our implementation, we apply a mixed measurement which can be partitioned as intra-layer and inter-layer criterions. For the inter-layer importance measurement, we introduce sensitivity as a criterion to determine the pruning strategy, i.e., the layer to be pruned and the number of filters to be pruned in

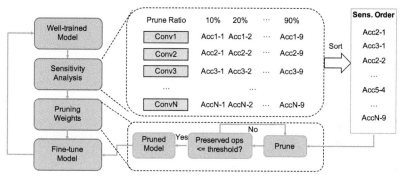

Figure 4.10 The schematic diagram of neural network pruning. To prune a well-trained model, we first analyze the sensitivity of different layers in the network. Then based on the sensitivity, corresponding weights will be removed. Then the pruned model will be fine-tuned to recover the accuracy. The above steps are iteratively conducted to gradually prune the model.

this layer. The sensitivity of a neural layer is defined as the accuracy gap with and without the pruned filters, and can be obtained in an easy manner. We apply a mask matrix to disable the weights in one layer and corresponding locations in the next layer, and evaluate the accuracy on a sample dataset. For the intra-layer importance measurement, we directly select the pruned filters based on their absolute weight sums. For example, to prune 30% of filters in a 64×64×3×3 convolutional kernel, we sort the weight sums in ascending order, then prune 19 channels from top to bottom. Then we test and record the accuracy drop as the sensitivity of this layer.

After evaluating the sensitivity, the next step is weight pruning. The sensitivity will be sorted in ascending order. A user may set a computation amount threshold, and the pruner will prune the weights from the least sensitive ones to the most sensitive ones, until the remaining computation amount becomes smaller than the preset threshold. The final step is finetuning, as it is unavoidable that the performance of pruned model would significantly drop from its original performance. Therefore, the pruned model needs to be finetuned to restore accuracy. Note that the mentioned process should be iteratively made to gradually prune the model.

4.3.1.2 Network Quantization

Network quantization is to quantize the weights and activation into lower-bit numbers to reduce the data size and computational complexity. We can classify the quantization methods from two aspects. First, classified by the application time of quantization, the quantizers can be classified into post-training quantization and quantization-aware training. Second, classified by the behavior of quantization, the quantizers can be classified into uniform and non-uniform quantization, respectively.

Post-training quantization is retrain-free. That is to say, no model re-training would be conducted. When obtaining a floating-point model, the quantizer can directly quantize the weights and activations only based on a small set of data to calibrate the quantization scale.

Quantization-aware training considers the quantization effects into training and is more likely to obtain a quantized model with better accuracy than post-training quantization, especially in lower-bit cases, e.g., 4-bit. Generally speaking, the process simulates the quantization behavior and encourages the model to adapt to the error introduced by the quantizer. However, such methods are slower than post-training quantization since it requires a complete training process and the full dataset. In all the competitions that we have taken part in, we adopt the quantization-aware training scheme instead of the post-training quantization to retain higher performances.

Uniform quantization quantizes the data into uniform-distributed levels. Assuming the data distribution lies in (x_{min}, x_{max}), to quantize the data into 8-bit integer, we may set the scale Δ as $\Delta = \max(abs(x_{min}), abs(x_{max}))/128$. The a floating-point number x can be quantized into an integer by:

$$x_{int8} = clip(round(\frac{x}{\Delta}), \min = -128, \max = 127). \qquad (4.7)$$

To recover the floating-point number, the process will be:

$$x_{rec} = x_{int8} * \Delta. \qquad (4.8)$$

As can be seen, if the scale Δ can be a power of 2, the data can be represented as an integer with a bit-shift operation. The selection of scale Δ should also be carefully designed. A simple way is to find the max absolute value of the data distribution, then we can obtain the scale by:

$$\Delta = pow(2, ceil(\log_2(\max(abs(x_{min}), abs(x_{max}))/128)). \qquad (4.9)$$

However, such method may be greatly impacted by some extreme outliers which lies far away from normal distribution. Therefore, an improved way to select the scale by minimizing the mean squared error between the quantized data and the floating-point data:

$$b = ceil(\log_2(\max(abs(x_{min}), abs(x_{max}))/128)) \qquad (4.10)$$

$$\hat{b} = \arg \min_{b_i} MSE((round(x/\Delta) * \Delta - x) \qquad (4.11)$$

$$\Delta = pow(2, b_i), b_i \in [b - 4, b] \qquad (4.12)$$

Non-uniform quantization applies a different scale Δ when quantizing the data in different ranges. The motivation behind this strategy is that the distribution of data is always non-uniform. Therefore, putting more levels in regions with more data would reduce the quantization errors for more values. However, non-uniform quantizers are not usually applied in practice because the processing of non-uniform data is harder.

4.3.1.3 Evaluation and Practices

DAC-SDC'18: As for the UAV application of the low-power object detection challenge [153], the dataset is provided by drone company DJI and comprises 12 main categories and 95 sub-categories. There are 100000 images for training and 50000 images as the hidden test set for contest evaluation. The metrics are IoU for accuracy evaluation, throughput (FPS) for processing speed, and power for energy consumption. The final score is the combination of IoU, FPS, and energy consumption.

We apply the quantization and pruning techniques on a modified VGG16+SSD network, where the last two convolutional layers of the VGG backbone are removed. We prune and fine-tune the floating-point model with 90% compression rate at first. Then, we utilize an 8-bit quantization-aware training process to achieve higher IoU. We obtain the final optimized model with 14.2× and 10× reduction in parameters and operations, respectively. The FPS is increased from 1.5 to 12, with the IoU decreased from 0.657 to 0.624, as illustrated in Table 4.4.

LPCVC'20: In the LPCVC competition 2020 object detection track, we adopt the aforementioned pruning method on MobileNetV2+SSD, and then finetuned the pruned model with adaptive distillation for the object detection task. Compared with the baseline model MobileNetV2+SSD

TABLE 4.4 DAC-SDC'18 results. Neural Network Pruning Brings 8× fps Improvement with 0.033 IoU Drop on a Modified SSD Network

Compression rate	IoU	Fps	Ops	Params
0	0.657	1.5	81.78G	32.60M
0.9	0.624	12	8.12G	2.29M

that has 0.23 mAP on COCO val2017 and 26.5ms latency on Pixel 4 single-core CPU, we achieved 0.227 mAP and 24.2ms latency. The evaluation score is calculated by equation 4.13 and indicates the improvements over the Pareto frontier. Our model obtained a score of 0.0326.

$$M(A, T) = A - k \log T - a_0 \qquad (4.13)$$

where A and T is the actual accuracy and latency of the model, and $k = 16.894553358968146, a0 = -34.42191514521174$.

We also apply knowledge distillation to improve the performance of the pruned model. In classification tasks, one just needs to add a cross-entropy loss function on the output of the teacher network and the student network [159]. While in object detection tasks, we implement the Adaptive Distillation [160] strategy that penalize more for hard examples. The soft loss can be writen as

$$ADL = (1 - e^{-(KL(p_t\|p_s)+\beta T(p_t))})^\gamma KL(p_t\|p_s) \qquad (4.14)$$

where $KL(p_t\|p_s)$ indicates the Kullback-Leibler divergence between the output distribution of the teacher network and the student network, and $T(p)$ is the entropy of the classification results of the teacher network. We can see that when $KL(p_t\|p_s)$ or $T(p_t)$ is large, the multiplier on $KL(p_t\|p_s)$ would be larger. The idea behind this loss reshaping is to penalize the hard-to-mimic (high $KL(p_t\|p_s)$) and hard-to-learn (high $T(p_t)$) more.

In our experiment, only the anchors assigned with ground-truth samples are used to calculate the soft loss, and we use the unpruned network as the teacher network. γ is set to 1.0 and β is set to 1.5. We use TFLite for quantization and deployment. The result is shown in Table 4.5.

TABLE 4.5 LPCVC'20 Results. Adaptive Distillation Brings 0.011 mAP Improvement on the Pruned Model

Model	latency(ms)	mAP
Baseline(float)	47.2	0.235
Baseline(quantized)	26.5	0.23
Pruned	24.2	0.216
Pruned + AD	24.2	0.227

4.3.2 Pruning with Hardware Cost Model

4.3.2.1 Iterative Search-based Pruning Methods

The previous pruning method only considers the task performance in the pruning process. In each iteration, it approximately evaluates the influence/sensitivity of each layer to the task performance and chooses the pruning ratios according to the sensitivity only. Finally, a threshold on the task performance is used to terminate the pruning process. This pruning process does not consider the hardware cost in the pruning ratio selection. Thus, the pruned models can be sub-optimal for hardware efficiency. Also, it cannot handle cost-related constraints, for example, one might require that the latency of the pruned model must be under some budget. To conduct pruning with a hardware cost model, many studies adopt various search methods to explicitly search for a pruning ratio configuration. These methods usually follow a two-level iterative scheme, and we summarize the workflow of these methods in Algorithm 1.

An early method, AMC [161], uses an RL-learned policy to decide the pruning ratios, and the objective is constructed using both the hardware cost and task performance. However, there exist controllability issues with the reinforcement learning-based pruning method. To get a pruned model under a certain budget, one needs to either impose a constraint on the action space or craft a specially-shaped cost function as a part of the objective. The first type requires much manual effort to design the sampling strategy to avoid out-of-budget samples and is not general. In the second type of strategy, the penalty is very high when the pruned model's resource is beyond budget. However, proper reward shaping for learning the RL policy requires manual hyperparameter tuning. Therefore, this type of black-box search strategy lacks controllability for practical usage.

In contrast to the RL-based black-box search strategy, another early study, NetAdapt [162], adopts an easy but interpretable search strategy.

Algorithm 1 The general workflow of iterative search-based pruning methods

---

1: *Model*: The model to be pruned
2: N_o: Number of outer iterations
3: N_i: Number of inner iterations
4: $\boldsymbol{a}$: Prune ratio configurations
5: Components: 1) *Controller*: Search strategy; 2) *Evaluate*: A fast evaluation of the task performance of the prune ratios

6: **for** $i = 1, \cdots, N_o$ **do**
7:     **for** $j = 1, \cdots, N_i$ **do**
8:         $\boldsymbol{a} = controller.sample\_as()$ ▷ Sample candidate prune ratios $\boldsymbol{a}$
9:         $\boldsymbol{accs} = Evaluate(\boldsymbol{a}, Model)$ ▷ Run fast sensitivity analyisis or evaluation of the candidate prune ratios
10:         $controller.fit(\boldsymbol{a}, \boldsymbol{accs})$ ▷ Update the controller using the evaluation results
11:     **end for**
12:     $\boldsymbol{a} = controller.decide()$ ▷ Decide a prune ratio decision for this outer iteration
13:     $Model = PruneAndFinetune(Model, \boldsymbol{a})$ ▷ Prune using the decided prune ratio, and finetune the model
14: **end for**

NetAdapt makes K proposals where K is the number of layers, and in each proposal, only one convolution layer is pruned to the maximum number of channels satisfying the budget constraint in this iteration. Other studies [163, 164] use heuristic local search methods, like simulated annealing (SA) [164] and evolutionary algorithm [163] that can directly search toward multiple objectives and get the Pareto curve. Explicit constraints could be easily incorporated into these search strategies.

4.3.2.2 Local Programming-based Pruning and the Practice in LPCVC'19

In the LPCVC competition at ICCV 2019, we developed another pruning method with very few hyperparameters that can utilize the hardware model and facilitate a controllable pruning process with various resource constraints. The task is to perform ImageNet classification using the Xilinx Ultra96-V2 board with the DPU IP. And the evaluation criteria is as follows: (1) Accuracy should be > 68.5%; (2) Latency should be < 10ms; (3) Score is calculated as $(acc \times 100 - 68.5) \times 0.08 + 6/latency$. We choose MNASNet-100 as our base network, which has an accuracy of 72.4% and a latency of 7.37ms after 8-bit quantization and finetuning. We only prune the inner (expansion) channels of each block, while respecting the hardware parallelism.

Similar to NetAdapt, our method adopts a progressive pruning scheme, in which the resource constraint tightens for $TargetDiff$ in each outer iteration. And in each outer iteration, we first conduct a block-wise analysis to profile the sensitivity of each block with respect to both the task performance (ImageNet accuracy) and the resource consumption (latency). Then, we fit linear functions on both the task performance (in a local region) and resource consumption for each layer independently to get $\boldsymbol{w}_{acc} \in \mathcal{R}^K$ and $\boldsymbol{w}_{lat} \in \mathcal{R}^K$. Finally, we solve the linear programming problem in Equation 4.15 to get the pruning ratios.

$$
\begin{aligned}
\arg\min_{a} \quad & \boldsymbol{w}_{acc}^T \boldsymbol{a} \\
\text{s.t.} \quad & \boldsymbol{L_a} \leq \boldsymbol{a} \leq \boldsymbol{U_a} \\
& \boldsymbol{w}_{lat}^T \boldsymbol{a} \geq TargetDiff
\end{aligned}
\qquad (4.15)
$$

where $\boldsymbol{a} \in (0,1]^K$ is the block-wise prune ratios, $TargetDiff$ is the target latency difference we'd like to achieve in this outer iteration. The upper bound of prune ratios $\boldsymbol{U_a}$ is determined by finding inflection-point on the accuracy curves heuristically, and the lower bound of prune ratios $\boldsymbol{L_a}$ is determined by the resource constraint of Ultra96. w_{acc}, w_{lat} are found by linear fitting the latency and accuracy profiling data in the linear region. In our experiment, we find that the latency curves follows the ladder linear pattern, and the ladder width is the channel parallelism.

In our solution, we conduct two outer iterations. In the first outer iteration, we want the latency to be reduced to 6ms from 7.37ms ($TargetDiff = 1.37$ms), and in the second iteration, we set $TargetDiff$ to be 0.5ms. After pruning, we round up the channel numbers to be aligned with the channel parallelism. We also fuse several small 1×1 convolutions into other convolutions, since DPU cannot process them efficiently despite their small FLOPs. Finally, our solution achieves an accuracy of 68.6% without using the validation data (70.4% accuracy without quantization), while the FLOPs is reduced from 587M to 456M, and the latency is reduced from 7.37ms to 5.26ms on DPU B1600. The

TABLE 4.6 Comparison of Iterative Pruning Methods: AMC [161], NetAdapt [162], AutoCompress [164], MetaPruning [163], and Our local LP-based Method

Method Interface	AMC	NetAdapt	Auto-Compress	MetaPruning	LP-based (Ours)
Controller. *sample_as*	MLP, per-K layer heuristic thresholding	single-layer proposals	SA mutation	EA mutation	grid sample
Evaluate	without tuning	with tuning	without tuning	HyperNetwork	without tuning
Controller. *fit*	RL learning	-	SA accept decision	EA population update	local model fitting
Controller. *decide*	same *sample_as*	pick highest acc	SA-discovered a	EA population Top-1 a	LP to get a
PruneAnd− *Finetune*	-	prune & long tuning	ADMM-based prune & tuning	-	prune & long tuning
N_o	1	multiple	multiple	1	multiple
N_i	multiple	1	multiple	multiple	1

evaluation score is 1.149. The profiling and pruning codes can be found at https://github.com/walkerning/LPCVC.

As we stated before, the iterative pruning methods comply with the general workflow shown in Algorithm 1. Table 4.6 summarizes the design choices adopted by various iterative methods (including our LP-based one) for interfaces in the general framework. As we can see, AMC, AutoCompress, and MetaPruning conduct black-box search processes using RL-learned policy, SA and EA, respectively. In contrast, NetAdapt develops a simple method, in which K naive proposals that only prune one layer are made in each outer iteration. Instead, our method assumes that the task performance follows a linear model in a local prune ratio region, and solves the best prune ratio configuration in each outer iteration. This local programming-based method has only one hyperparameter (the target resource difference in each round $TargetDiff$), and various constraints can be incorporated effortlessly.

As a possible extension of our method, we can use any local-convex function instead of the linear function to fit the task performance. And the resulting optimization problem is still a convex optimization problem that can be easily solved. This extension might be useful for reducing the outer iteration number N_o, since the local region in which the fitting is appropriate can be enlarged. However, if we want to extend our method to use a non-linear resource model like a quadratic function, which is the case if we'd like to prune the outer channels of each block too. The optimization problem is no longer a convex optimization, and one should resort to iterative methods to find a solution to the optimization problem. Even in this case, the optimization process would be very fast since each evaluation only involves a function calculation instead of evaluating a pruned network on the validation dataset.

4.3.3 Architecture Search Framework

Recently, Neural Architecture Search (NAS) has received extensive attention due to its capability to discover competitive neural network architectures in an automated manner. To enrich our software tool stack for efficient NN inference, we build a NAS framework aw_nas that implements various NAS algorithms in a modularized manner. And we use this framework to reproduce the results of mainstream NAS algorithms of various types. Also, due to the modularized design, one can experiment with different NAS algorithms with aw_nas for various application scenarios (e.g., classification, detection, fault tolerance, adversarial robustness, hardware efficiency, and etc.). The codes can be found at https://github.com/walkerning/aw_nas. In this section, we first give a simple introduction to the modularization and general workflow of a NAS system. Then, we present a case study of running hardware-aware NAS for efficient NN inference on a Xilinx DPU.

4.3.3.1 Framework Design

The main design principle lying behind aw_nas is modularization. There are multiple actors that are working together in a NAS algorithm, and they can be categorized into well-defined components based on their roles. The list of components and the aw_nas supported choices for each component are summarized in Table 4.7.

The interface between these components is well-defined. We use a "rollout" (class *awnas.rollout.base.BaseRollout*) to represent the interface object between all these components. Usually, a search space defines one or more rollout types (a subclass of *BaseRollout*). For example, the basic cell-based search space **cnn** corresponds to two rollout types: (1) **discrete** rollouts that are used in reinforcement learning (RL)-based, evolutionary-based controllers, etc., and (2) **differentiable** rollouts that are used in gradient-based NAS.

TABLE 4.7 aw_nas Supported Component Types. Different Component Choices Can Be Combined to Construct Different NAS Algorithms. See the "examples" Directory in the Code Repository for More Examples

Component	Description	Current supported types
Dataset	define the dataset	Cifar-10/100, SVHN, (Tiny-) ImageNet, PTB, VOC, COCO, TT100k
Objective	the rewards to learn the controller, and (optionally) the objectives to update the evaluator	classification, detection, language, fault tolerance, adversarial robustness, hardware (latency, energy ...)
Search space	define what architectural decision to be made	cell-based CNN, dense cell-based CNN, cell-based RNN, NasBench-101/201, blockwise with mnasnet/mobilenet backbones
Controller	select architectures to be evaluated	random sample, simulated annealing, evolutionary, RL-learned sampler, differentiable, predictor-based
Weights manager	fill the architectures with weights	supernet, differentiable supernet, morphism-based
Evaluator	how to evaluate an architecture	parameter-sharing evaluator (mepa), separately tune and evaluate (tune)
Trainer	the orchestration of the overall NAS search flow	a general workflow (simple), parallelized evaluation and async update of controller (async)

The search workflow of a NAS algorithm and the important interface methods are illustrated in Figure 4.11. Specifically, one iteration of the search flow goes as follows:

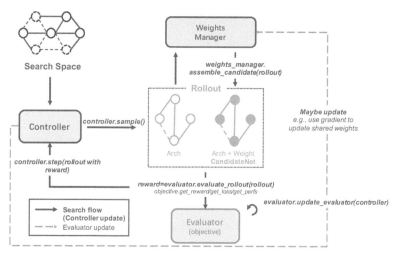

Figure 4.11 Search workflow and interfaces. The solid arrows denote the update loop of the controller (i.e., search strategy): *controller.sample* samples candidate architectures from the search space, *weights_manager.assemble_candidate* construct parametrized models of the candidate architectures, *evaluator.evaluate_rollout* evaluate the candidate models using specific objectives, and then the evaluation results are used by *controller.step* to update the controller state. The dashed arrows denote the optional update loop of the evaluator, for example, when using the parameter-sharing evaluator, the supernet weights manager in the evaluator should be updated.

1. *rollout = controller.sample()*: The **controller** is responsible for sampling candidate architectures from the search space.

2. *weights_manager.assemble_candidate(rollout)*: The weights manager fills the sampled architecture with weights.

3. *evaluator.evaluate_rollout(rollout)*: The evaluator evaluate the rollout that contains the architecture and weights information.

4. *controler.step(rollout)*: The rollout that contains the reward information is used to update the controller.

5. Optionally, some types of evaluator might need to be updated periodically by calling *evaluator.update_evaluator(controller)*, which might issue calls to *controller.sample weights_manager.assemble_candidate* too.

TABLE 4.8 Comparison of Fixed and the Discovered Search Spaces with ResNet-based Blocks on CIFAR-10

Method	Fix Search Space			Discovered Search Space			
	Acc. (%)	Lat. (ms)	Cost	λ	Acc. (%)	Lat. (ms)	Cost
DNAS	86	1.287	0.9872	0.01	86.7	1.299	**0.9838**
			0.9173	0.1	86	1.28	**0.9169**
			0.2180	1	85.8	1.255	**0.1992**
RL-NAS	83.5	1.385	1.0003	0.01	86.5	1.292	**0.9848**
			0.9353	0.1	85.4	1.27	**0.9193**
			0.2855	1	84.7	1.25	**0.2015**

Taking the ENAS [165] method as an example, the dataset and objective are of type **cifar10** and **classification**, respectively. The search space type **cnn** defines a cell-based CNN search space. And the controller **rl** is a RL-learned RNN network. The weights_manager **supernet** is a parameter-sharing based supernet. As for the evaluator **mepa**, with its most basic configuration, just forward batches sampled from the dataset and call *objective.get_reward* to get the rollout's reward.

As a case study of applying the aw_nas framework for hardware-aware NAS, we introduce our work on black-box search space profiling and selection for specific hardware accelerators [166]. We'd like to craft a compact layer-wise search space for a given NN hardware accelerator in a black-box way. Specifically, given a NN hardware accelerator without a white-box simulator, we aim to select proper operation primitives for each layer to construct a smaller search space that is suitable for the accelerator. This search space tuning process is used as a preprocessing step of the vanilla NAS process. And we demonstrate that this search space tuning process can enable the NAS process to find better architectures with smaller search costs for the given NN accelerator. For this case study, all the documentation and codes can be found at https://github.com/walkerning/aw_nas/tree/master/examples/research/bbssp.

4.3.3.2 Case Study Using the aw_nas Framework: Black-box Search Space Tuning for Hardware-aware NAS

The workflow is illustrated in Figure 4.12. For each candidate search space, we first generate search space base networks (SSBNs) and evaluate

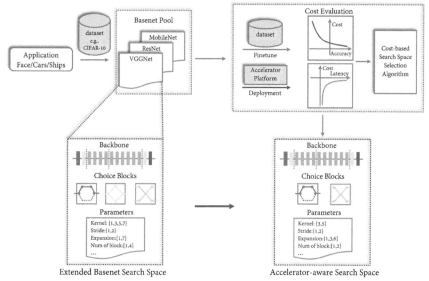

Figure 4.12 The workflow of the black-box search space profiling and selection method. The first step is to generate the basenet pool consisting of all the original search space. The second step is to profiling the extended basenet search space based on the cost evaluation. Finally, an accelerator-aware search space can be generated using a heuristic algorithm according to the evaluated costs.

their accuracy and latency by training and testing the networks on the targeted accelerator. Then the cost of each SSBN is calculated as

$$\text{Cost} = \exp(-\frac{\text{acc} - \text{thres}}{\text{scale}}) - \frac{\lambda}{\text{latency}}. \qquad (4.16)$$

Then, we define the cost of a search space as the average of all its SSBNs' costs. After the above profiling process, we search for the primitive choices of each layer. And the discovered search space has a much smaller controllable size and can provide a tradeoff between accuracy and latency by adjusting the user-defined parameter λ.

After the search space profiling and selection stage, we run parameter-sharing NAS on the search space with two types of controllers: differentiable (DNAS), reinforcement learning (RL-NAS). The reward used in the RL-NAS search process is calculated as $Reward(a) = Acc(a) \cdot (1/Lat(a))^{\beta}$, and the loss used in the DNAS search process is calculated as $Loss(a) = CE(a) \cdot log(LAT(a))^{\beta}$, where $Acc(a)$, $CE(a)$, and $Lat(a)$ are the accuracy, cross-entropy loss, and latency of architecture α, respectively. The latency $Lat(a)$ is estimated using the latency LUT. β is a coefficient to trade-off between latency and task performance.

In this case study, the hardware accelerator is a Xilinx DPU with the parallelism of 4096 operations per cycle, synthesized with 333 MHz on Xilinx ZCU102 FPGA. This DPU has a similar hardware design to the previously described Angel-eye accelerator. Table 4.8 shows the performance comparison between the discovered search space and a fixed search space with ResNet-based network blocks on CIFAR10. Compared with the fixed search space, the discovered one achieves lower cost under three different λ settings. In other words, a better tradeoff result can be found using the same search strategy in the fixed SS, which means the search space profiling method enables the search process to converge to a better result by pruning away redundant information in SS while preserving the useful information of the original SS [166].

4.4 SUMMARY

In this chapter, we introduce the hardware design of our ISA-based CNN accelerator, which is the winning solution for the DAC-SDC 2018 challenge. Then, we introduce our pruning and quantization practices in DAC-SDC 2018, LPCVC 2019/2020 solutions. Finally, we introduce our NAS framework together with a case study of running hardware-aware NAS on Xilinx DPU.

CHAPTER 5

Progressive Automatic Design of Search Space for One-Shot Neural Architecture Search

Xin Xia

Bytedance Inc

Xuefeng Xiao

Bytedance Inc

Xing Wang

Bytedance AI Lab

CONTENTS

DOI: 10.1201/9781003162810-5

5.1   ABSTRACT

Neural architecture search (NAS) has attracted growing interest. To reduce the search cost, recent work has explored weight sharing across models and made major progress in One-Shot NAS. However, it has been observed that a model with higher one-shot model accuracy does not necessarily perform better when stand-alone trained. To address this issue, in this paper, we propose **P**rogressive **A**utomatic **D**esign of search space, named PAD-NAS. Unlike previous approaches where the same operation search space is shared by all the layers in the supernet, we formulate a progressive search strategy based on operation pruning and build a layer-wise operation search space. In this way, PAD-NAS can automatically design the operations for each layer. During the search, we also take the hardware platform constraints into consideration for efficient neural network model deployment. Extensive experiments on ImageNet show that our method can achieve state-of-the-art performance.

5.2   INTRODUCTION

Neural Architecture Search (NAS) has received increasing attention in both industry and academia, and demonstrated much success in various computer vision tasks, such as image recognition [167, 168, 169, 170, 171, 172], object detection [173, 174], and image segmentation [175, 176, 177]. Early work [171, 172] is mainly built on top of reinforcement learning (RL) [178], where tremendous amount of time is needed to evaluate candidate models by training them from scratch. In [179], the authors use evolutionary algorithm (EA) and achieve comparable result to RL. Recently, more and more researchers have adopted weight sharing approaches [180, 169, 165, 181, 182] to reduce the computation. They utilize an over-parameterized network, which is defined to subsume all architectures and needs to be trained only once.

Weight-sharing approaches can be mainly divided into two categories. In the first category, researchers use a continuous relaxation of search space, e.g., ProxylessNAS [183], DARTS [169], and FBNet [184]. The architecture distribution is continuously parameterized. Supernet training and architecture search are deeply coupled into single stage and jointly optimized by gradient-based methods. Deep coupling between the architecture parameters and supernet weights introduces bias and instability to the search process. One-Shot NAS, e.g., SPOS [167], SMASH [180], and others [185, 186], belongs to the other category. The optimization of the supernet weights and architecture parameters are decoupled into two sequential steps. The fairness among all architectures is ensured by sampling architecture or dropping out operators uniformly. During the architecture search, the validation accuracy of a model is predicted by inheriting the weights from the trained supernet. Unfortunately, as stated in [187, 185, 186], weight coupling issue exists in one-shot methods that one-shot model accuracy cannot truly reflect the relative performance of architectures.

In this paper, we aim to address weight coupling issue in one-shot approaches from the view of operation search space. Current one-shot approaches [185, 180, 186, 167] use the same operation search space for all the layers. However, in practice, we observe some operations will never be selected by certain layers in the final architectures. The reason lies in that sub-networks that contain these redundant operations either violate the hardware platform constraints, or perform poorly on the validation dataset and are excluded during the architecture search step. So a natural question arises: if an operation will never be selected, why spend the effort on training it at the very beginning? Keeping these operations will degrade the performance, since the more operations in the supernet, the more severe the interference between the operations. Thus, an effective way to mitigate the weight coupling is to remove these operations before training. A follow-up question is how we can identify if an operations is redundant or not, before the training of a supernet.

To answer the questions above, we propose a simple yet effective approach named PAD-NAS. Our algorithm can automatically design its own operations for each layer and build a layer-wise search space through a progressive search strategy. The flow of our algorithm is illustrated in Figure 5.2. In the first stage, we perform the training of an initial supernet, where the same operation search space is shared by all the layers. For the next $M-2$ stages, we start from the supernet coming from the previous stage, and estimate the operation probability distribution for each layer from the architectures that reside on the Pareto frontier [188]. Next, we prune the operations layer by layer and remove the ones whose probabilities are below certain threshold to build the pruned supernet for the next stage. Finally, we finetune the pruned supernet. We repeat the process above until stopping criterion is satisfied. In the final stage, we search the architectures from the supernet and return the architectures with highest accuracy.

The effectiveness of PAD-NAS is demonstrated on ImageNet. We name the models discovered by PAD-NAS as PAD-NASs. PAD-NASs

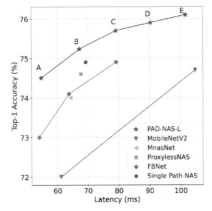

Figure 5.1 The trade-off between Pixel 3 latency and top-1 ImageNet accuracy. All models use the input resolution 224 and the latency is measured on a single large core of the same device using TFLite[189].

achieves state-of-the-art (SOTA) performance on ImageNet and outperforms efficient networks designed manually and automatically, such as MobileNetV2 [190], FBNet [184], ProxylessNAS [183], and SPOS [167]. As shown in Figure 5.1, PAD-NAS-L-A achieves 74.5% top-1 accuracy

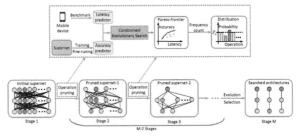

Figure 5.2 The overall framework of PAD-NAS. **Bottom:** PAD-NAS is divided into M stages: the first stage is initial supernet training, and the next $M-2$ stages repeat the process of pruning the operations and training the pruned supernet. The last stage is the architecture evolution and selection. **Top:** The procedures of operation pruning. Supernet is used as an accuracy predictor after training, and a latency predictor is built for target mobile devices. The constrained evolutionary search algorithm is used to evolve the network architecture, and the probability distribution of the operations for each layer is estimated from Pareto frontier.

with 271M FLOPs and 54.7 ms latency on an Pixel 3 phone, 1.5% higher than FBNet-A [184]. Top-1 accuracy of PAD-NAS-L-B is 0.6% higher than Proxyless-R [183], while PAD-NAS-L-C achieves 0.8% absolute gain in top-1 accuracy compared with FBNet-C [184].

Our main contributions are summarized as follows:

- We present PAD-NAS, an efficient neural architecture search framework for One-Shot NAS. It automatically designs the search space for each layer through a progressive search strategy.

- we propose a new search space pruning method for One-Shot NAS. We estimate the probability distribution of operations for each layer by counting their frequencies in the searched architectures and remove the one whose probabilities are below certain threshold.

- Extensive experiments demonstrate the advantage of PAD-NAS. It achieves SOTA performance on ImageNet and significantly mitigates weight coupling.

5.3 RELATED WORK

Recently, in order to reduce the computation cost of NAS, some researchers propose weight sharing approaches to speed up the architecture search, such as ENAS [165], DARTS [169], and One-Shot NAS [185, 167]. All sub-network architectures inherit the weights from the trained supernet without training from scratch. DARTS softens the discrete search space into a continuous search space and directly optimizes it by the gradient method. [180] and [185] propose a method which decouples supernet training and architecture search into two sequential stages, including supernet training and architecture search. However, due to the weight coupling in the supernet, the accuracy of the supernet prediction has a certain deviation from the ground truth, which result in inaccurate ranking of the architectures. The authors of SPOS [167] further propose uniform sampling and single-path training to overcome weight coupling in One-Shot NAS. ProxylessNAS [183] binarize entire paths and keep only one path when training the over-parameterized supernet to reduce memory footprint. FBNet [184] use a proxy dataset (subset of ImageNet) to train the continuously parameterized architecture distribution. FairNas [186] is based on [167] and proposes a new fairness sampling and training strategy for supernet training. GreedyNAS [191] propose to greedily focus on training potentially-good paths, which implemented by multi-path sampling strategy with rejection. ABS [192] propose an angle-based search space shrinking method by adopting a novel angle-based metric to evaluate capability of child models and guide the shrinking procedure. RegNet [193] propose to design network design spaces, which parametrize populations of networks, and present a new network design paradigm.

Different from the methods above, we propose a progressive search strategy to reduce the number of operations in each stage and build a layer-wise operation search space for One-Shot NAS automatically. Our work is also closely related to P-DARTS[194] and HM-NAS[195]. [194] proposes a progressive version of DARTS to bridge the depth gap between search and evaluation scenarios. Its core idea is to gradually increase the depth of candidate architectures. [195] incorporates a multi-level architecture encoding scheme to enable an architecture candidate to have arbitrary numbers of edges and operations with different importance. [194, 195] belong to differential NAS, which prune the operations according to the value of architecture parameters. However, there are no architecture parameters in One-Shot NAS.

Another relevant topic is network pruning [196, 197, 198] that aim to reduce the network complexity by removing redundant, non-informative connections in a pre-trained network. Similar to this work, we start from an over-parameterized supernet and then prune the redundant operations to get the optimized architecture. The distinction is that they aim to prune the connections in a pre-trained network, while we focus on improving One-Shot NAS performance through operation pruning.

5.4 METHOD

5.4.1 Problem Formulation and Motivation

One-Shot NAS [185, 180] contains two stages. The first stage is supernet training, which is formulated as:

$$W_S = \arg\min_W Loss_{train}(N(S, W)), \qquad (5.1)$$

where $Loss_{train}(\cdot)$ is the loss function on the training set, $N(S, W)$ is the supernet, represented by the search space S and its weights W, and W_S are the learned weights of the supernet.

The second stage is the architecture search. It aims to find the architectures from the supernet that has the best one-shot accuracy on the validation set under mobile latency constrain, expressed as:

$$s^* = \arg\max_{s \in S} Acc_{val}(N(s, W_S(s)))$$

$$s.t. \text{Lat}_{min} \le \text{Latency}(s^*) \le \text{Lat}_{max}, \tag{5.2}$$

where Lat_{min} and Lat_{max} are lower and upper mobile latency constraint. Each sampled sub-network inherits its weights from W_S as $W_S(s)$. Therefore, one-shot accuracy $Acc_{val}(\cdot)$ only requires inference on validation dataset. This completes the search phase. In the evaluation phase, without violating the mobile latency constraint, the architectures with the highest one-shot accuracy will be selected to do the stand-alone training from scratch.

However, as stated by many researchers [185, 180, 167], weight coupling issue exists in One-Shot NAS, which results in ranking inconsistency between supernet predicted accuracies and that of ground-truth ones by stand-alone training from scratch. Architectures with a higher supernet predicted accuracy during the search phase may perform worse in the evaluation phase.

In this paper, we aim to mitigate weight coupling and improve the performance of One-Shot NAS. We start with the analysis of operation search space. In practice, we observe that redundant operations exist in the supernet. These redundant operations will never be selected by the searched architectures. As shown in Figure 5.3, where x-axis represents the layer name and y-axis represents the operation name, the operation distribution in each layer is sparse. For example, in layer1, the distribution concentrates on the operation IBconv_K3_E1, while all the other operations never appear in this layer. The reason lies in the architectures that contain the redundant operations either violate the constrains in equation 5.2, or has a low validation accuracy and are excluded during the architecture search.

Motivated by [196] where uncritical connections in deep networks can be removed without affecting the performance, we claim that existing operation search space is redundant, and needs to be pruned. However, different from network pruning [196], whose main purpose is to decrease the network complexity, our main goal is to mitigate weight coupling and improve One-Shot NAS performance by pruning unnecessary operations. These operations will hurt the supernet training since more operations means more intense interference between the operations in the supernet,

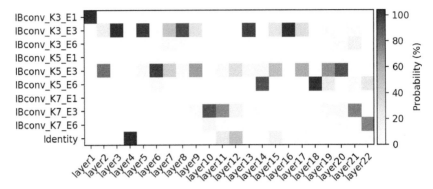

Figure 5.3 Probability distribution of each layer's operation in the searched architectures.

resulting in more severe ranking inconsistency between supernet predicted accuracy and ground-truth one. However, large number of operations is demanded for NAS to discover promising models. Due to this conflict, we present PAD-NAS which follows a coarse-to-fine manner to refine the operation search by progressively pruning the operations and build the layer-wise operation search space automatically.

5.4.2 Progressive Automatic Design of Search Space

The whole process of PAD-NAS is elaborated in Algorithm 2. $s^{i,j}$ denotes the ith operation for layer j in the supernet, $p^{i,j}$ represents the probability of operation $s^{i,j}$. We start with an initial operation search space, which is shared by all the layers in the supernet. Next, we construct the supernet following Table 5.1 with this initial search space and train it with path-wise manner proposed in [167]. This is the first stage of PAD-NAS, served as the initialization. For the next $M-2$ stages, we first use constrained evolutionary algorithm to search the architectures that belong to Pareto frontier on the input supernet. The output supernet of the previous stage is fed as the input of the current stage. Then, we estimate the probability distribution of each operation layer-by-layer from Pareto frontier and remove the operations whose probabilities are below certain threshold P_{th}. To this end, we build the layer-wise operation search space for the current stage. The last part is to finetune the pruned supernet and feed it into the next stage. The process above is repeated until we reach the targeted number of search stages. When it comes to the last stage of PAD-NAS, constrained evolutionary algorithm is used to search the architectures on

TABLE 5.1 Architecture of the Supernet. Output Channels Denotes the Output Channel Number of a Block. Repeat Denotes the Number of Blocks in a Stage. Stride Denotes the Stride of the First Block in a Stage. The Output Channel Size of the First Three Stage Is 32-16-32 in Basic Search Space and 16-16-24 in Large Search Space

Input shape	Block	Output channels	Repeat	Stride
$224^2 \times 3$	3×3 conv	32 (16)	1	2
$112^2 \times 32(16)$	SBS	16 (16)	1	1
$112^2 \times 16(16)$	SBS	32 (24)	4	2
$56^2 \times 32(24)$	SBS	40	4	2
$28^2 \times 40$	SBS	80	4	2
$14^2 \times 80$	SBS	96	4	1
$14^2 \times 96$	SBS	192	4	2
$7^2 \times 192$	SBS	320	1	1
$7^2 \times 320$	1×1 conv	1280	1	1
$7^2 \times 320$	global avgpool	-	1	1
$7^2 \times 1280$	fc	1000	1	-

the supernet, which is the output of the $(M-1)$th stage, and return the architectures with highest accuracy. We list more details as follows.

Initial Supernet Training. For the training of initial supernet, we adopt a path-wise [167] manner, where supernet training and architecture search are decoupled into two sequential steps, to ensure the trained supernet can reflect the relative performance of sub-networks. However, since weight coupling is inevitable in weight-sharing approaches [167, 169], supernet predicted accuracy can only coarsely indicate the relative ranking of sub-networks. The trained initial supernet is the starting point of PAD-NAS and served as the initialization.

Constrained Evolutionary Search. After initial supernet training, we need to search the architectures. Evolutionary search performs better than random search in previous one-shot works [185, 180, 167]. The evolutionary algorithm is flexible in dealing with hardware platform constraints in equation 5.2, since the mutation and crossover processes can be manipulated to generate proper candidates to satisfy the constraints. We aim to estimate the probability distribution of operations by counting their frequencies in the searched architectures. The estimated operation

---

Algorithm 2 PAD-NAS

Input: operation search space $S = \{s^{i,j}\}$, probability threshold P_{th}, the number of progressive search stages M, latency constraints $Lat_{\min}$ and $Lat_{\max}$, and the number of layers J in the supernet

 1: Initial supernet training: construct and train supernet $N(S, W)$
 2: **for** $k = 2$ to $M - 1$ **do**
 3:     Architecture search: search the architectures on the supernet $N(S, W)$
 4:     **for** $j = 1$ to J **do**
 5:         Distribution estimation: count and normalize the frequency of $s^{i,j}$ to get $p^{i,j}$
 6:     **end for**
 7:     Pruning: remove $s^{i,j}$ from S if $p^{i,j} \leq P_{th}$
 8:     Pruned supernet training: construct and finetune supernet $N(S, W)$
 9: **end for**
10: Repeat step 3
11: **return** the architectures with highest accuracy

---

distribution is unstable and exhibits a large variance if the evolutionary search algorithm in [167] is used.

To deal with this problem, we propose Constrained Evolutionary Search (CES), which is built on top of Non-Dominated Sorting Genetic Algorithm II (NSGA-II) [199]. Note that the authors of [200] are the first to introduce NSGA-II into NAS. Architecture search can be formulated as a multi-objective optimization problem to balance between mobile latency and architecture accuracy. We add latency constraint into the iteration of NSGA-II to formulate CES. More specifically, we discard the architectures that violate the constraint in the crossover and mutation of CES. This change gives us better result. Different from the evolutionary search in [167] that architectures with poor accuracy are discarded in each iteration, no architectures are discarded but all of them are sorted by its latency and accuracy in each iteration. The estimated operation distribution by CES is stable and its corresponding variance is small.

During the search, the architecture accuracy is predicted by the supernet through weight inheritance, and its corresponding latency on mobile devices is estimated by the latency predictor. We build a lookup table of mobile latency for operations in the supernet and the architecture latency is predicted by summing up the latency of all of its operations.

Operation Pruning. We prune the operations based on its corresponding probability distribution. Since there is no way to get the ground-truth distribution, the sampling method is applied here. We count the frequency of operations in the architectures after architecture search and normalize it to get the approximate probability distribution. However, not all the architectures are equally important. Only the architectures that belong to Pareto frontier are used. These architectures dominate all the other ones. Here, we say one architecture dominates the others if and only if its latency is no bigger than others while its accuracy is no lower than others. Next, we remove the operations whose probability are below pre-defined threshold.

During CES, the architecture accuracy is predicted by supernet, which introduces error into the sorting part of CES, due to weight coupling. This implies Pareto frontier returned by CES is noisy, which may cause inaccurate estimation of operation probability distribution. To mitigate this effect, we count the frequency of operations from the architectures whose nondomination rank [199] is smaller than 10, where the nondomination rank of Pareto frontier is 1.

Pruned Supernet Training. After operation pruning, we get the pruned supernet. The training of pruned supernet is the same as initial supernet training. The only distinction is weights are randomly initialized in initial supernet training, while weights in the pruned supernet are inherited from the supernet in the previous stage. We only need to fine-tune the pruned supernet instead of training it from scratch. This reduces the search cost without affecting the performance.

5.5 EXPERIMENTS

5.5.1 Dataset and Implement Details

Dataset. Throughout the paper, we use the ILSVRC2012 dataset [201]. To be consistent with previous works, 50 images are randomly sampled from each class of the training set, and a total of 50,000 images are used as the validation set. The original validation set is used as the test set.

Search Space. For fair comparison, our basic search space is the same as ProxylessNAS [183], similar to FBNet [184], and shown in the left column of Table 5.2. The optional operations in each SBS are 6 types of operation (with a kernel size 3×3 , 5×5 or 7×7 and an expansion factor of 3 or 6), plus one identity operation. In addition, the expansion

TABLE 5.2 Operations Table. IBConv_KX_EY Represents The Specific
Operator IBConv with Expansion Y and Kernel size X. IBConv Denotes
Inverted Bottleneck in MobilenetV2

Basic search space	Operators exclusively in large search space	
IBConv_K3_E3	IBConv_K3_E1	IBConv_K5_E4
IBConv_K3_E6	IBConv_K3_E2	IBConv_K5_E5
IBConv_K5_E3	IBConv_K3_E4	IBConv_K7_E1
IBConv_K5_E6	IBConv_K3_E5	IBConv_K7_E2
IBConv_K7_E3	IBConv_K5_E1	IBConv_K7_E4
IBConv_K7_E6	IBConv_K5_E2	IBConv_K7_E5
Identity		

factor in the first SBS block is fixed as 1, and the identity operation is
forbidden in the first layer of every block. The basic search space size is
$3 \times 6^6 \times 7^{15} \approx 6.64 \times 10^{17}$.

Moreover, we enlarge the search space by adding 12 more operations and formulate the large search space, as shown in the right column of right column of Table 5.2. We do not add any special operations, but a fine-grained version of the basic search space. In the basic search space, the expansion is either 3 or 6, while in the large search space, the expansion ranges from 1 to 6. The large search space size is $3 \times 18^6 \times 19^{15} \approx 1.55 \times 10^{27}$. The primary reason to use the enlarged search space in this paper is to show PAD-NAS can automatically design the search space, mitigate the weight coupling issue, and bring better results, compared to the small search space.

Implementation Details. For the stand-alone training of the searched architecture (after evolutionary search) from scratch, we use the same settings (including data augmentation strategy, learning rate schedule, dropout rate, etc.) as [167]. The network weights are optimized by momentum SGD, with an initial learning rate of 0.5, a momentum of 0.9, and a weight decay of 4×10^{-5}. A linear learning rate decay strategy is applied for 240 epochs and the batch size is 1024. It is worth noting that we have not used neither squeeze-and-excitation [202] nor swish activation functions [203], which are some tricks, not related to the weight coupling problem in this paper. Extra data augmentations such as mixup [204] and autoaugment [205] are not used as well.

For the training of the supernet, we first train the initial supernet for 120 epochs with an initial learning rate of 0.5. All other training settings are the same as stand-alone training. Next, we apply operation pruning for the first time and finetune the pruned supernet for 80 epochs with an initial learning rate of 0.1. Finally, we prune the operations for the second time and finetune the pruned supernet-2 for another 40 epochs with an initial learning rate of 0.1. The total number of supernet training epochs is 240, exactly the same as stand-alone training.

For pruning the operation of the supernet, the pruning threshold P_{th} is set to 1% for each layer. And, the latency lower Lat_{min} and upper bound Lat_{max} is set to 60ms and 70ms for basic search space and 50ms and 100ms for large search space. The latency is measured on Pixel 3 using TFLite. The number of progressive search stages M is set to 4. In the constrained evolution search, the initial population size is set to 64, the max evolution iteration is set to 40, and the polynomial mutation and two-point crossover is adopted.

Notations. Two baseline methods are frequently mentioned in the following part. In the first baseline method, we fully train the initial supernet with single path and uniform sampling in [167] and search the architectures with CES proposed in this paper. We name it I-Supernet (Initial Supernet). I-Supernet is a special case of PAD-NAS with $M = 2$ in Figure 5.2 and Algorithm 2. There is no operation pruning inside. The second baseline method is called P-Supernet (Pruned Supernet). P-Supernet corresponds to PAD-NAS with $M = 3$, where we only prune the operations once. PAD-NAS denotes our main result, corresponding to PAD-NAS with $M = 4$, where we prune the operations twice. I-Supernet-S, P-Supernet-S, and PAD-NAS-S denote the corresponding method on the basic search space while I-Supernet-L, P-Supernet-L, and PAD-NAS-L represent the one on the large search space.

To make a fair comparison, for the three methods above, all the experimental setups are exactly the same, e.g., the total training epochs, learning rate, and hyperparameters in the CES. The only difference between these three methods is the number of progressive search stages M.

5.5.2 Comparison with State-of-the-art Methods

ImageNet Results. The experimental results are shown in Table 5.3. We compare our searched architectures with SOTA efficient architectures

TABLE 5.3 Comparison with the State-of-the-arts on ImageNet under Mobile Setting. The FLOPs and Latency Are Calculated with 224×224 Input

No.	Model	Params	FLOPs	Latency	Top-1 Acc (%)
1	MobileNetV1 [206]	4.2M	569M	89.97ms	70.6
2	MobileNetV2 [190]	3.4M	300M	61.13ms	72.0
3	MobileNetV2 (×1.4) [190]	6.9M	585M	104.65ms	74.7
4	ShuffleNetV2 [207]	3.4M	299M	-	72.6
5	NASNet-mobile [172]	5.3M	564M	144.05ms	74.0
6	DARTS [169]	4.7M	574M	-	73.3
7	P-DARTS [208]	4.9M	574M	-	75.6
8	SPOS [167]	3.5M	319M	-	74.3
9	REGNETX-400MF [193]	5.2M	400M	-	72.7
10	MnasNet [209]	4.2M	317M	64.24ms	74.0
11	DenseNAS-B [210]	-	314M	66.62ms	74.6
12	ABS [192]	3.8M	325M	68.21ms	74.4
13	Proxyless-R (mobile) [183]	4.1M	320M	67.66ms	74.6
14	FBNet-A [184]	4.3M	249M	54.05ms	73.0
15	FBNet-B [184]	4.5M	295M	63.79ms	74.1
16	FBNet-C [184]	5.5M	375M	79.18ms	74.9
17	Single Path NAS [211]	4.4M	334M	68.67ms	74.9
18	FairNAS-C [186]	4.4M	321M	67.32ms	74.7
19	PAD-NAS-S-A	3.7M	310M	61.31ms	74.6
20	PAD-NAS-S-B	4.1M	315M	64.71ms	74.9
21	PAD-NAS-S-C	4.2M	347M	69.85ms	75.2
22	PAD-NAS-L-A	4.0M	271M	54.72ms	74.5
23	PAD-NAS-L-B	4.2M	334M	67.23ms	75.2
24	PAD-NAS-L-C	4.6M	385M	79.08ms	75.7
25	PAD-NAS-L-D	4.6M	401M	90.21ms	75.9
26	PAD-NAS-L-E	4.7M	444M	101.53ms	**76.1**

both designed automatically and manually. The primary metrics we care about are top-1 accuracy on the test set and mobile latency. If the latency is not available, the FLOPs is used as the secondary efficiency metric.

For the basic search space, we present three models PAD-NAS-S-A, PAD-NAS-S-B, and PAD-NAS-S-C according to their latency. Compared to manually designed network, as shown the first four rows in Table 5.3, it can reach significantly higher accuracy with lower latency(or FLOPs). In particular, PAD-NAS-S-C achieves 75.2% with 69.85ms latency, which

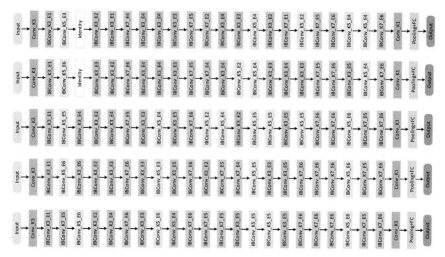

Figure 5.4 Architectures of PAD-NAS-L-A,B,C,D,E in Figure 5.1 (from top to bottom).

surpasses MobileNetV2(1.4X) top-1 accuracy by 0.4% while much faster (from 104.65ms to 69.85ms). Comapared with Proxyless-R [183] which shares exactly the same search space, ours PAD-NAS-S-B improves the accuracy by 0.3% but still faster (2.95ms less for processing single image on Pixel 3). Finally, in comparison with other SOTA NAS algorithms, the three models all achieve higher accuracy with lower latency.

For the large search space, we list five models according to their latency, named PAD-NAS-L-A, PAD-NAS-L-B, PAD-NAS-L-C, PAD-NAS-L-D, and PAD-NAS-L-E, whose detailed architectures are illustrated in Figure 5.4. The top-1 accuracy of PAD-NAS-L-A is 74.5%, which is 1.5% higher than its counterpart FBNet-A [184]. PAD-NAS-L-B achieves 75.2% top-1 accuracy, which is 0.6% higher than Proxyless-R [183], and exhibits the same accuracy as PAD-NAS-S-C with lower latency. PAD-NAS-L-C achieves 75.7% top-1 accuracy, 0.8% higher than FBNet-C [184]. These results demonstrate our proposed PAD-NAS can be applied to larger search space, and achieve even better performance.

Transfer Learning Results We transfer the backbone searched by PAD-NAS to semantic segmentation task on cityscapes. The results are shown in Table 5.4. The result for MobilenetV2 is reported in [212]. We can see PAD-NAS model outperforms efficient networks automatically and manually

TABLE 5.4 Semantic Segmentation Results on Cityscapes val Set. The FLOPs and Latency Are Calculated with 512×1024 Input

Model	FLOPs	Latency	mIOU(%)
MobileNetV2	12.6B	773ms	72.7
Proxyless-R (mobile)	13.5B	831ms	73.1
PAD-NAS-L-A	11.3B	680ms	73.3

5.5.3 Automatically Designed Search Space

Impact of Search Space. Theoretically, NAS should give better result when larger search space is used. However, as stated in Table 5.5, I-Supernet-L achieves even lower stand-alone top-1 accuracy than I-Supernet-S. There exists a gap between theory and practice in current one-shot approaches. More operations brings more intense interference between operations during the training of supernet, which makes the weight coupling more severe. Thanks to the proposed progressive search strategy, PAD-NAS-L performs better than PAD-NAS-S, which implies PAD-NAS can bridge this gap. Besides, we can see top-1 accuracy achieved by PAD-NAS-L is 1.02% higher than I-Supernet-L while this number is only 0.42% for basic search space. This shows the advantage of PAD-NAS is more obvious in larger search space. We can also notice big gap exists between supernet accuracy and stand-alone one in the baseline method I-Supernet, as large as 21.99%, while this gap is much lower in PAD-NAS, only 2.94%.

We also present the corresponding supernet accuracy curve on validation dataset for PAD-NAS and I-Supernet in Figure 5.5. The supernet accuracy of I-Supernet-L is always below I-Supernet-S. Similar trend also exists in the first 120 epochs of PAD-NAS. However, once operation

TABLE 5.5 Performance Comparison between I-Supernet and PAD-NAS on Different Search Spaces. Stand-alone Top-1 Acc Means the Best Top-1 Accuracy of Searched Architecture when Stand-alone Trained from Scratch under the Same Latency(69ms)

Algorithm	Supernet Top-1 Acc (%)	Stand-alone Top-1 Acc (%)
I-Supernet-L	52.32±0.51	74.31±0.14
PAD-NAS-L	72.39±0.08	75.33±0.09
I-Supernet-S	65.51±0.13	74.70±0.09
PAD-NAS-S	71.42±0.07	75.12±0.08

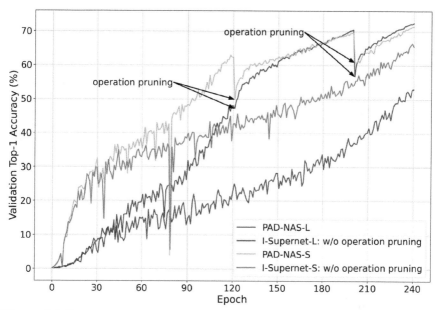

Figure 5.5 Supernet accuracy curve comparison between I-Supernet and PAD-NAS.

pruning is introduced, the gap between PAD-NAS-L and PAD-NAS-S is becoming much smaller. And PAD-NAS-L eventually surpasses PAD-NAS-S at epoch 240 after we prune the operations twice. Here, we repeat the experiments 5 times with different random seeds.

Random Search Baseline. Under the same latency constrain, we randomly select 5 models from the basic small and large search space and denote their results as Random-S and Random-L. Moreover, we also randomly select 5 models from the Auto-Designed search space (AD) by PAD-NAS, whose results are named as Random-S-AD and Random-L-AD. From the results of Table 5.6, the random search results corresponding to

TABLE 5.6 Random Search Results under the Same Latency(69ms)

Algorithms	Top-1 Acc(%)
Random-L	72.71±0.46
Random-L-AD	74.91±0.11
Random-S	73.88±0.19
Random-S-AD	74.72±0.10

AD are greatly improved compared to the basic small search space(74.91% vs. 72.71% and 74.72% vs. 73.88%), which indicates that PAD-NAS can automatically design the search space while maintaining model diversity. Similar findings are also mentioned in the recent RegNet[193].

Kendall Rank Analysis. The ranking consistency of stand-alone and one-shot model accuracy is an important problem in One-Shot NAS algorithm. It indicates whether supernet can reflect the relative performance of models. Due to high training cost, we sample 30 models at approximately equal distances on the Pareto frontier and do the stand-alone training from scratch to get the ranking, as shown in Figure 5.6. We can observe in PAD-NAS-L, one-shot accuracy is more relevant to stand-alone accuracy, compared to the baseline method I-Supernet-L where there is no operation pruning involved.

We also use Kendall's rank correlation coefficient τ [213] to measure this consistency. τ ranges from -1 to 1, meaning the rankings are totally reversed or completely preserved, whereas 0 means there is no correlation at all. As shown in Table 5.7, as the number of progressive search stage M increases, the spaces size is reduced dramatically by orders-of-magnitude, and Kendall's ranking coefficient τ has significantly increased from 0.4977 to 0.8879. Here, we repeat the experiments 3 times with different random seeds.

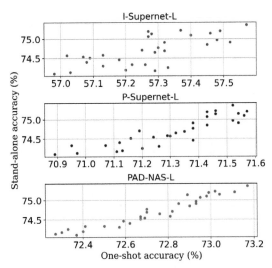

Figure 5.6 Top-1 accuracy of 30 stand-alone trained architectures vs. one-shot models.

TABLE 5.7 Ranking Consistency Comparison with Baseline Methods. Size Represents Search Space Size

Algorithm	τ	Size	M
I-Supernet-L	0.4977 ± 0.0573	10^{27}	2
P-Supernet-L	0.7451 ± 0.0332	10^{15}	3
PAD-NAS-L	0.8879 ± 0.0201	10^{9}	4

5.5.4 Ablation Studies

Search Cost Analysis. Evolution search is performed twice more than the baseline method, which costs 7 more hours. All experiments are trained with 8 Tesla V100 GPUs. The analysis is shown in Table 5.8.

Impact of Evolutionary Algorithm. We study the impact of different evolutionary algorithms on the estimated probability distribution of operations. Two algorithms are compared here: one is our CES while the other is the one used in SPOS [167]. Each algorithm is repeated 5 times with different random seeds to get the mean and variance of estimated distribution. As illustrated in Figure 5.7, the result of our CES exhibits a much lower variance than SPOS [167], which indicates CES gives us more stable result. Besides, the distribution corresponding to CES is sparser than SPOS, which makes operation pruning more efficient.

MobileNetV3-based Comparisons. The experimental results with MobileNetV3-based search space (PAD-NAS-V3) are shown in Table 5.9, where SE and swish are used. We can see PAD-NAS model achieves better performance.

TABLE 5.8 The Comparison of Search Cost

Method	Baseline[167]	PAD-NAS
Supernet training	32hours	39hours
Evolution search	3.5hours	3.5hours
Retrain	24hours	24hours
Total	59.5hours	66.5hours

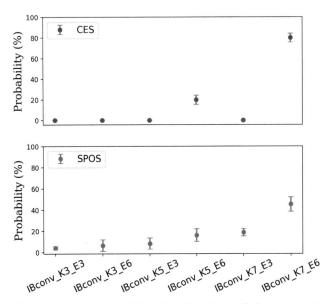

Figure 5.7 Estimated probability distribution of the operations in the last layer by CES and SPOS [167], respectively.

TABLE 5.9 Comparison with the SOTA under MobileNetV3-based Search Space

Model	FLOPs	Latency	Top-1 Acc (%)
MobileNetV3 (×0.75) [212]	155M	40.4ms	73.3
MobileNetV3[212]	219M	52.6ms	75.2
GreedyNAS-C[191]	284M	58.1ms	76.2
PAD-NAS-V3-A	182M	40.8ms	75.0
PAD-NAS-V3-B	222M	49.2ms	75.9
PAD-NAS-V3-C	266M	55.1ms	**76.5**

5.6 CONCLUSION

In this paper, we present PAD-NAS, a progressive search strategy for One-Shot NAS. It automatically designs a layer-wise operation search space through operation pruning. PAD-NAS mitigates weight coupling issue and significantly improves the ranking consistency between supernet predicted accuracy and stand-alone trained accuracy. Experimental results demonstrate the effectiveness of our method, which achieves SOTA performance on ImageNet.

CHAPTER 6

Fast Adjustable Threshold for Uniform Neural Network Quantization

Alexander Goncharenko

Novosibirsk State University, ENOT.AI

Andrey Denisov

Expasoft

Sergey Alyamkin

ENOT.AI

CONTENTS

DOI: 10.1201/9781003162810-6

A neural network quantization is a highly desired procedure to perform before running neural networks on mobile devices. Quantization without fine-tuning leads to accuracy drop of the model, whereas commonly used quantization aware training is done on the full set of the labeled data and therefore is both time- and resource-consuming. Real-life applications require simplification and acceleration of quantization procedure that will maintain the accuracy of full-precision neural network, especially for modern mobile neural network architectures like Mobilenet-v1, MobileNet-v2, and MNAS.

Here we present two methods to significantly optimize the training with quantization procedure. The first one is introducing the trained scale factors for quantization thresholds that are separate for each filter. The second one is based on mutual rescaling of consequent depth-wise separable convolution and convolution layers. Using the proposed techniques, we quantize the modern mobile architectures of neural networks with the set of train data of only $\sim 10\%$ of the total ImageNet 2012 sample. Such reduction of train dataset size and small number of trainable parameters allow to fine-tune the network for several hours while maintaining the high accuracy of quantized model (accuracy drop was less than 0.5%).

6.1 INTRODUCTION

Mobile neural network architectures [214, 215, 216] allow running AI solutions on mobile devices due to the small size of models, low memory consumption, and high processing speed while providing a relatively high level of accuracy in image recognition tasks. In spite of their high computational efficiency, these networks continuously undergo further optimization to meet the requirements of edge devices. One of the promising optimization directions is to use quantization to int8, which is natively supported by mobile processors, either with or without training. Both methods have certain advantages and disadvantages.

Quantization of the neural network without training is a fast process as in this case a pre-trained model is used. However, the accuracy of the resultant network is particularly low compared to the one typically obtained in commonly used mobile architectures of neural networks [217]. On the other hand, quantization with training is a resource-intensive task which results in low applicability of this approach.

Current article suggests a method which allows speeding up the procedure of training with quantization and at the same time preserves a high accuracy of results for 8-bit discretization. The developed method was winning solution (the first place) of Low-Power Image Recognition Challenge II in two tracks: image classification with low latency (within 30 ms) and image classification for interaction (up to 100 ms).

6.2 RELATED WORK

In general case the procedure of neural network quantization implies discretization of weights and input values of each layer. Mapping from the space of float32 values to the space of signed integer values with n significant digits is defined by the following formulae:

$$S_w = \frac{2^n - 1}{T_w} \tag{6.1}$$

$$T_w = max|W| \tag{6.2}$$

$$W_{int} = \lfloor S_w \cdot W \rceil \tag{6.3}$$

$$W_q = clip(W_{int}, -(2^{n-1} - 1), 2^{n-1} - 1) = \\ = min(max(W_{int}, -(2^{n-1} - 1)), 2^{n-1} - 1) \tag{6.4}$$

Here $\lfloor \rceil$ is rounding to the nearest integer number, W is the weights of some layer of neural network, T_w is the quantization threshold for weights, max calculates the maximum value across all axes of the tensor, max calculates the minimum value across all axes of the tensor and W_q is quantized weights. Input values can be quantized both to signed and unsigned integer numbers depending on the activation function on the previous layer.

$$S_I = \frac{2^n - 1}{T_I} \qquad (6.5)$$

$$T_I = max|I| \qquad (6.6)$$

$$I_{int} = \lfloor S_I \cdot I \rceil \qquad (6.7)$$

$$I_q^{signed} = clip(I_{int}, -(2^{n-1} - 1), 2^{n-1} - 1) \qquad (6.8)$$

$$I_q^{unsigned} = clip(I_{int}, 0, 2^n - 1) \qquad (6.9)$$

Here $\lfloor \rceil$ is rounding to the nearest integer number, I is the input tensor of some layer of neural network, T_I is the quantization threshold for input tensor, max calculates the maximum value across all axes of the tensor and max calculates the minimum value across all axes of the tensor. I_q^{signed}, $I_q^{unsigned}$ are signed and unsigned quantized inputs correspondingly. After all inputs and weights of the neural network are quantized, the procedure of convolution is performed in a usual way. It is necessary to mention that the result of operation must be in higher bit capacity than operands. For example, in [218] authors use a scheme where weights and activations are quantized to 8-bits while accumulators are 32-bit values.

It is important to note that we implicitly assume the distribution of weights and activations is symmetric around zero in the quantization approach described above. We will call the quantization of such a distributions *symmetric*. The asymmetric case (when the left or right border of the distribution is shifted from zero) will be described in Section 6.3.1.4 ("Training of asymmetric thresholds").

Potentially quantization threshold can be calculated on the fly, which, however, can significantly slow down the processing speed on a device with low system resources. It is one of the reasons why quantization thresholds are usually calculated beforehand in calibration procedure. A set of data is provided to the network input to find desired thresholds (in the example above—the maximum absolute value) of each layer. Calibration dataset contains the most typical data for the certain network and this data does not have to be labeled according to procedure described above.

6.2.1 Quantization with Knowledge Distillation

Knowledge distillation method was proposed by G. Hinton [219] as an approach to neural network quality improvement. Its main idea is training of neural networks with the help of pre-trained network. In [220, 221] this method was successfully used in the following form: a full-precision model was used as a model-teacher, and quantized neural network—as a model-student. Such paradigm of learning gives not only a higher quality of the quantized network inference, but also allows reducing the bit capacity of quantized data while keeping an acceptable level of accuracy.

6.2.2 Quantization without Fine-tuning

Some frameworks allow using the quantization of neural networks without fine-tuning. The most known examples are TensorRT [222], Tensorflow [223], and Distiller framework from Nervana Systems [224]. However (at the moment of the research), in the last two models calculation of quantization coefficients is done on the fly, which can potentially slow down the operation speed of neural networks on mobile devices. In addition, to the best of our knowledge, TensorRT framework does not support quantization of neural networks with the architectures like MobileNet.

6.2.3 Quantization with Training/Fine-tuning

One of the main focus points of research publications over the last years is the development of methods that allow to minimize the accuracy drop after neural network quantization. The first results in this field were obtained in [225, 226, 227, 228]. The authors used the Straight Through Estimator (STE) [229] for training the weights of neural networks into 2 or 3 bit integer representation. Nevertheless, such networks had substantially lower accuracy than their full-precision analogs.

The most recent achievements in this field are presented in [230, 231] where the accuracy of trained models is almost the same as for original architectures. Moreover, in [231] the authors emphasize the importance of the quantized networks ensembling which can potentially be used for binary quantized networks. In [218] authors report the whole framework for modification of network architecture allowing further launch of learned quantized models on mobile devices.

In [232] the authors use the procedure of threshold training which is similar to the method suggested in our work. However, the reported approach has substantial shortcomings and cannot be used for fast conversion of pre-trained neural networks on mobile devices. First of all it has a requirement to train threshold on the full ImageNet dataset [233]. Besides, it has no examples demonstrating the accuracy of networks used as standards for mobile platforms.

In the current article we propose a novel approach to set the quantization threshold with fast fine-tuning procedure on a small set of unlabeled data that allows to overcome the main drawbacks of known methods. We demonstrate performance of our approach on modern mobile neural network architectures (MobileNet-v2, MNAS).

6.3 METHOD DESCRIPTION

Under certain conditions (see Figures 6.1–6.2) the processed model can significantly degrade during the quantization process. The presence of outliers for weights distribution shown in Figure 6.1 forces to choose a high value for thresholds that leads to accuracy degradation of quantized model.

Outliers can appear due to several reasons, namely specific features of calibration dataset such as class imbalance or non-typical input data. They also can be a natural feature of the neural network, that are, for example, weight outliers formed during training or reaction of some neurons on features with the maximum value.

Overall it is impossible to avoid outliers completely because they are closely associated with the fundamental features of neural networks. However, it is possible to find a compromise between the value of threshold and distortion of other values during quantization and get a better quality of the quantized neural network.

6.3.1 Quantization with Threshold Fine-tuning

6.3.1.1 Differentiable Quantization Threshold

In [226, 228, 229] it is shown that the Straight Through Estimator (STE) can be used to define a derivative of a function which is non-differentiable in the usual sense (*round*, *sign*, *clip*, etc). Therefore, the value which is an argument of this function becomes differentiable and can be trained with the method of steepest descent, also called the gradient descent method. Such variable is a quantization threshold and its training can directly lead to the optimal quality of the quantized network. This approach can be further optimized through some modifications as described below.

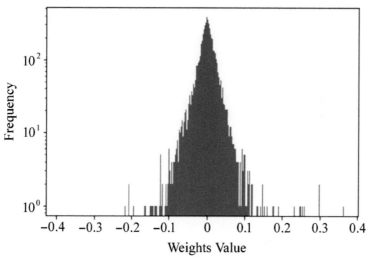

Figure 6.1 Distribution of weights of ResNet-50 neural network before the quantization procedure.

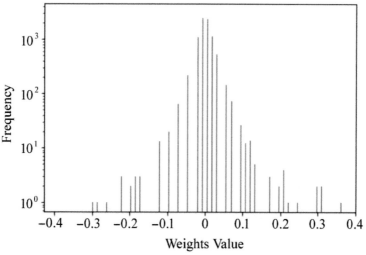

Figure 6.2 Distribution of weights of ResNet-50 neural network after the quantization procedure. The number of values appeared in bins near zero increased significantly.

6.3.1.2 *Batch Normalization Folding*

Batch normalization (BN) layers play an important role in training of neural networks because they speed up train procedure convergence [234]. Before making quantization of neural network weights, we suggest to perform BN folding with the network weights similar to method described in [218]. As a result we obtain the new weights calculated by the following formulae:

$$W_{fold} = \frac{\gamma W}{\sqrt{\sigma^2 + \varepsilon}} \tag{6.10}$$

$$b_{fold} = \beta - \frac{\gamma \mu}{\sqrt{\sigma^2 + \varepsilon}} \tag{6.11}$$

Here μ and σ are running mean and running variance of full precision network, γ and β are trainable parameters of batch normalization layer, and W is weights of preceeding layer (convolutional or fully connected). We apply quantization to weights which were fused with the BN layers because it simplifies discretization and speeds up the neural network inference. Further in this article the folded weights will be implied (unless specified otherwise).

6.3.1.3 *Threshold Scale*

All network parameters except quantization thresholds are fixed. The initial value of thresholds for activations is the value calculated during calibration. For weights it is the maximum absolute value. Quantization threshold T is calculated as

$$T = clip(\alpha, min_\alpha, max_\alpha) \cdot T_{max} \tag{6.12}$$

where α is a trainable parameter which takes values from min_α to max_α with saturation and T_{max} stands for T_w or T_I as in basic quantization procedure description in Related Work section. The typical values of min_α and max_α parameters are found empirically, which are equal to 0.5 and 1.0 correspondingly. Introducing the scale factor simplifies the network training since the update of thresholds is done with different learning rates for different layers of neural network as they can have various orders of values. For example, values on the intermediate layers of VGG network may increase up to 7 times in comparison with the values on the first layers.

Therefore the quantization procedure can be formalized as follows:

$$T_{adj} = clip(\alpha, 0.5, 1) \cdot T_I \tag{6.13}$$

$$S_I = \frac{2^n - 1}{T_{adj}} \tag{6.14}$$

$$I_q = \lfloor I \cdot S_I \rceil \tag{6.15}$$

The similar procedure is performed for weights. The current quantization scheme has two non-differentiable functions, namely *round* and *clip*. Derivatives of these functions can be defined as:

$$I_q = \lfloor I \rceil \tag{6.16}$$

$$\frac{dI_q}{dI} = 1 \tag{6.17}$$

$$X_c = clip(X, a, b) \tag{6.18}$$

$$\frac{dX_c}{dX} = \begin{cases} 1, if X \in [a, b] \\ 0, otherwise \end{cases} \tag{6.19}$$

Bias quantization is performed similar to [218]:

$$b_q = clip(\lfloor S_i \cdot S_w \cdot b \rceil, -(2^{31} - 1), 2^{31} - 1) \tag{6.20}$$

6.3.1.4 Training of Asymmetric Thresholds

Quantization with symmetric thresholds described in the previous sections is easy to implement on certain devices; however it uses an available range of integer values inefficiently which significantly decreases the accuracy of quantized models. Authors in [218] effectively implemented quantization with asymmetric thresholds for mobile devices, so it was decided to adapt the described above training procedure for asymmetric thresholds.

T_l and T_r are left and right limits of asymmetric thresholds. However, it is more convenient to use other two values for quantization procedure: left limit and width, and train these parameters. If the left limit is equal to 0, then scaling of this value has no effect. That is why a shift for the left limit is introduced. It is calculated as:

$$R = T_r - T_l \tag{6.21}$$

$$T_{adj} = T_l + clip(\alpha_T, min_{\alpha_T}, max_{\alpha_T}) \cdot R \tag{6.22}$$

The coefficients min_{α_T} and, max_{α_T} are set empirically. They are equal to -0.2 and 0.4 in the case of signed variables, and to 0 and 0.4 in the case of unsigned. Range width is selected in a similar way. The values of min_{α_R} and, max_{α_R} are also empiric and equal to 0.5 and 1.

$$R_{adj} = clip(\alpha_R, min_{\alpha_R}, max_{\alpha_R}) \cdot R \tag{6.23}$$

6.3.1.5 Vector Quantization

Sometimes due to high range of weight values it is possible to perform the discretization procedure more softly, using different thresholds for different filters of the convolutional layer. Therefore, instead of a single quantization factor for the whole convolutional layer (scalar quantization) there is a group of factors for each output channel (vector quantization or channel-wise quantization as well). This procedure does not complicate the realization on devices, however it allows increasing the accuracy of the quantized model significantly. Considerable improvement of accuracy is observed for models with the architecture using the depth-wise separable convolutions. The most known networks of this type are MobileNet-v1 [214] and MobileNet-v2 [215].

6.3.2 Training on the Unlabeled Data

Most articles related to neural network quantization use the labeled dataset for training discretization thresholds or directly the network weights. In the proposed approach it is recommended to discard initial labels of train data which significantly speeds up transition from a trained non-quantized network to a quantized one as it reduces the requirements to the train dataset. We also suggest to optimize root-mean-square error (RMSE) between outputs of quantized and original networks before applying the softmax function, while leaving the parameters of the original network unchanged.

The suggested above technique can be considered as a special type of quantization with distillation [220] where all components related to the labeled data are absent.

The total loss function L is calculated by the following formula:

$$L(x; W_T, W_A) = \alpha H(y, z^T) + \beta H(y, z^A) + \gamma H(z^T, z^A) \qquad (6.24)$$

In our case α and β are equal to 0, and

$$H(z^T, z^A) = \sqrt{\sum_{i=1}^{N} \frac{(z_i^T - z_i^A)^2}{N}} \qquad (6.25)$$

where:

- z^T is the output of non-quantized neural network,

- z^A is the output of quantized neural network,

- N is batch size,

- y is the label of x example.

6.3.3 Quantization of Depth-wise Separable Convolution

During quantization of models having the depth-wise separable convolution layers (or DWS-layers) it was noticed that for some models (MobileNet-v2, MNasNet with the lower resolution of input images) vector quantization gives much higher accuracy than the scalar quantization. Besides, the usage of vector quantization instead of scalar only for DWS-layers gives the accuracy improvement.

In contrast to the scalar quantization, vector quantization takes into account the distribution of weights for each filter separately—each filter has its own quantization threshold. If we perform rescaling of values so that the quantization thresholds become identical for each filter, then procedures of scalar and vector quantization of the scaled data become equivalent.

For some models this approach may be inapplicable because any non-linear operations (like swish, sigmoid or tanh activation functions) on the scaled data as well as addition of the data with different scaling factors are not allowed. Scaling the data can be made for the particular case $DWS \rightarrow [ReLU] \rightarrow Conv$ (see Figure 6.3). In this case only the weights of the model change.

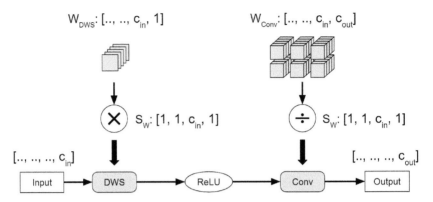

Figure 6.3 Scaling the filters of DWS + Convolution layers where the output of DWS + Convolution remains unchanged. Numbers in square brackets denote the dimension of the scaling factors. W_{DWS} represents the weights of the DWS layer, and W_{Conv} the weights of the convolution layer. Note that the scaling factor $S_W > 0$.

6.3.3.1 Scaling the Weights for MobileNet-V2 (with ReLU6)

As mentioned above, the described method is not applicable for models which use the non-linear activation functions. In case of MobileNet, there is ReLU6 activation function ($ReLU6(x) = min(max(x, 0), 6)$) between the convolutional operations. When scaling the weights of a DWS-filter, the output of the DWS-layer is also scaled. One way to keep the result of the neural network inference unchanged is to modify the ReLU6 function, so that the saturation threshold for the k-th channel is equal to $6 \cdot S_W[k]$. However, it is not suitable for the scalar quantization technique.

In practice, the output data for some channels of a DWS-layer X_k may be less than 6.0 on a large amount of input data of the network. It is possible to make rescaling for these channels, but with the certain restrictions. The scaling factor for each of these channels must be taken so that the output data for channels X_k does not exceed the value 6.0.

If $X_k < 6$ and $X_k \cdot S_W[k] < 6$, then

$$min(6, X_k \cdot S_W[k]) = S_W[k] \cdot min(6, X_k) \qquad (6.26)$$

Consequently:

$$ReLU6(X_k \cdot S_W[k]) = S_W[k] \cdot ReLU6(X_k) \qquad (6.27)$$

We propose the following scheme of scaling the DWS-filter weights.

1. Find the maximum absolute value of weights for each filter of a DWS-layer.

2. Using the set of calibration data, determine the maximum values each channel of the output of the DWS-layer reaches (before applying ReLU6).

3. Mark the channels where the output values exceed 6.0 or are close to it as "locked". Let N_{fixed} is number of such channels. The corresponding filters of the DWS layer must stay unchanged. We propose to lock the channels where the output data is close to the value 6.0, because it could reach this value if we use a different calibration dataset. In this article we consider 5.9 as the upper limit.

4. Calculate the maximum absolute value of weights for each of the locked filters $T_{w_i^{fixed}} = max(|w_i^{fixed}|)$ where i is channel index. The average of these maximum values $T_a = \frac{1}{N_{fixed}} \sum_{i=0}^{N_{fixed}} T_{w_i^{fixed}}$ becomes a control value that is used for scaling the weights of non-locked filters. The main purpose of such choice is to minimize the difference between the thresholds of different filters of the DWS-layer.

5. Find the appropriate scaling factors for non-locked channels.

6. Limit these scaling factors so that the maximum values on the DWS-layer output for non-locked channels do not exceed the value 6.0.

6.4 EXPERIMENTS AND RESULTS

6.4.1 Experiments Description

6.4.1.1 *Researched Architectures*

The procedure of quantization for architectures with high redundancy is practically irrelevant because such neural networks are hardly applicable for mobile devices. Current work is focused on experiments on the architectures which are actually considered to be a standard for mobile devices (MobileNet-v2 [215]), as well as on more recent ones (MNasNet [216]). All architectures are tested using 224 x 224 spatial resolution.

6.4.1.2 Training Procedure

As it is mentioned above in the Section 6.3.2 ("Training on the unlabeled data"), we use RMSE between the original and quantized networks as a loss function. Adam optimizer [235] is used for training, and cosine annealing with the reset of optimizer parameters—for learning rate. Training is carried out on approximately 10% part of ImageNet dataset [233]. All images in train dataset were sampled in uniform way. Testing is done on the validation set. Note that 100 images from the training set are used as calibration data. Training takes 6–8 epochs depending on the network.

6.4.2 Results

The quality of network quantization is represented in the Tables 6.1–6.2.

Experimental results show that the scalar quantization of MobileNet-v2 has very poor accuracy. A possible reason of such quality degradation is the usage of ReLU6 activation function in the full-precision network. Negative influence of this function on the process of network quantization is mentioned in [236]. In case of using vector procedure of thresholds calculation, the accuracy of quantized MobileNet-v2 network and other researched neural networks is almost the same as the original one.

The Tensorflow framework [223] is chosen for implementation because it is rather flexible and convenient for further porting to mobile devices. Pre-trained networks are taken from Tensorflow repository [237]. To verify the results, the program code and quantized scalar models in the .lite format, ready to run on mobile phones, are presented in the repository [238].

The algorithm described in the section 6.3.3 ("Quantization of depthwise separable convolution") gives the following results. After performing the scalar quantization of the original MobileNetV2 model, its accuracy becomes low (the top-1 value is about 1.6%). Applying the weights rescaling before the quantization increases the accuracy of the quantized

TABLE 6.1 Quantization in the 8-bit Scalar Mode

Architecture	Symmetric thresholds	Asymmetric thresholds	Original accuracy
MobileNet v2	8.1%	19.86%	71.55%
MNas-1.0	72.42%	73.46%	74.34%
MNas-1.3	74.92%	75.30%	75.79%

TABLE 6.2 Quantization in the 8-bit Vector Mode

Architecture	Symmetric thresholds	Asymmetric thresholds	Original accuracy
MobileNet v2	71.11%	71.39%	71.55%
MNas-1.0	73.96%	74.25%	74.34%
MNas-1.3	75.56%	75.72%	75.79%

model up to 67% (the accuracy of the original model is 71.55% [1]). To improve the accuracy of the quantized model we use fine-tuning of weights for all filters and biases. Fine-tuning is implemented via trainable point-wise scale factors where each value can vary from 0.75 to 1.25. The intuition behind this approach is to compensate the disadvantages of the linear quantization by slight modification of weights and biases, so some values can change their quantized state. As a result, fine-tuning improves the accuracy of the quantized model up to 71% (without training the quantization thresholds). Fine-tuning procedures are the same as described in the Section 6.4.1.

6.5 CONCLUSION

This article demonstrates the methodology of neural network quantization with fine-tuning. Quantized networks obtained with the help of our method demonstrate a high accuracy that is proved experimentally. Our work shows that setting a quantization threshold as multiplication of the maximum threshold value and trained scaling factor, and also training on a small set of unlabeled data allow using the described method of quantization for fast conversion of pre-trained models to mobile devices.

---

[1]The network accuracy is measured on a full validation set ImageNet2012 which includes single-channel images.

CHAPTER 7

Power-efficient Neural Network Scheduling

Ying Wang, Xuyi Cai, and Xiandong Zhao

Institute of Computing Technology, Chinese Academy of Sciences

CONTENTS

DOI: 10.1201/9781003162810-7

7.1 INTRODUCTION TO NEURAL NETWORK SCHEDULING ON HETEROGENEOUS SoCs

The thrive of deep neural networks (DNN) are propelling the rapid development of computer vision, speech, and language processing technology. Particularly, the confluence of IoT (Internet of Things) and neural networks (NNs) are enabling ubiquitous intelligent computing capability on edge and embedded devices. To squeeze sufficient neural network processing power from the resource-limited edge devices, researchers are delving into the deep optimization stack of neural networks for energy efficient system-level solutions, such as smart vehicles, robotics, and vision-based surveillance systems. The past years have witnessed the intensive researches on how to improve the energy-efficiency of neural network related tasks, including the model-level optimization such as lightweight network architectures and hardware-aware Network Architecture Search (NAS) [239, 209, 240, 111, 241], network compression and quantization techniques [196, 242], approximate computing technology [243], specialized neural network hardware [244, 245, 246, 247], and even cross-stack co-design approaches [248]. Amongst these techniques, network scheduling is an essential interfacing middleware between the network model and the underlying hardware. Thus how to efficiently schedule and map the neural network tasks onto the underlying power-constrained hardware makes fundamental impacts to the system, especially the mobile computing devices that are made of System-on-Chip (SoC). When the algorithm-level optimization has been finalized and the hardware design is frozen, the network task scheduling policy will play an important role in system-level efficiency of neural network application. Before delving into the details of neural network task implementation on real devices, we must clarify some of the important notations and definitions about scheduling.

7.1.1 Heterogeneous SoC

Before the scheduler deploys neural network tasks onto the hardware, let us revisit the typical System-on-Chip (SoC) architecture that is widely used on power-efficient embedded and mobile devices. Figure 7.1 shows a typical architectural diagram of the mobile SoC released by [249], which is integrated with a central general-purpose processor core, a Specialized Unit (SU) to accelerate a certain or a domain of application kernels, the low and high speed peripherals, and the system bus or Network-on-Chip (NoC) that couples them up. For instance, Nvidia Tegra Jetson X1 is also a typical heterogeneous SoC, in which the GPGPU is integrated as the SU to accelerate parallel computing workload.

In a heterogeneous SoC, the CPU core and the SU are thought working in a collaborative fashion. In accordance with its specialized purpose, a SU can be referred to as an accelerator, e.g., GPU, Neural Processing Unit (NPU), Image Processing Processor (ISP) or other IPs. It is also referred to as a fused system. Sometimes, the SUs and the CPU cores have shared memory on-chip, and they support explicit data transfer via the on-chip bus or NoC, which can be optimized by the programmer to enhance data locality on the chip. Otherwise, they must exchange data and communicate by the off-chip shared main memory. When the CPU

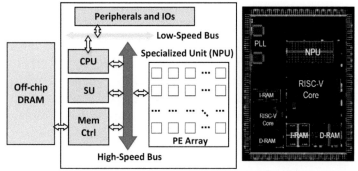

Figure 7.1 A diagram of heterogeneous SoC and its physical layout [249].

and SU cores cooperate to fulfill one workload, the data exchange can be enabled by bus or memory operations. As it illustrates in Figure 7.1, the NPU have separate on-chip memory spaces from the CPU, data transfer either happens over the NoC, and through the main memory, which incurs large overhead and needs to be addressed by scheduling.

7.1.2 Network Scheduling

In a heterogeneous SoC, the job of neural network scheduling is to partition the target neural network task into stages, reorder, map the divided stage on the CPU and SUs. Moreover, using asynchronous pipelining, the SU will start the computation kernel mapped on it while the CPU is completing the next stage. Depending on when the scheduling is conducted, at the neural network compile time or at runtime, there will be dynamic scheduling and static scheduling. In dynamic scheduling, the mapping of different application stages is made at runtime, while in static scheduling the mapping of workload stages onto SUs and CPU are already fixed before every execution round. No matter the scheduler is static or dynamic, the goal is to maximize the workload performance or increase power and resource utility of the hardware. In general, there are several important observations that help the designer to efficiently schedule the workload on heterogeneous SoCs: First, the scheduler must take advantages of the architectural merits of SUs and CPUs. For example, CPUs are considered as good at kernels with complicated control paths and intensive random memory instructions, while GPU is powerful in processing highly-parallel matrix and streams. Thus, the workload partitioning and mapping must be carefully conducted to match the features of SUs and CPUs; Second, the scheduler must try best to reduce or hide the data exchange penalty between CPU and SU; Last of all, the scheduler must improve the resource of SU and CPU through load-balance, so that no power will be wasted by the idling hardware. The three rules are also very important for scheduling the neural network task on power-efficient SoCs.

Coarse-Grained NN Tasks Scheduling on Heterogeneous SoCs: Sometimes, a neural network task is made of multiple interdepended program phases: the processing stages that deal with the operations to the original input, e.g., the optical-related image processing and time-and-frequency transformation of audio samples, the neural network backbone that is well supported on SUs like GPU and NPU, and some other operations or post-processing steps related to the task output. In this paper, we refer the high-level task partitioning and mapping as coarse-grained scheduling, which is more of the traditional scheduling problem on typical heterogeneous SoC. In coarse-grained scheduling, the three rules above will be major guidelines that we observed in the implementation of neural network tasks on energy-efficient SoCs.

Fine-Grained Neural Network Scheduling on SUs: After coarse-grained scheduling, if the network backbone or the divided network blocks are mapped onto the SUs, there will be a further step of fine-grained model mapping in order to improve the execution efficiency on the according accelerator. Basically, a DNN algorithm can be represented as coarse-computational directed acyclic graphs (coarse-grained CDAG), which is made of interconnected matrix operators. How such a coarse-grained CDAG and its operators are implemented leave a potential space for the fine-grained DNN scheduling to exploit. Unlike coarse-grained scheduling, this step is often conducted by the SU compiler. The details of fine-grained scheduling will be introduced in Section 7.3.

Conclusively, scheduling a compound neural network task onto a heterogeneous SoC is a non-trivial job and makes an important complement to the other performance improvement stacks, and it must be fully aware of the application and the underlying hardware to extract the most processing efficiency out the system.

7.2 COARSE-GRAINED SCHEDULING FOR NEURAL NET-WORK TASKS: A CASE STUDY OF CHAMPION SOLUTION IN LPIRC2016

In this section, we will exemplify the manual scheduling process on a typical heterogeneous SoC with a DNN-based objection detection task, which is also the contest mission in Low-Power Image Recognition Challenge 2016. Since the exemplary system was constructed more than half a decade ago, many of useful SOTA techniques, such as lightweight DNN backbone, Network Architecture Search, novel single-stage detection frameworks, etc., are absent. However, from the perspective of neural network task scheduling, it still makes a proper example of software/hardware co-design for neural network tasks.

7.2.1 Introduction to the LPIRC2016 Mission and the Solutions

Although convolutional neural networks (CNNs) are showing increasing accuracy on a variety of image datasets, the growth of network scale and depth, agitated by the complexity of image content, still poses a significant challenge to the consolidation of CNN-based image recognition solutions on power-constrained computing devices, which struggle to offer real-time CNN inference capability with limited computational and memory resources.

For low-end mobile devices, an appropriate CNN-based detection framework should be able to fully exploit the capability of hardware and increase the power utility using architectural and algorithmic optimization. Low-Power Image Recognition Challenge (LPIRC) is initiated to promote such research on energy-efficient object detection technology, and it particularly highlights the trade-off between accuracy and power consumption in the detection solutions [250]. In contrast to state-of-the-art CNN-based detection frameworks focused on accuracy improvement, detection solutions that are aware of the network computational overhead are expected to have a better score in terms of mAP/Wh (Watt-hour) in this contest. In LPIRC2016, we implemented a low-power image recognition system by manually scheduling the neural network based detector on NVDIA Jetson TX1.

In the on-site contest of LPIRC2016, two different object detection architectures: BING+FAST-RCNN and Faster-RCNN are generally employed to complete the image recognition missions of track-1 and track-3, respectively. These two frameworks are all based on the popular CNN architecture and implemented on the NVIDIA Jetson-TX1 development board. The design philosophy in building this low-power but cost-effective object detection system is quite simple: to trade-off between model accuracy and computational complexity through design space exploration, and to maximize the utility of the heterogeneous CPU+GPU SoC through scheduling. For example, we can adjust the hyper-parameters of the employed CNN model, sacrifice accuracy for speed-up by means of dimension reduction techniques and other input-level approximation methods [251]. It helps create a massive space of design parameters for the target CNN-based object detection framework. With a sufficiently large design space, we can evaluate the accuracy and power consumption of different implementations, and search for a proper design point that yields the highest score measured in mAP/WH. With this method, we successfully boost the energy-efficiency of the Fast R-CNN and Faster R-CNN implementations on NVIDIA TX1, which are respectively the winners in track-1 and track-3 of LPIRC2016 [252].

7.2.2 Static Scheduling for the Image Recognition Task

Early region-based object detection systems include two stages: hypothesizing bounding boxes, extracting features of each box and applying a classifier to them. The first stage often relies on inexpensive coarse-grained proposal search methods, such as Selective Search, EdgeBox, BING, etc. [253, 254]. With the region proposal results, the CNN mainly plays as a classifier, and it does not need to predict the object bound coordinates in the image all by itself. These flows are defined as two-stage object detection methods. The original Region CNN (R-CNN) is a typical two-stage object detector and very computationally expensive because it has to perform a ConvNet forward pass for every single object proposal [255]. To accelerate this procedure, later work such as OverFeat computes the shared convolutional features from an image pyramid for detection results [256]. Similarly, SPPnets were proposed to accelerate the R-CNN by sharing some convolutional feature maps across the regions [257]. Fast R-CNN [258] also trains the detector to work on shared convolutional features, and it even claims to classify object proposals and refine their bounding-box coordinates simultaneously. In general, the two-stage CNN-based object detectors are thought as expensive and less efficient because it requires additional computation stage of bounding-box search, however, they can be used to construct power-efficient object detection solutions by properly balancing the two stages in heterogeneous multi-core SoC when the accuracy is not the sole optimization target of the system [252].

Task Partitioning and Mapping: As shown in Figure 7.2, they include three steps: (1) coarse-grained region proposal extraction; (2) CNN feature extraction and object classification; and (3) fine-grained bounding box regression. Fast R-CNN is an enhancement over the famous framework of object detection called region-based CNN (R-CNN). An important feature is that Fast R-CNN needs a coarse-grained region proposal extraction method, such as Selective Search, EdgeBox, BING, and so on. Among them, BING is a very fast method for extracting object proposals. According to the paper, BING generates a small set of category-independent, high quality windows. Such coarse-grained proposals generated by BING or Selective Search extract the possible coordinates for Fast R-CNN to generate region proposals, which are called "bounding box". Intuitively, the Fast R-CNN can be partitioned into two separate steps by the scheduler: proposal search and R-CNN. To better exploit the strength of CPU and GPU, Selective Search or Bing can be scheduled on CPU while the neural network backbone of the Fast R-CNN can be

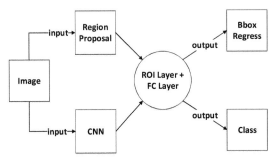

Figure 7.2 Framework of Fast R-CNN and Faster R-CNN.

mapped to GPU, so that we can pipeline the two tasks and hide the performance overhead of proposal searching with CNN inference.

7.2.3 Manual Load Balancing for Pipelined Fast R-CNN

After task partitioning, the next important step of scheduling is to load balance the stages of detector between CPU and GPU. Table 7.1 summarizes the running speed of the entire object detection system. Selective search takes 1–2 seconds on CPU while the Fast R-CNN with VGG-16 backbone takes 320ms on the 2000 SS proposals. Faster R-CNN with VGG-16 takes 198ms in total for both proposal and detection, in which RPN only takes 10ms to complete the additional layers. However, all these experiments are run on a K40 GPU. When we run it on Nvidia Jetson TX1, the conclusion is very different in Table 7.1. Because of the relatively weaker GPU and CPU in TX1, RPN alone consumes a large proportion of execution time. The difference of the speed for RPN between Xeon+K40 and TX1 is mainly because RPN has to be divided and mapped separately onto CPU and GPU by the scheduler. Once deployed onto CPU, the wimpy ARM cores in TX1 will spend a plenty of time in processing the results of RPN and becomes the bottleneck of system. Therefore, though as an improvement to Faster R-CNN, the

TABLE 7.1 Timing (ms) on a K40 GPU, Except SS Proposal Is Evaluated in a Xeon CPU [259]

model	system	conv	proposal	total	rate
VGG	SS+Fast R-CNN	146	1510	1830	0.5fps
VGG	RPN+Fast R-CNN	141	10	198	5fps
ZF	RPN+Fast R-CNN	31	3	59	17fps

performance advantages of Faster R-CNN will not be that evident in TX1 when compared to Fast R-CNN.

For Fast R-CNN, if we use BING as the region proposal method, it only takes about 15 percent of total time when we run BING on CPU and run Fast R-CNN backbone on GPU at the same. In contrast, RPN of Faster R-CNN is sharing results of some convolution layers with R-CNN and it is supposed to run serially on GPU cores. Thus, RPN cannot run with R-CNN at the same time. Otherwise, we have to put it to CPU cores for pipelined execution, the speed will become even less attractive after scheduling because of RPN is closely coupled with the other part of network, which will induce additional data exchange overhead between CPU and GPU.

Therefore, once Fast R-CNN and Faster-RCNN are adopted as the potential tasks to schedule on the TX1 platform, the next step is to fine-tune the design parameters of the detection frameworks to ensure we reach a balanced task scheduling result on the SoC. As we have mentioned, neural networks have many design parameters to tune by trading-off excessive recognition accuracy for energy-efficiency boost, which is the general principle of approximate computing. There are many design choices in the implementation of Fast R-CNN in the target hardware NVIDIA Jetson TX1 SoC, including the hyper-parameter of CNN, data representation precision, region proposal methods and input scaling factor. For efficient implementation and power efficient scheduling, we decided to adjust the stages divided from the Fast R-CNN framework through approximate computing, until it reaches two essential goal after scheduling:

1. Perfect load-balance; To load-balance in between CPU and GPU, which means the pipelined stages of proposal search and CNN inference are completely equal in computation time when they are assigned to the CPU and GPU cores respectively. In this case, neither the CPU nor the CPU core will be forced to wait for the response from the other side, so that no static power will be wasted on core idling.

2. Best-effort tuning for each stage; To make sure that the marginal profit of tuning each parameter knob is equivalent to the others', which means changing any of the parameters will decrease the contest score measured in mAP/MH. In ideally balanced cases, neither the accuracy nor the power consumption is over-emphasized, so that the mAP/Wh will be maximized.

Since adjusting any of the approximate computing knobs will change the balance of the pipeline, the tuning will be conducted manually and iteratively until a best-effort load-balance is almost reached. In the challenge, we mainly adjust the available design parameters as follows:

- Bounding-box Number: Besides options of region proposal extractor, the number of output box and the input image size are also two important parameters in the pipeline of object detection. For example, when we directly crop the image into half size, the speed of BING is almost increased more than 100 percent. Moreover, if BING is configured to generate the top K scored proposals for Fast R-CNN instead of the top-scored 1000 boxes (K<<1000), the latency of the detection pipeline stage will also be shrunk significantly at a minor cost of accuracy loss. In scenarios when the application is more sensitive to speed, it is to be avoided to pass too many of proposals to the stage of region-based CNN classification. Last of all, Faster R-CNN can also be accelerated by reducing the proposals. Conclusively, the option and the detailed parameters of region proposal methods can lead to many tuning knobs that could be adjusted to reach trade-off between accuracy and speed.

- CNN Backbone Selection: Different CNNs models can be fitted into the framework of Fast R-CNN for the step of feature extraction, classification, and box regression, after proposal search. A deep convolutional neural network have many tunable hyper-parameters used to configure the strength and computational cost of the network even for the same task [260]. Changing such hyper-parameters has a direct impact on the memory access and computation performance of object detection on the SoC. In this work, we assume the hyper-parameter of CNNs, i.e., depth, kernel size, activation functions, and other features, as adjustable parameters in the stage of design space exploration for energy-efficient object detector on TX1. The choice of replacing the backbone CNNs in Fast R-CNN into a possible customized network can also create an ample space of design points with varied detection accuracy and detection speed.

- Input-level Approximation: Cropping the under-test image into different sizes also directly changes the number of operations in the entire procedure of CNN inference for object detection. The speed-up effect brought by image down-sizing is evident because the volume of convolutional operations drops almost at the same ratio as the resizing factor. We see in experiments that when image size changes from 600*600 to 300*300, the inference speed almost increase $3\times$ and the detection accuracy of the original Fast R-CNN decrease only about 3 percent.

- Representation Precision: GPGPUs like TX1 support half-precision operation to accelerate matrix multiplication. The computation throughput can be elevated at a minor cost of accuracy loss when full-precision operation is replaced with half-precision operation [252]. The recently released TensorRT framework even support the execution mode of 8-bit fixed point representation, which could potentially improve the memory performance without the need to re-train the model. This mechanism also poses an option of speedup and accuracy trade-off.

- Truncated Singular Value Decomposition: The convolutional, fully-connected and RoI layers that dominate the inference time of the detector [261], are basically decomposed into matrix multiplication in GPGPU execution. Particularly, for Fast R-CNN, localized RoI layers account for nearly half of the inference time when the proposal number and the number of object classes are huge. Singular value decomposition (SVD) is a linear-algebra technique that factorizes a matrix and eventually compresses it for less intensive computation on GPUs. Supposing the linear operation in an FC layer is deemed as a product of Wx, x is the feature vector and W is the matrix of weight. SVD is to decompose the W to let $W = UDV$, where D is the diagonal matrix with singular value on the diagonal, and U and V are orthogonal matrices. To reduce the size of W, W can be approximated by keeping only the d largest elements in D diagonals, so that the approximated W becomes the product of smaller matrices U, D, and V. Such truncated SVD is proved to effectively reduce the scale of FC and convolutional weight, also with a controllable amount of accuracy loss. Depending on the proposal number, truncated SVD is proved to reduce detection time by 30%–70% with only a small (0.3 percentage point) drop in mAP and without additional fine-tuning after model compression [252].

There are also other non-approximate design parameters like Batch Sizing. GPU is a high-throughput architecture that supports the exploitation of batch-level parallelism in object detection. However, the best batch size for high throughput processing is depending on the scale of CNN model and also the memory capacity of the computing platform.

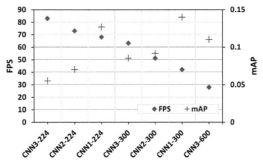

Figure 7.3 Accuracy and speed of several implementations.

Thus, we can try different batch sizes for a potential detection framework and find a proper one that leads to better throughput on the GPU.

The adjustable parameters listed above do not cover all the acceleration techniques for CNN-based detection methods, but they are chosen as the basic tuning knobs in our design exploration phase for LPIRC-score maximization, for their simplicity and generality. Figure 7.3 shows the varying detection accuracy and speed when some of the parameters are adjusted. It reveals that a larger exploration space exists for speed-accuracy trade-off. In this figure, the model size of CNN1, CNN2, and CNN3 are decreasing while the images under test are cropped into 224*224, 300*300, and 600*600 inputs. As the rise of Network Architecture Search (NAS) and automl, all the design parameter of the neural network tasks can be potentially evaluated and selected through the automated method, e.g., Bayesian[262], ADMM[263, 264], evolutionary algorithm[265], reinforcement learning, and other methods[266, 248].

7.2.4 The Result of Static Scheduling

For track-1 of 2nd Low-Power Image Recognition Challenge (LPIRC) [250], we prepare two solutions, which are designed based on Fast R-CNN. Track-1 have 10000 images test set provided by a local server, TX-1 can fetch the image through Ethernet and process them in a total of 10 minutes.

The evaluation metric of track-1 is mAP/Wh. We formalized this task as an optimization problem whose object is mAP/Wh. mAP is the accuracy metric of object detection methods, and it is ratioed against the energy consumption it takes to accomplish the detection task. The accuracy is constrained by the accuracy of detection framework and it

TABLE 7.2 Results in Track-1 of 2nd LPIRC with TX1

Solution	Accuracy	Energy(WH)	Score
Fast R-CNN(C)	0.0347	0.79	0.044
Fast R-CNN(P)	0.0260	0.94	0.028

takes a lot of opportunity cost to pursue if we use powerful but large CNN models to improve it. However, the energy consumption the task needs can be easily reduced by speeding-up Fast R-CNN and finding a design point in our configuration space. In short, if we start from a powerful object detector, we can keep on accelerate it and compromise its accuracy with the method introduced in last sections. The accuracy is sacrificed until the marginal benefits of enhancing detector accuracy and relaxing detector accuracy, $\Delta mAP/\Delta Wh$ and $\Delta Wh/\Delta mAP$, are equalized.

In addition, the constraint of such a problem is that multi-threaded BING executed onto ARM cores has the same speed as the flowing Fast R-CNN, and also the image fetching time through the Ethernet, so that the task can be parallelized without bottleneck and the resources/energy on TX1 are fully utilized

Finally, we choose the fastest configuration, in which the number of hypothesizing boxes p is 50, the cropping size s is 224, the input batch size b is 50, and the number of diagonals t is set to 256 in the SVD process. For the H parameter, we try two networks, Fast R-CNN(C) and Fast R-CNN(P). The results of track 1 shows in Table 7.2. In the competition, the Fast R-CNN(P) solution was ranked the 1st in all 9 solutions. Further, 20120 images from ILSVRC 2013 training dataset were used for fine-tuning and the accuracy was calculated by mAP. The gap between the offline test results (higher mAP) and the competition results is because the result of contest is discounted by a factor.

As the development of object detection technology and hardware architecture, the results of the LPIRC2016 contest cannot be compared to state-of-the-art (SOTA) solutions in either performance or power evaluation. However, through the result analysis, we can see that detection objection system is multi-layered framework in which the single-aspect improvement such as better DNN models or SoC devices cannot guarantee the system-level optimality. It is encouraged to customize the whole system and scheduling the workloads properly to achieve the design goal for a particular application scenario. In this chapter, we show manually selecting the design parameter of the model is able to achieve perfect balance between CPU and the SUs. However, when the design parameter space becomes larger and larger, automated solutions will be more efficient [267].

7.3 FINE-GRAINED NEURAL NETWORK SCHEDULING ON POWER-EFFICIENT PROCESSORS

In the last section, we use the reward-wining solution of LPIRC2016 to illustrate how to manually schedule the neural network based task onto SoCs in a coarse grained way. However, the mapping of neural network backbone, for example, the VGG model of Fast R-CNN in the last chapter, on the SUs, e.g., GPGPU, is also very important for the power efficiency of the system.

7.3.1 Network Scheduling on SUs: Compiler-Level Techniques

Neural network scheduling aims to translate the arithmetic operations of the model into the low-level code for the given hardware platform. Popular DL frameworks, such as TensorFlow [268], PyTorch [269], MXNet [270], and CNTK [271], have been released to simplify the scheduling of different NN models. However, most of these frameworks support only a limited number of hardware devices, and their performance also rely on the manually engineered and vendor-specific operator libraries (e.g., MKL-DNN, cuBLAS, and cuDNN). When novel operators emerge as the rapid development of NN architectures, these libraries require intensive labor work of manual scheduling to realise performance boost across various hardware devices.

The issue of NN scheduling on diverse hardware devices has stimulated the research on DL compilers such as TVM [272], Tensor Comprehension (TC) [273], Glow [274], nGraph [275], and XLA [276]. To decouple software and hardware, DL compilers transform the NN models into multilevel intermediate representations (IRs) and conduct network scheduling optimizations. Specifically, the compiler frontend takes a NN model from existing DL frameworks as input and then transforms the model into high-level IR (as known as graph IR) via hardware-independent scheduling-level optimization, and afterward the compiler backend will transform the high-level IR into low-level IR and performs hardware-dependent scheduling-level optimization, code generation, and compilation.

The high-level IRs (e.g., DAG-based IR for TC, Glow, nGraph, and XLA, and Let-binding-based IR for TVM) of DL compilers are hardware-independent and thus can be applied to different hardware backend. The hardware-independent scheduling optimizations in the compiler frontends include strength reduction (i.e., replace expensive operators by inexpensive ones), constant folding (i.e., convert the constant expressions), static memory planning (i.e., pre-allocate memory space to hold each intermediate tensor), and data layout transformations (i.e., try to find the best data layouts to store tensors). These scheduling techniques will reduce the computation and data communication redundancy as well as to improve the performance of NN models at the computation graph level.

After high-level scheduling, the low-level IRs are employed to perform hardware-dependent scheduling and code generation. They can be divided into Halide-based IRs (e.g., TVM), polyhedral-based IRs (e.g., TC and nGraph), and other customized IRs (e.g., Glow, MLIR, and HLO IR of XLA). Hardware-dependent scheduling techniques, include hardware intrinsic mapping (i.e. map the certain low-level programs onto the kernels that are highly optimized for the target hardware), memory allocation and fetching (i.e., efficient memory management to improve data locality), memory latency hiding and loop transformation (including loop/layer fusion, loop tiling, loop reordering, loop unrolling, etc.). Because of the enormous DNN scheduling space, it is necessary to leverage the auto-tuning or auto-schedule functionality in the compiler backend to derive performance optimization. Thereby, different auto-tuning (or auto-schedule) search strategies, e.g., genetic algorithm (GA), reinforcement learning (RL), have been tried by the compilers to achieve efficient scheduling on SUs. In addition to its massive search space, NN scheduling can also bear different optimization goals. For power-constrained devices, network scheduling can also be intentionally geared toward the goal of energy saving. In this chapter, we will study the issue of power-efficient network scheduling on SUs for DNN model inference.

7.3.2 Memory-Efficient Network Scheduling

It has been reported that the off-chip memory dissipates the biggest proportion of chip power in most of the neural network processors [277], thus power efficient network scheduling must pay more attention to the cross-layer scheduling, which is well-known for its efficacy of mitigating memory accesses. Cross-layer scheduling, i.e., layer fusion, was proposed to reduce the feature-induced off-chip traffic. [278, 279, 280] propose to exhaustively evaluated the network fusion options on their specific designs and seek to find the optimal one for simple DNN models without branches. DNNVM [281] designed a heuristic subgraph isomorphism algorithm to find layer fusion opportunities and adopted a heuristic shortest-path algorithm to choose network fusion schemes. [282] proposed a fusion algorithm along with a heuristic cost function, however, its scheduling algorithm complexity is too high to be used in very-deep network architectures and on-line deployment cases. [283] reduced the problem by transforming the complicated network into a linear proxy model and employed the Dynamic Programming (DP) algorithm to find the optimal network fusion result for that proxy, which however may not be the optimal result of the original network model.

In short, these network fusion schemes are far from optimal since their approaches do not fully explore the entire scheduling space of the valid fused layer-groups, or they do not consider sophisticated DNN architectures and different accelerator architectures. Moreover, there is lack of an accurate performance model for them to evaluate the memory access cost of potential fusion schemes, so that their derived fusion solutions are sometimes non-optimum. Based on the above analysis, we need a new network fusion framework with an accurate memory-cost model which can reach the optimal network fusion solutions for any combination of accelerators and models.

7.3.3 The Formulation of the Layer-Fusion Problem by Computational Graphs

Before jumping to the design of optimal network fuser, we must formalize the network fusion problem, which is not formally or comprehensively defined in prior works. The formulation will guide us to exhaustively explore the entire scheduling space and evaluate the best solution. In general, DNN algorithms can be represented as computational directed acyclic graphs (CDAGs) $\mathcal{G} = (\mathcal{V}, \mathcal{E})$, which provide a global view of the interconnected operators, but they do not specify how each operator must be implemented as shown in Figure 7.4. For a CDAG $\mathcal{G} = (\mathcal{V}, \mathcal{E})$ of a DNN model, let $L = \{G_1, G_2, \ldots\}$ ($L \in \mathcal{L}$, $\mathcal{L}$ is the layer fusion space) be a layer fusion scheme that divides the network $\mathcal{V}$ into disjointed fused layer-groups G_i ($\mathcal{V} = \bigcup G_i$) which have no cyclic dependencies in them, as shown in Figure 7.4. Thus the optimal layer fusion problem is to find an optimal layer fusion scheme that has the minimum off-chip memory accesses:

$$\min_{L \in \mathcal{L}} \sum_{G_i \in L} MinCost\,(G_i) \tag{7.1}$$

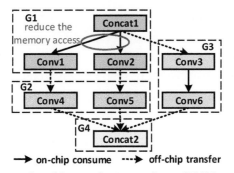

on-chip consume ---> off-chip transfer

Figure 7.4 An example of layer fusion, where DNN model is partition into serveral fused layer-groups and the intermediate feature maps within the fused layer-groups will not be stored to the off-chip memory.

wherein $MinCost\,(\cdot)$ is the cost function that models the memory overhead of the fused layer-groups, which will be detailed in Section 7.3.4.

In a fused layer-group, the intermediate results generated by one operator can be consumed immediately without being evicted to the off-chip memory. Figure 7.5 exemplifies the execution process of a fused layer-group with two convolution (Conv) layers, e.g., G3 in Figure 7.4.

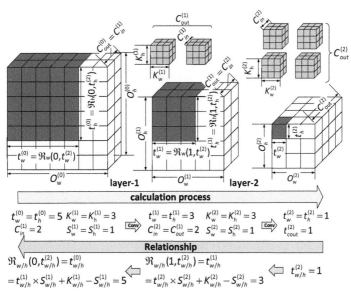

Figure 7.5 The execution process of a fused layer-group with two convolution layers.

For the l-th Conv layer of a certain fused layer-group, the height/width of the output feature map (ofmap) is $O_h^{(l)} \times O_w^{(l)}$ and the size of the weight kernel is $K_h^{(l)} \times K_w^{(l)}$. Additionally, C_{in}, C_{out} are the channel numbers of input and output feature map, respectively. The convolution strides in the h and w directions are $S_h^{(l)}$ and $S_w^{(l)}$, respectively. Besides the Conv layers, other layers can also be expressed by these factors. For example, a fully-connected (FC) layer can be considered as a special Conv layer by setting $S_h^{(l)}$, $S_w^{(l)}$, $O_w^{(l)}$, $O_h^{(l)}$, $K_w^{(l)}$, and $K_h^{(l)}$ to 1.

As illustrated in Figure 7.5, a fmap tile $t_h^{(l)} \times t_w^{(l)}$ is a partition from one fmap and it is the minimum basic unit that can be processed by the DNN processors. In a fused layer-group, the fmap tiles are directly passed to the subsequent layers without being evicted to the off-chip memory. The height/width of the ofmap tiles between two layers satisfy the equation $t^{(l-1)} = t^{(l)} \cdot S^{(l)} + K^{(l)} - S^{(l)}$ [278], wherein layer-$(l-1)$ is the input producer of layer-l. Thus, according to the producer-and-consumer relationship, the required minimum size of the of map tile for layer-l in the fused-group, which is referred to as $\mathcal{R}(l, t^{(n)})$, can be deduced and determined backwardly from the last layer (layer-n) in the group. For example, as shown in in Figure 7.5, to output one single pixel in layer-2 depends on $3 \times 3 \times 2$ output pixels from layer-1, which further relies on $5 \times 5 \times 2$ output pixels from layer-0 (input layer).

Previous works on layer fusion assume that all the required parameters and the feature pixels with the size of at least $\sum_{l=0}^{n-1} C_{out}^{(l)} \cdot \mathcal{R}_h(l,1) \cdot \mathcal{R}_w(l,1) + 1$ in total must be kept in the on-chip buffer at the same time, only to generate one output pixel for layer-n. For example, all the orange data (parameters) and all the blue data (feature pixels) in Figure 7.5 must stay in the on-chip buffer simultaneously to trigger the processor in their assumption. At this point, the parameters are loaded and kept on-chip, so that the computation of the entire fused layer-group can be started. Thereby, as assumed in prior works, the off-chip memory traffic volume equals to the sum of the input fmap size of layer-1, the output fmap size of layer-n, and the parameter size of the whole layer-group, i.e., $\left| ifmap^{(1)} \right| + \left| ofmap^{(n)} \right| + \sum_{l=1}^{n} \left| param^{(l)} \right|$. Otherwise, this fused-group is deemed invalid by the scheduler due to the lack of on-chip memory space for parameters.

However, allowing parameter-induced memory access during the computation of a fused-layer group will possibly gain benefits in terms of the total memory access overhead, since it is very common that the buffer space of a given off-the-shelf DNN processor cannot accommodate all the parameters of the fused layer-group. In addition, allowing parameter refill during the inference process of the fused layer-group exposes the potential benefit of reducing the on-chip buffer space required by the fmap pixels, which can further reduce off-chip memory accesses. Consequently, in contrast to prior work, we will present a memory-cost model (Section 7.3.4) to precisely evaluate the achievable minimum memory-traffic achieved by each layer-group, which takes into account the parameter-induced memory accesses.

7.3.4 Cost Estimation of Fused Layer-Groups

Prior work on layer fusion usually put some restrictions on the configuration of the on-chip memory, so they cannot reach the true memory cost of the grouped layers in a network. In this section, we will remodel the memory cost by removing such assumptions.

Allowing Parameters-Induced Access: First of all, unlike prior works, we discard their assumption that all the parameters associated to the fused layers must be kept in the on-chip memory, and then present how to find the true minimum off-chip memory access number for layer-group G_i with n layers. As illustrated in Figure 7.6, when the on-chip buffer of the processor is large enough to keep all the parameters of the fused layer-group, parameters are loaded outside the loop nest. However, when

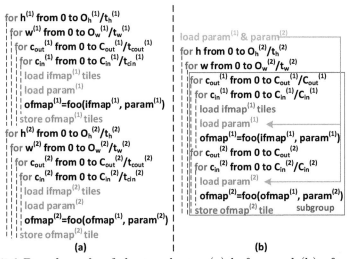

Figure 7.6 Pseudo code of the two layers (a) before and (b) after fusing as Figure 7.5 (`foo` represents the execution of the target DNN processor).

the parameters of the fused layer-group cannot fit into the on-chip buffer of the target processors, all parameters are loaded in each sub-group iteration. In this case, the parameters in the loop nest do not need to be loaded onto the chip all at once. Instead, only the part of parameters required by the operations are loaded according to the dataflow of the target DNN processors, and the loaded parameters are reused as much as possible by the fmap tiles before releasing the buffer space they occupy. In our cost model, such a reuse mechanism in the sub-groups are faithfully accounted for, so that we further push down the lower-bound of memory cost in our model. Therefore, we have that the total number of subgroups is $Q = \left\lceil \frac{O_h^{(n)}}{t_h^{(n)}} \right\rceil \cdot \left\lceil \frac{O_w^{(n)}}{t_w^{(n)}} \right\rceil$. Accordingly in our cost model, the involved off-chip parameter-induced memory traffic for the layer-group is measured as $p\_access = Q \sum_{l=1}^{n} \left| param^{(l)} \right|$. Thus, minimizing p_access is equivalent to maximizing $t_h^{(n)} \cdot t_w^{(n)}$, which is limited by the on-chip buffer size of a given processor.

Reducing the Buffer Space Requirement: Since the memory cost is directly correlated to the on-chip memory space, we further present how we reduce the on-chip buffer space requirement to the greatest extent for the fmap pixels to approach the true memory lower-bound. In a fused layer-group, we allow the subsequent layer to be invoked immediately once a subset of fmap tiles produced by the predecessor layer are ready. For example, in Figure 7.7, layer-1 starts processing T_c ifmap tiles as soon as they are provided, which means a buffer space of $T_c{}^1$ is sufficient for processing the tiles smoothly without encountering memory stalls.

Although the on-chip buffer space requirement for the fmap pixels can be reduced significantly, the mentioned buffer requirement reduction (**BRR**) method cannot be applied to two successive layers simultaneously, because it assumes that all channels of ifmap must be reduced to produce one channel of ofmap. As the example illustrated in Figure 7.7, buffering only T_c ifmap and ofmap tiles of layer-1 causes the partial sums to be evicted to the off-chip memory and then reloaded to the on-chip buffer again as required by the next accumulation operation. In order to deduce

---

[1]Herein, T_* is the **t**hroughput of the processors determined by the PE-array size and the dataflow [284, 285], since we realized that there are impacts of tile size on the on-chip resources utilization of the target processor. It implies that the tile size needs to be greater than the throughput determined by the dataflow and PE-array of a given DNN processor; otherwise, the processor datapath and the corresponding on-chip resources cannot be fully utilized.

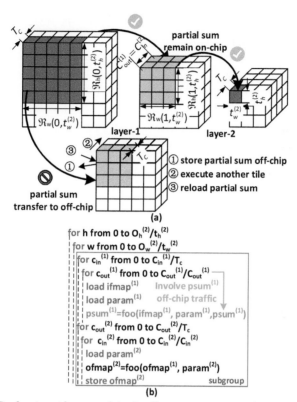

Figure 7.7 Reducing the on-chip buffer space requirement of the fused layer-group in Figure 7.5.

the minimum on-chip buffer requirement of the fmap pixels, we present an algorithm to determine which layers in the fused layer-group are compatible to the BRR method.

As formulated in Algorithm 3, the layer group is traversed in reversed topological order (line 1), and we define $dp[l][1]$ as the case when layer-l adopts the BRR method, and $dp[l][0]$ as the case when BRR does not apply. If the layer-l adopts the BRR method, its successor layers must not use the BRR method (line 5); otherwise, its successor layers are allowed to use the BRR method (line 6). Then, we have the BRR decision $\phi(\cdot)$ for each layer and so obtain the minimum on-chip buffer requirement (lines 9–10).

Minimizing Off-Chip Memory Access: Having removing the prior memory restrictions, we can maximize $t_h^{(n)} \cdot t_w^{(n)}$ under the premise that the buffer requirement of fmaps cannot exceed the on-chip buffer

---

Algorithm 3 BRR Method Applying Decision

1: **In:** fused layer-group G_i
2: **Out:** Binary indicates whether a layer uses methods of reduce buffer requirements ϕ
3: **for** $l \in$ reversed topological order of G_i **do**
4: $\quad dp[l][1] = T_c \cdot \mathcal{R}_h(l, 1) \cdot \mathcal{R}_w(l, 1)$
5: $\quad dp[l][0] = C_{out}^{(l)} \cdot \mathcal{R}_h(l, 1) \cdot \mathcal{R}_w(l, 1)$
6: $\quad$ **for** $v \in succ(l)$ **do**
7: $\quad\quad dp[l][1] += dp[v][0]$
8: $\quad\quad dp[l][0] += \min(dp[v][1], dp[v][0])$
9: $\quad$ **end for**
10: $\quad minCost = \min(dp[0][1], dp[0][0])$
11: $\quad \phi = Backtracking(dp, minCost)$
12: **end for**
13: ϕ

---

capacity B_c:

$$maxT = \max \ t_h^{(n)} \cdot t_w^{(n)}$$
$$s.t. \ \sum_{l=0}^{n} t_c^{(l)} \cdot \mathcal{R}_h(l, t_h^{(n)}) \cdot \mathcal{R}_w(l, t_w^{(n)}) \leq B_c$$
$$t_c^{(l)} = \begin{cases} \min(T_c, C_{out}^{(l)}), & \phi(l) = 1 \\ C_{out}^{(l)}, & \phi(l) = 0 \end{cases} \tag{7.2}$$
$$\min(T_h, O_h^{(n)}), \min(T_w, O_w^{(n)}) \leq t_h^{(n)}, t_w^{(n)} \leq O_h^{(n)}, O_w^{(n)}$$

$t_c^{(l)}$ is the channel tile of the ofmap and $\phi(l)$ is the BRR method applying decision from Algorithm 3. Therefore, the achievable minimum off-chip memory access volume is[2]:

$$MinCost(G_i) = |ifmap^{(1)}| + |ofmap^{(n)}| + \frac{O_h^{(n)} O_w^{(n)}}{maxT} \sum_{l=1}^{n} |param^{(l)}| \tag{7.3}$$

Besides the Conv layer, the cost model can be applied to fully-connected layers, deconvolution layers, dilated convolution layers, etc., by reconfiguring the $S_h^{(l)}$, $S_w^{(l)}$, $O_w^{(l)}$, $O_h^{(l)}$, $K_w^{(l)}$, and $K_h^{(l)}$.

---

[2] For simple illustration, we assume that there is only one input layer and only one output layer in the layer-group. In fact, our derivation is also applicable to more complicated layer-group with multiple input and output layers, e.g., {v1,v2,v3,v4} in Figure 7.8.

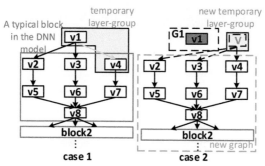

Figure 7.8 Two cases of DP sub-structure.

7.3.5 Hardware-Aware Network Fusion Algorithm (HaNF)

So far, based on the computing graph, we have ensured that the potential layer fusion schemes can be correctly evaluated, and then we need to evaluate each fusion scheme in the massive layer fusion space.

Given a CDAG of NN model $\mathcal{G} = (\mathcal{V}, \mathcal{E})$, $\mathcal{V} = \{v_1, ..., v_N\}$, as shown in Figure 7.8, we observe that for any profitable layer-group (which can contain one or multiple layers), it can grow by including more nodes on the graph or remain unchanged. These two choices are mutually exclusive, and the decisions can be made by comparing the cost of them. However, for complicated NN models, it will make the decision space grow exponentially because there are $2^{|succ(v_i)|}$ combination choices for each multi-output node to be included[3], so the total combination space is $\prod_i 2^{|succ(v_i)|}$. Fortunately, in layer fusion, layer-groups that result in cyclic data dependencies, e.g., layer-group {v1, v3, v6, v8} in Figure 7.8, are invalid; otherwise, the operation on fused layer-groups must be interrupted to load from off-chip memory, e.g., the memory request occurs after v1 to fetch the data from v5 that v8 needs, which stalls the continuous execution of the fused layer-groups. Under this constraint, the total number of available combinations of layer-groups after space growth becomes $\sum_i 2^{|succ(v_i)|}$ $(O(2^{\max_i |succ(v_i)|} |\mathcal{V}|))$.

Based on this observation, we present the HaNF algorithm that searches for the optimal fusion scheme recursively. We start from the input node (v_1) of $\mathcal{V}$, and a virtual "routing" node of zero memory-overhead will be created as the ancestor node of the original input nodes when there are more than one input nodes. We have a temporary layer-group $\{v_1\}$ that can only be extended but not be reduced, and there are

---

[3] $|succ(v_i)|$ is the number of successor nodes of v_i.

two choices for $\{v_1\}$ as shown in Figure 7.8 : (1) attach it with one of its external successor node v_i if there is no cycle dependencies after graph growth, and the new temporary layer-group becomes $\{v_1, v_i\}$; (2) use it as a valid fused layer-group, i.e., $G_1 = \{v_1\}$, the new graph becomes $\mathcal{V} = \mathcal{V} - \{v_1\}$, and the temporary layer-group changes accordingly. In both cases, the algorithm works recursively, and thus the original problem is reduced into smaller sub-problems.

By summarizing over these two cases, we have the general form of sub-problems as: minimizing the memory cost of the graph $\mathcal{V}'$ which starts from a non-reducible layer-group G_{temp}. And we can represent the optimal value (minimum memory access number) of the sub-problems as $\mathcal{M}(G_{temp}, \mathcal{V}')$. Therefore, we obtain the optimal sub-structure property:

$$\mathcal{M}(\{v_1\}, \mathcal{V}) = \min \begin{cases} \min\limits_{v_i \in succ(\{v_1\})} \mathcal{M}(\{v_1, v_i\}, \mathcal{V}) \\ MinCost(\{v_1\}) + \mathcal{M}(\{v'\}, \mathcal{V} - \{v_1\}) \end{cases} \tag{7.4}$$

where $v_1 = input(\mathcal{V})$, $v' = input(\mathcal{V} - \{v_1\})$ are the input node of the graph $\mathcal{V}$, $\mathcal{V} - \{v_1\}$, respectively. $succ(v_i)$ returns the successor nodes of v_i.

7.3.6 Implementation of the Network Fusion Algorithm

With the optimal sub-structure property, the HaNF procedure is shown in Algorithm 4. The **Fuser** first checks the temporary layer-group G_{temp}, and then initializes it using the input node of the graph (lines 2–5) if it is null. Then, **Fuser** practices the basic steps of DP: 1) Lines 6–8 checks whether the sub-problem had already been solved and stored in the global DP table, which implements the "memorizing" mechanism of DP; 2) Lines 9–12 returns the result of the base case, i.e., when G_{temp} has no successor node. After that, the **Fuser** extends G_{temp} with one of its successor nodes and chooses the optimal one that has minimum off-chip memory traffic (lines 14–21) according to the first case of $\mathcal{M}(\cdot)$. Line 15 directly filters the invalid layer-groups that form a cyclic data dependency in the graph. Lines 22–29 separates the G_{temp} from the other nodes and start again from the remaining graph according to the second case of $\mathcal{M}(\cdot)$. Line 22 avoids the invalid fused layer groups over the virtual nodes. The other layer groups are valid and will be evaluated whether they are profitable or not. Finally, the algorithm returns the optimal layer fusion solutions and the minimum off-chip memory access volume (line 31).

In this way, HaNF explore the whole valid layer fusion space $\mathcal{L}$ and evaluated the memory cost $\sum_{G_i \in L} MinCost(G_i)$ of each layer fusion scheme $L = \{G_1, G_2, \ldots\} (\forall L \in \mathcal{L})$. The optimal layer fusion scheme and the cost of each sub-graph $\mathcal{V}'$ are stored after being evaluated by the memory-cost model, to avoid repetitively solving the same sub-problems.

Algorithm 4 HaNF

1: **In:** network graph $\mathcal{V}$; temporary layer-group G_{temp}; processor configuration acc
2: **Out:** optimal layer fusion solution S_{opt}; minimum off-chip memory access volume $minAccess$
3: **if** $G_{temp} = \varnothing$ **then**
4:     $v_1 \leftarrow input(\mathcal{V})$
5:     $G_{temp} = \{v_1\}$
6:     **if** $(G_{temp}, \mathcal{V}) \in table$ **then**
7:         **return** $table[(G_{temp}, \mathcal{V})]$
8:     **end if**
9:     **if** $succ(G_{temp}) = \varnothing$ **then**
10:         $table[(G_{temp}, \mathcal{V})] \leftarrow (\{G_{temp}\}, MinCost(G_{temp}, acc))$
11:         **return** $table[(G_{temp}, \mathcal{V})]$
12:     **end if**
13:     $(S_{opt}, minAccess) \leftarrow (\varnothing, \infty)$
14:     **for all** $v_i \in succ(G_{temp})$ **do**
15:         **if not** $isCycle(G_{temp}, \mathcal{V})$ **then**
16:             $(s, access) \leftarrow \text{Fuser}\mathcal{V}, G_{temp} + \{v_i\}, acc$
17:         **end if**
18:         **if** $access < minAccess$ **then**
19:             $(S_{opt}, minAccess) \leftarrow (s, access)$
20:         **end if**
21:         **if** $not\_only\_contain\_virtual\_node(G_{temp})$ **then**
22:             $(s, access) \leftarrow \text{Fuser}\mathcal{V} - G_{temp}, \varnothing, acc$
23:             $s \leftarrow \{G_{temp}\} \cup s$
24:             $access \leftarrow access + MinCost(G_{temp}, acc)$
25:             **if** $access < minAccess$ **then**
26:                 $(S_{opt}, minAccess) \leftarrow (s, access)$
27:             **end if**
28:         **end if**
29:     **end for**
30:     $table[(G_{temp}, \mathcal{V})] \leftarrow (S_{opt}, minAccess)$
31:     $(S_{opt}, minAccess)$
32: **end if**

TABLE 7.3 NPU Configurations

on-chip buffer	PE-array	register files/PE	dataflow	arithmetic units
128KB	32×32	64B	$X\|K^*$	16-bit fixed-point

* parallel the loop X and K

7.3.7 Evaluation of Memory Overhead

To fairly evaluate HaNF, we use the latest network schedulers, the work of Zheng et al. [283] (**baseline1**) and DNNVM [281] (**baseline2**) as the comparison baselines. We compare the end-to-end inference latency and the main memory access volume of HaNF with that of baselines using several state-of-the-art CNN models, including the classic AlexNet [286], VGG16 [287], GoogleNet [143], SqueezeNet [288], MobileNet [206], ResNet [140], and the latest NasNet designed via NAS [172]. Unless otherwise specified, the batch size of these workloads is set to 4. In evaluation, HaNF and the baselines are applied to the NPU as described in Figure 7.1, which are implemented and synthesized with Design Compiler using 65nm process technology. The NPU configurations are shown in Table 7.3. Figure 7.9 shows the latency and DRAM access comparison between the baselines and HaNF. Compared to baseline1 and baseline2, HaNF reduces and 15.0% – 52.4% and 16.8% – 72.2% DRAM accesses, respectively. HaNF also achieves significant performance gains when the DRAM cost dominates the network inference latency. Compared with baseline1 and baseline2, the performance of the HaNF scheduler shows an average improvement of 1.1x and 1.4x, respectively.

The analysis given in Section 7.3.4 indicates that the on-chip buffer capacity is an important factor that influences the main memory access traffic. We test the memory access volume of HaNF under different

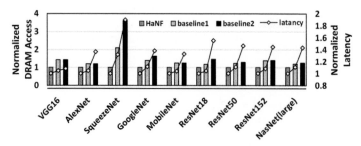

Figure 7.9 Latency and DRAM access comparison between baselines and HaNF.

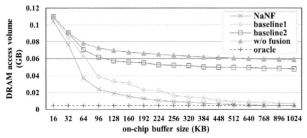

Figure 7.10 The impact of the on-chip buffer size on the performance of SqueezeNet.

on-chip buffer capacities, as shown in Figure 7.10. Oracle refers to the ideal case when the NPU has a on-chip buffer of unlimited size. W/o fusion refers to the implementation without network fusion optimization. It is seen that HaNF outperforms baseline1 and baseline2 in all cases with different on-chip buffer sizes. All these solutions under comparison achieve relatively more memory access reduction on larger on-chip buffers, while the marginal benefits brought by buffer up-sizing are higher when the on-chip buffer size is small (smaller than 96KB). When the on-chip buffer size is larger than 96KB, the restricted layer fusion space stops baseline2 from further reducing the main memory accesses, since there is no more layer-fusion opportunities available for baseline2 to exploit. In contrast, when baseline2 is no longer gaining benefits from buffer-size increase, HaNF and baseline1 still produce performance gains even though at a much lower rate than that in smaller buffers. The performance growth continues until all intermediate feature maps can be consumed on-chip and no longer evicted out to the main memory. Even in cases of smaller buffers, baseline1 does not perform as well as HaNF, because its simplified search strategy will omit some of the profitable layer-groups by default, and it does not have an accurate cost estimation to drive their search algorithm. However, when the on-chip buffer is large enough (>1MB), baseline1 can also approach Oracle.

7.3.8 Performance on Different Processors

HaNF can easily be applied to most of the deep learning processors and minimize their off-chip memory accesses. As a case study, we integrate HaNF into three DNN processors, ShiDianNao, Eyeriss, and [289], as shown in Table 7.4. Note that the buffer of ShiDianNao is split into three units: an input buffer, an output buffer, and a synapse buffer. [289]

TABLE 7.4 Reduction in Memory Access (%) Compared to the Processor Baselines

	AlexNet	VGG16	GoogleNet	ResNet18	SqueezeNet
Eyeriss	33.25	31.75	49.56	34.44	87.59
ShiDianNao	37.86	52.45	49.45	24.39	85.26
[289]	32.43	27.41	47.31	16.86	84.29

is designed to achieve the communication lower bound for the Conv layers, but do not consider the optimization of layer fusion. HaNF even achieves better results on Eyeriss without using data compression when compared to baseline Eyeriss with sparse compression, and it outperforms the baseline ShiDianNao when applied to the ShiDianNao accelerator. HaNF also further drops the off-chip communication lower bound for the convolutional accelerator proposed in [289], because HaNF can eliminate the unnecessary off-chip traffic induced by intermediate feature maps.

7.4 SCHEDULER-FRIENDLY NETWORK QUANTIZATIONS

7.4.1 The Problem of Layer Pipelining between CPU and Integer SUs

With the popularity of specialized neural accelerators as SUs in SoCs, quantizing the neural network models from high precision floating points to integer representation becomes a very important step. Many research works have been done to enable high accuracy and low bit-width quantization technology to shrink the model size or reduce the computation complexity of neural networks. However, not many of them consider that quantization is also an indispensable stage to deploy models on integer SUs, and sometimes they will leave certain layers such as the bottom and top layers unquantized or still keeps some of the parameters in a high precision state. This is not a big issue for floating-point engines such as GPGPU since such unquantized parameters often contributes to marginal memory footprint, but it makes some performance impacts on the SoCs that are integrated with weak CPU cores and integer SUs such as Diannao[244]. If some of the neural layers go unquantized, they will have to be scheduled onto the CPU cores because they are not operable on pure integer SUs. As we discussed in previous sections, partitioning the neural networks between CPUs and SUs will raise data communication overhead especially when they have to synchronize data through the power-consuming shared off-chip memory. For example, for some of the generative networks, the instance normalization layers, will not be quantized into low-precision for the sake of model accuracy and must be executed on a general-purpose processor, which will degrades the inference performance significantly. In our simulation with real-world content generation tasks [290, 291, 292], when leaving the unquantized Instance Normalization (IN) layers on the CPU while the other layers on an integer SU, there will be considerable runtime latency growth due to IN outsourcing [293]. The explanation for the poor performance is two-fold: first, data exchange through the off-chip memory is very expensive when we partition the network into fine-grained blocks and separate them between SUs and CPUs. Second, general-purpose processors are much slower than specialized processors when dealing with neural layers.

Therefore, partitioning the neural networks across the CPU and integer SUs due to inconsistent quantization will harm the end-to-end latency of network inference, even though it may not make a huge difference in the throughput if balanced pipelining is achieved. Thus, for the latency and energy critical scenarios, scheduler-friendly quantization will try to convert all of the neural network parameters into hardware-consistent integers so that the CPU cores run only the control-intensive and high precision scalar operations that we have to move out of the SUs.

7.4.2 Introduction to Neural Network Quantization for Integer Neural Accelerators

Quantization technique is not only widely used to shrink the size of large-scale DNN models, but also closely related to the implementation of specialized hardware that maps the procedure of network inference onto energy-efficient low-precision integer or fixed-point arithmetic circuits. From the hardware perspective, low-precision integer accelerators or processors are dominating the solutions targeted on neural network inference, especially for mobile and embedded scenarios. Google's Tensor Processing Unit 1.0 (TPU) [294], Unified Deep Neural Network Accelerator (UNPU) [295], Eyeriss [245], Stripes [296], Pragmatic[297], and many other newly proposed hardware implementations are generally reliant on the effectiveness of the underlying quantization techniques, which are crucial for the low-precision integer hardware designed to process binary, ternary, 4-bit or 8-bit networks. In other words, quantization is not only a method to reduce the memory footprint as in prior work on model compression, but also a mandatory step to make the network deployable on integer hardware.

Though there is a lot of prior work that investigates low-precision quantization, they mainly target on reducing the memory overhead caused by floating or high precision data representation in the networks, but not focus on specialized integer hardware for network inference. To enable neural network processors to work with low-precision integer operands and minimize the accuracy losses, a good network quantizer must satisfy the constraints as enlisted in Table 7.5.

First, all the parameters, including weights, bias, activation, partial results that eventually accumulate to an activation, and even the scaling factors, which are indispensable for low-precision networks like binary and ternary representation, must be quantized into low bitwidth integers as

TABLE 7.5 Comparison between Different Quantizers: *All-Layer (AL)* denotes quantizing all the parameters of all the operators in networks, including weights, bias, activations, and the scaling factor for low-precision networks; *(BatchNorm) BN* donates that the BN operation is only invoked in training but merged into weights and induces no overhead in integer inference; *Linear-Symmetric (LS)* denotes linear symmetric quantization; *Activation Functions (AF)* donates the support of *Leaky ReLU* and activation functions besides *ReLU*. *Structure-Intact (SI)* indicates the network structure is unmodified.

Method	AL	BN	LS	AF[e]	SI
Deep Compression [261][b]					✓
WQ [298][b]					✓
LQ-Nets [299]					✓
Min-Max Linear Quantization[a]				✓	✓
DoReFa [300][c]				✓	✓
RQ [301]			✓	✓	✓
WRPN [302]		✓	✓	✓	
PACT [303][d]			✓	✓	✓
LLSQ(ours)	✓	✓	✓	✓	✓

[a] Naive linear quantization, which finds min-max value at runtime.
[b] Clustering-based approaches to quantize weights.
[c] DoReFa falls into linear asymmetric quantizer due to the need for offset.
[d] In [303], they use PACT to quantize activations, and DoReFa to quantize weights.
[e] DoReFa, RQ, WRPN, and PACT are designed for *ReLU*, but they can be extended to support other activation functions in theory.

required by the underlying specialized hardware. In some prior work [300, 304, 299, 302, 303], they either leave bias and scaling factors unquantized or keep the first and last layer in full or high precision. Besides, some designs rely on high-precision internal register or ALUs to support high-precision partial results that are generated during the computation stage before the final output of activations or features. For example, [305], which quantizes the weights and activations to 8-bit, directly use 32-bit accumulators to cache the intermediate values or partial results to avoid register overflows. However, for 4-bit and lower bitwidth, the integer accelerators cannot afford high bitwidth accumulators, which indicates higher silicon area and power cost. For integer-only-arithmetic, we quantize the bias to fixed-point numbers by using a straight-forward method. The value range of these numbers is wide, resulting in overflows of the low bitwidth accumulators. To overcome this problem, we quantize the bias to 8-bit and finetune the bias of the model. As shown in Figure 7.11, the bitwidth of accumulators can be reduced to 16-bit.

Second, the *BatchNorm (BN)* layer does not necessarily need to be processed on the NPU or the CPU during inference. For most of the convolutional neural networks, *BN* layers are often after the *Conv* or *FC* layers. In these situations, *BN* can be merged into the weights and biases of the corresponding *Conv* or *FC* layers. However, in [300, 299],

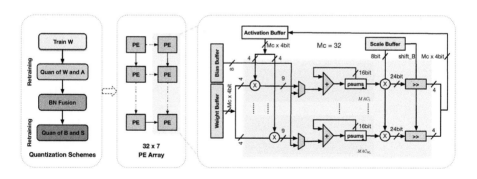

Figure 7.11 An overview of LLSQ: using pre-trained weights for fast convergence; Retraining of the network with quantized weights and activations; *BN* fusion for efficient inference; Quantization of bias and scaling factors; Deployment of the quantized model to our accelerator. As shown in this figure, weights, activations, bias, and scaling factors are quantized to low-bit integers. And the bandwidth of accumulator can be set to lower (e.g., 16-bit in our experiments).

they use asymmetric or non-linear quantization, causing barriers to BN fusion. There are two ways to overcome this obstacle. One is "BN folded training"[305], which adopts BN fusion before weights quantization in every training step; the other is to use symmetric linear quantization. However, the first method doubles the training time, while the second one has no additional computational overhead, which will be introduced in Section 7.4.7.

Third, linear quantization is necessary for state-of-the-art accelerators. There are many non-linear quantization methods which achieve excellent bitwidth reduction efficacy and accuracy tradeoffs. In these cases, it requires additional transformation to have correct arithmetic results after quantizing the value into non-linear distribution. For example, as in [261, 298], it necessitates the operation of table lookup to have correct multiplication between quantized values. However, the linear quantization can make full use of the low-precision arithmetic components in off-the-shelf accelerators. Further, linear quantization can be divided into symmetric mode and asymmetric mode. Asymmetric quantization has one more parameter (e.g., zero-point [305]) than symmetric quantization, and it requires additional subtraction or linear-operation before multiplication. As a result, the symmetrical mode is compatible with the mainstream integer accelerator chip design and does not require the redesign of datapath in these hardware.

Fourth, different CNNs or applications usually use a variety of activation functions. For instance, the object detection model [306] typically uses *Leaky ReLU*. And the bottleneck of ResNet block does not use any activation function. The quantization methods are expected to be adapted to these situations. However, [299, 298] only focus on the quantization of activations after *ReLU*. In this paper, we demonstrate our method is friendly to different activation methods such as *Leaky ReLU*.

Some of the previous researches change the network structure for better quantization performance, e.g., [302] double or even triple the convolutional filters to reduce accuracy degradation. For the energy-efficient integer neural network chips, it needs to remap the changed network architecture to hardware and adds to computational and memory access overhead due to the increased filters and parameters. As a result, keeping the network structure intact is important. Concerning all the factors above, in this section, we present a learned linear symmetric quantization (LLSQ) method that enable the quantized models to be completely executed on the integer neural accelerator, which reduces the overhead of scheduling the network seperately onto high-precision CPU and integer SU.

7.4.3 Related Work of Neural Network Quantization

Edge or embedded neural network accelerators generally have three primary design goals— small-footprint, high-throughput/low-latency, and low-power. For different applications and scenarios, the prior researches on specialized deep learning processors are often falling into different categories: cloud-oriented hardware for warehouse machines, low-power mobile processors and ultra-low power accelerators for IoT or cyber-physical devices.

For mobile and embedded usage, specialized neural network processors are becoming increasingly popular as an efficient hardware solution of inference. DianNao [244] is proposed for fast inference of DNNs and it uses 16-bit fixed-point multipliers for small silicon area and low-energy. Later, ShiDianNao [307] is introduced and it burns extremely low energy consumption by putting all weights onto the SRAM to eliminate considerable DRAM accesses. Besides, DeepBurning [308] simplifies the design flow of accelerator for different NN models. Eyeriss [245] is also another representative of low-power accelerators. And it presents a row-stationary (RS) dataflow to minimize data movement energy consumption on a spatial architecture. To further reduce computation overhead, EIE [309] exploits the sparsity and low-bit compression of the NNs and achieves better throughput, energy and area efficiency. These typical edge neural network processors are accepting fixed-point data input and using fixed-point processing elements to reduce the power and chip area overhead caused by floating-point arithmetic components and memory. For the cloud scenarios, specialized architectures like TPU [294] and FPGA-based accelerator cards are also replacing conventional GPGPU and CPU for high-throughput inference tasks. Even for cloud-oriented inference architectures, fixed-point processing architectures like TPU are favored because they are able to deliver much higher throughput for the given power budget and silicon area overhead.

However, for the fixed-point or integer hardware targeted on neural network acceleration, quantization is prerequisite to convert the floating-point network model into the fixed-point format compatible with the specialized hardware, and it is also a critical step to ensure the accuracy of the network after conversion. Many prior quantization methods are intended to reduce the running overhead of networks but ignore the architecture and working mechanism of integer neural network processors, as illustrated in Table 7.5, and they sometimes face considerable accuracy losses, or performance penalty or even fail to be supported on the realistic integer datapath due to the unconsciousness of the underlying hardware. This problem becomes particularly important for the hardware that is designed to run low bitwidth networks such as binary, ternary, and 2/4-bit models. For instance, Deep compression and WQ are clustering-based quantization methods, and they still need high-precision values to represent the weights, bias, and activations. As a result, they are not compatible with the hardware that only supports low-precision computing. LQ-Nets uses non-linear quantization based on the binary code and basis vector, and it can theoretically calculate the inner products between quantized weights and activations by bitwise operations only. However, it requires intensive modifications to the design of current processors by adding a lot of look-up tables in the datapath. Further, bias and scaling factors are not quantized in PACT and WRPN, resulting in performance penalty when employing additional high-precision or float-point ALUs to deal with them. In contrast, our LLSQ is designed to ease the model quantization flow for the specialized integer neural network processors by conforming to the constraints specified in Table 7.5. To validate the importance of hardware-aware quantizer and software/hardware co-design, we also design a specialized CNN accelerator for wearable applications. And the specialized accelerator supports 2/4/8 integer operation and adopts the dataflow of low latency and energy design.

7.4.4 Linear Symmetric Quantization for Low-Precision Integer Hardware

In this section, we present a learned linear symmetric quantizer (LLSQ) [242] for integer neural network processors, which is fully aware of the integer hardware that has only the low-precision arithmetic units such as the partial-sum accumulators (e.g., 16-bit) and multipliers (e.g., 4-bit). Accordingly, all the model parameters, including weights, bias, activations, partial results that eventually accumulate to an activation, and even the scaling factors, which are indispensable for low-precision networks like binary and ternary representation, are quantized into low bitwidth integers. After that, the quantized models can be deployed onto specialized accelerators with integer-only-arithmetic and can avoid the costly pipelining between SUs and CPU.

LLSQ [242] adopts a unified way to quantize weights and activations. Figure 7.11 shows an overview of the quantization scheme. Compared with prior work, the LLSQ scheme pays more attention to the constraints imposed by real hardware. Thereby it employs a unified learned linear symmetric quantizer to quantize weights and activations. And the quantizer has only one parameter, known as the scaling factor. Linear symmetric quantization consumes little additional resources based on the mainstream integer accelerator designs while achieving state-of-the-art accuracy in various networks. After that, LLSQ adopts *BN* fusion for fast inference on hardware. As for bias and scaling factors, LLSQ also quantizes them to low-bitwidth integers. The integer accelerator illustrated in Figure 7.11 is an illustrative case of 4-bit quantization and hardware acceleration.

7.4.5 Making Full Use of the Pre-Trained Parameters

LLSQ shows that it is more efficient to start with the pre-trained full-precision parameters before quantization. [301, 310] also use pre-trained weights for fast convergence and deployment, while [299, 303, 311] train quantized network from scratch to show the robustness of the algorithm. However, for some object detection models, the backbone models and pre-trained weights are essential to the detection performance. [306] shows that the pre-trained high-resolution classification network gives an increase of almost 4% *mAP*. To have better performance in classification, object detection, and other CNN-based tasks, state of the art quantization approaches use pre-trained parameters to initialize the networks.

7.4.6 Low-Precision Representation and Quantization Algorithm

LLSQ uses channel-wise quantization for *Conv* layers and layer-wise quantization for *FC* layers and activations. And LLSQ adopt the symmetric linear quantization to quantize weights or activations into k bits words(e.g., 4-bit), which can be defined as

$$\boldsymbol{x}^q = Quantize_k(\boldsymbol{x}^r \mid \alpha)$$
$$\boldsymbol{q} = \frac{\boldsymbol{x}^q}{\alpha} = clamp(\lfloor \frac{\boldsymbol{x}^r}{\alpha} \rceil, -2^{k-1}, 2^{k-1} - 1) \quad (7.5)$$

where $\boldsymbol{x}^r \in \mathbb{R}$ is one kernel of weights or one layer of activations, the variable $\alpha \in \mathbb{R}^+$ is the quantization parameter, known as the scaling factor, while $\boldsymbol{q} \in \{-2^{k-1}, \ldots, 0, 1, \ldots, 2^{k-1} - 1\}$ is the integer values flowing in the integer accelerator and $\boldsymbol{x}^q \in \{-2^{k-1}\alpha, \ldots, 0, \alpha, \ldots, (2^{k-1} -$

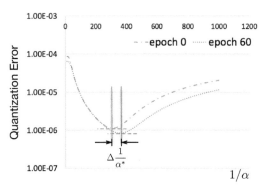

Figure 7.12 L2 distance of quantization. The data is from weights of the first FC layer in AlexNet. As shown in the figure, the optimal α^* changes with the updating of weights.

$1)\alpha\}$ is the quantized weights or activations. Note that for activations, which are non-negative values if the ActFun is $ReLU$, LLSQ clamps them to $[0, 2^k - 1]$, resulting in $q \in \{0, 1, \ldots, 2^k - 1\}$ and $x^q \in \{0, \alpha, \ldots, (2^{k-1} - 1)\alpha\}$, respectively.

As defined above, LLSQ uses α as the quantization parameter. And LLSQ optimizes it with:

$$\alpha^* = \arg\min_{\alpha} \int p(x^r)|x^q - x^r|^2 \qquad (7.6)$$

where x^r, x^q are the same factors defined in equation 7.5, $p(x^r)$ is the probability density distribution of x^r. In Figure 7.12, LLSQ presents the relationship between quantization error and α. When fixing weights x^r, we can find the optimal α^* by using the brute-force search approach, which induces high computation cost. Besides, the weights are updated during the re-training phase and the optimal value α^* changes accordingly. In other words, the optimal value for the factors is not fixed and it takes considerable computational overhead to find the dynamic optimal value.

LLSQ shows that there is no need to find the optimal value α^*, and it works well enough to find a near-optimal value $\tilde{\alpha}^*$. Generally, quantization can be considered as a regularization of the networks, and the quantization parameter α needs only to be adjusted to a near-optimal value to preserve the network capacity.

Then the problem becomes how to find a $\tilde{\alpha}^*$ in the forward pass of the network training phase. At the beginning of training, LLSQ assigns α an initial value. Then in every training iteration, LLSQ explore between

2α and $\alpha/2$ to find a better search direction $d_{better} \in \{-1, 0, 1\}$, and use $-\alpha^2 d_{better}$ as the simulated gradient (SG) of α which is detailed in equation 7.7. The gradients of other parameters are still obtained by backpropagation algorithm. After that, LLSQ update all parameters with the gradients or simulated gradients.

$$E_m = \sum_i (x_i^r - quantize_k(x_i^r \mid \alpha))^2$$

$$E_l = \sum_i (x_i^r - quantize_k(x_i^r \mid \frac{\alpha}{2}))^2$$

$$E_r = \sum_i (x_i^r - quantize_k(x_i^r \mid 2\alpha))^2 \qquad (7.7)$$

$$d_{better} = \arg\min([E_l, E_m, E_r]) - 1$$

$$\Delta G_\alpha = -\alpha^2 d_{better}$$

where x^r is one kernel of weights or one layer of activations. $\arg\min([E_l, E_m, E_r]) \in \{0, 1, 2\}$ selects the index of the smallest number in the array $[E_l, E_m, E_r]$.

7.4.7 BN Layer Fusion of Quantized Networks

LLSQ shows that merging the BN layers into convolutional layers can reduce the latency of network inference by removing additional computation overhead. The operator of quantized $Conv^4$ and FC layers can be expressed as

$$o = \alpha_a q_a \alpha_w q_w + b \qquad (7.8)$$

where α, q are the same as equation 7.5, $\alpha_a q_a$, $\alpha_w q_w$ donate the quantized activations and weights, while b is the bias and o is the output feature vector. Note that α_a, α_w, and b are full precision values. And the BN layer can be formulated as follows:

$$y = \frac{o - \mu}{\sqrt{\sigma^2 + \epsilon}} \gamma + \beta \qquad (7.9)$$

where μ and σ^2 are EMA statistics, γ and β are learned parameters in BN layers.

---

[4]For brevity, LLSQ only considers the operation of one channel.

Obviously, we can merge *BN* layers and figure out the corrected parameters:

$$\hat{\alpha}_w = \zeta\alpha_w; \quad \hat{b} = (b-\mu)\zeta + \beta; \quad \zeta = \frac{\gamma}{\sqrt{\sigma^2 + \epsilon}} \tag{7.10}$$

7.4.8 Bias and Scaling Factor Quantization for Low-Precision Integer Operation

Further, the outputs of layers are quantized according to equation 7.5. For integer-only-arithmetic, the bias use $\alpha_a\alpha_w$ as its scaling factor. And for the multiplier $\frac{\alpha_a\alpha_w}{\alpha_o}$, we use bit-shift quantization (See equation 7.12) so that no multiplication but bit-shift operation is needed in hardware.

$$\alpha_o \boldsymbol{q}_o = \alpha_a \boldsymbol{q}_a \hat{\alpha}_w \boldsymbol{q}_w + \hat{b}$$

$$\boldsymbol{q}_o = \frac{\alpha_a \hat{\alpha}_w}{\alpha_o}(\boldsymbol{q}_a\boldsymbol{q}_w + q_b) \tag{7.11}$$

$$where \; q_b = clamp(\lfloor \frac{\hat{b}}{\alpha_a\hat{\alpha}_w} \rceil, -2^{k_b-1}, 2^{k_b-1} - 1)$$

Note that $\alpha_a\alpha_w$ is a very small number, resulting in large quantization noise when adopting the clamp operation. In addition, the quantization of the scaling factors α can also raise the quantization noise of weights and activations. As a result, parameter re-training is required.

$$\boldsymbol{\alpha}^q = SQ_k(\boldsymbol{\alpha})$$

$$= \frac{clamp(round(2^{qcode} \cdot \boldsymbol{\alpha}), -2^{k-1}, 2^{k-1} - 1)}{2^{qcode}} \tag{7.12}$$

where $\boldsymbol{\alpha} \in \mathbb{R}^{+len(\boldsymbol{\alpha})}$ is the scaling factors to be quantized, $k \in \mathbb{Z}$ is the bitwidth, and $qcode \in \mathbb{Z}$ is the parameter of the bit-shift quantizer simply obtained by:

$$qcode = k - ceil(\log_2(\max(\boldsymbol{\alpha})) + 1 - 10^{-5}) \tag{7.13}$$

In the re-training phase, we adopt STE [312] to realize the non-differentiable quantization function.

For weights and bias, we have

$$\frac{\partial y}{\partial w^q} \simeq \frac{\partial y}{\partial w^r}; \quad \frac{\partial y}{\partial b^q} \simeq \frac{\partial y}{\partial b^r} \tag{7.14}$$

For activations, we have

$$\frac{\partial y}{\partial a^q} \simeq \begin{cases} \frac{\partial y}{\partial a^r} & if\ 0 \leq a^r \leq (2^k - 1)\alpha \\ 0 & otherwise \end{cases} \qquad (7.15)$$

7.4.9 Evaluation Results

In this section, three sets of experiments on Cifar10, ImageNet, and Pascal VOC datasets are presented. First, we conduct our proposed learned linear symmetric quantization (LLSQ) on weights and activations, leaving the first and last layers in full precision for a fair comparison with [299]. Second, we quantize the whole networks including the first and last layers, which is referred as LLSQF (LLSQ for Full network). Finally, we quantize the remaining bias and scaling factors. LLSQ is implemented in PyTorch [313], and most of the baselines it uses in evaluation are from PyTorch Model Zoo[5].

Quantization of Weights and Activations

We first employ the VGG-Small network on Cifar10 to verify the LLSQ method. After that, we use AlexNet [286], ResNet18, and ResNet34 [140], particularly the light-weight and hard-to-compress network architecture of MobileNet [206, 314] etc., to conduct more detailed experiments on the ImageNet dataset. Finally, we also quantize YOLOv2 [315] to demonstrate that LLSQ also works well for complicated applications and especially the task adopting the activation functions like *Leaky ReLU* other than *ReLU* used in previous work.

VGG-Small on Cifar10

The VGG-Small architecture is the same with [301, 299], consisting of six *Conv* layers, three *MaxPool* layers, and one *FC* layer. We adopt a cosine learn rate scheduler to train the VGG-Small reference and the quantized models. Specifically, we train the reference network for 400 epochs using an initial learning rate of 2e-2. And for the training of the quantized network, we use a warmup learning rate scheduler in the first ten epochs with an initial learning rate of 2e-3. In all quantization experiments, the total training epochs are 100. The VGG-Small quantization results are provided in Table 7.6. With 3-bit weights and 3-bit activations, the

---

[5]https://pytorch.org/docs/stable/torchvision/models.html

TABLE 7.6 Comparison with the State-of-the-art Low-bit Quantization Methods on CIFAR-10. The Bitwidth for Weights(w), Activations(a), Bias(b) and Scaling factor(α) are given.

Method	# Bits $w/a/b/\alpha$	Acc(%)	Degradation(%)
LQ-Nets[*]	Reference	93.8	
[299]	3/3	93.8	0.0
	2/2	93.5	0.3
RQ	Reference	93.05	
[301]	8/8	93.30	-0.25
	4/4	91.57	1.48
	2/2	90.92	2.31
LLSQ[*](ours)	Reference	93.34	
	4/4	**94.34**	-1.00
	3/3	**94.02**	-0.68
	2/2	93.31	0.03
LLSQF(ours)	4/4	94.30	-0.96
	3/3	94.07	-0.73
	2/2	93.12	0.22
	4/4/8/8	93.84	-0.50

[*] first and last layer in full precision

accuracy using our method is better than state-of-the-art method, LQ-Nets. And even when the first and last layers are all quantized in the same way, the loss of accuracy is minimal.

ImageNet Dataset

We then quantize AlexNet, ResNet18, ResNet34, and MobileNetv2 on ILSVRC2012 dataset with different bitwidth configuration to demonstrate the effectiveness of the method. All of the pre-trained float-point weights except MobileNetv2[6] are downloaded from the PyTorch Model Zoo, and they are trained for 90 epochs with a step learning rate scheduler. After loading the pre-trained weights, we employ a warmup learning scheduler in the first three epochs and the cosine scheduler in the remained 57 epochs with an initial learning rate of 2e-2.

As shown in Figure 7.13, when quantizing both weights and activations, our degradation of accuracy is significantly smaller than

---

[6]https://github.com/tonylins/pytorch-mobilenet-v2

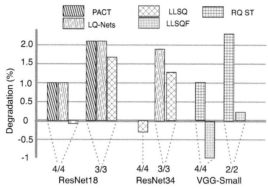

Figure 7.13 Quantization results on different networks. Comparison with other state-of-art methods. Lower is better.

LQ-Nets[299], PACT[303], and RQ[301]. Especially, LLSQ outperforms the baselines when quantizing weights and activations into 4-bit. And it also outperforms other non-linear quantization methods with different bitwidth. In overall, the accuracy drop is less than 0.4% in ResNet18, ResNet34, and AlexNet when quantizing the whole network. We also quantize MobileNetv2, a more compact network, and obtain results that are significantly better than RQ.

Object Detection on Pascal VOC

We also apply the proposed LLSQ to YOLOv2. The backbone of YOLOv2 is Darknet-19, and its activation function is *Leaky ReLU*, so that the activations contain negative values. For YOLOv2 on Pascal VOC, we adopt the same quantization configuration (see Section 7.4.6) of the weights to the activations. Results are listed in Table 7.7. As shown in the table, LLSQ induces minor losses of *mAP* in different bitwidth presentation. For comparison, we also quantize the activations into signed 5-bit integers using PACT, and consequently face considerable mAP losses ($54.8mAP$). Please note that we use the open-source PyTorch implementation of YOLOv2 [7] as the baseline. We train the quantized model for 170 epochs (2/3 of baseline) with an initial learning rate of 1e-4 (1/10 of baseline).

---

[7]https://github.com/marvis/pytorch-yolo2

TABLE 7.7 LLSQ on YOLOv2 Detector

bitwidth	FP32	w4a5	w32a5	w4a8	w4a32
mAP	73.2	70.3	71.2	73.4	74.2

BN Fusion and Quantization of Bias and the Scaling Factor

We adopt *BN* fusion in the PostAct (*Conv→BN→ReLU*) networks according to the formula in Section 7.4.7. And the scaling factor of bias is the product of the corresponding scaling factors belonging to the activations and the weights, respectively. After that, we visualize the bias value distribution of VGG-Small. Figure 7.14 shows b/α_b is distributed between a vast range $(-1000, 1000)$, resulting in overflows of low bitwidth accumulators. And the overflow phenomena have a significantly harmful impact on the network performance. To deal with this issue, we quantize the bias and the scaling factors to 8-bit words, and then fine-tune the networks to restore the original performance. Generally, we need fine-tuning for one epoch only. After the quantization of bias and scaling factor, we have a fully quantized model and have it deployed onto our integer-only accelerator with 16 bitwidth accumulators. Table 7.6 show, that the accuracy loss is negligible with $\boldsymbol{w4a4b8\alpha8}$ quantization on both VGG-Small.

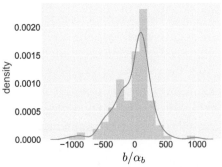

Figure 7.14 Distribution of the bias/scaling factor. The data is from VGG-Small with w4a4 quantization.

Deployment onto Realistic Hardware

The introduced linear symmetric quantization is intended to deploy the quantized networks to specialized integer-only-arithmetic CNN accelerators or other integer-only hardware. Our accelerators adopt the typical 2D systolic array architecture [245], but they are featured with 4-bit or 2-bit low-precision operation. As shown in Figure 7.11, the 8/4/2-bit accelerator has a 32x7 array of processing elements (PE). And the MAC unit in each PE consists of a 4-bit multiplier and a 16-bit accumulator. For the 4-bit accelerator, we use INT4 representation for weights, UINT4 for activations, INT8 for the bias and scaling factors, respectively. For the 2-bit accelerator, we use INT2 for weights, UINT2 for activations, INT8 for the bias and scaling factors, respectively. Through the quantization process described in the paper, we can have a fully quantized network that works directly on the CNN accelerator. In addition, as we use linear symmetric quantization, we can use a straight-forward way to conduct multiply-accumulate operations without introducing shifters or lookup tables, which means the quantized models can run on state-of-the-art integer accelerators and ensures that their output accuracy degradation is minimal as presented in the above sections. Finally, we implement the 8/4/2-bit integer neural network processors with Synopsys Design Compiler (DC) under the 40nm technology, clocked at 800MHz. Table 7.8 shows that the 4/2-bit implementation achieves up to 2.56x lower silicon area and 5.56x lower power compared to that of the 8-bit baseline.

TABLE 7.8 Comparison of our Low-Precision Integer Neural Network Processors

Bitwidth	#MAC Unit	Throughput (GOps/sec)	Silicon Area (mm^2)	Power (mW)
8-bit	224	179.2	4.71	228
4-bit	224	179.2	**2.80**	**93**
2-bit	224	179.2	**1.84**	**41**

Implemented and synthesized with Synopsys Design Compiler (DC) under the 40nm technology.

7.5 SUMMARY

As AI begins to enter into many different fields, the implementation of AI algorithms on different hardware devices is also the major concerns for the system designers who want to unleash the power of AI algorithms in real-world problems. Especially for the widely-used low power and mobile devices, which embraces heterogeneous processor architectures and hardware specialization to increase system energy efficiency, how to analyze, decompose and map the AI algorithms onto the underlying hardware significantly impacts the implementation costs of the task. In this chapter, by investigating into our LPIRC solutions, we show how to manually schedule the neural network based computer vision tasks onto the typical mobile SoCs and achieve the high utility of computing resources and energy. In addition to the manual scheduling method, we also introduce and model the fine-grained neural network scheduling problem on the processors. It is proved that the model and the task-graph based scheduler will lead to the memory-level optimality on the popular neural network processors. Finally, besides the scheduler itself, a scheduler-friendly network quantizer is described and it reveals that the algorithm-level optimization take to the neural networks must be coordinated with the architecture of the hardware. We believe that there will be a constant evolution of novel hardware architectures and AI models, however, as the critical bridge between the algorithms and the hardware, the scheduling and mapping layer should be carefully designed and optimized to make the best use of the system energy, computing resources and finally the power of AI techniques.

FURTHER READING

[316] Zhao, X., Wang, Y., Cai, X., Liu, C., and Zhang, L. (2019, September). Linear symmetric quantization of neural networks for low-precision integer hardware. *In International Conference on Learning Representations.*

[317] Wang, Y., Quan, Z., Li, J., Han, Y., Li, H., and Li, X. (2018, March). A retrospective evaluation of energy-efficient object detection solutions on embedded devices. *In 2018 Design, Automation and Test in Europe Conference and Exhibition (DATE)* (pp. 709-714). IEEE.

[318] Zhao, X., Wang, Y., Liu, C., Shi, C., Tu, K., and Zhang, L. (2020, July). BitPruner: network pruning for bit-serial accelerators. *In 2020 57th ACM/IEEE Design Automation Conference (DAC)* (pp. 1-6). IEEE.

[252] Wang, C., Wang, Y., Han, Y., Song, L., Quan, Z., Li, J., and Li, X. (2017, January). CNN-based object detection solutions for embedded heterogeneous multicore SoCs. *In 2017 22nd Asia and South Pacific Design Automation Conference (ASP-DAC)* (pp. 105-110). IEEE.

[293] Xu, H., Wang, Y., Wang, Y., Li, J., Liu, B., and Han, Y. (2019, November). ACG-Engine: An Inference Accelerator for Content Generative Neural Networks. In 2019 IEEE/ACM International Conference on Computer-Aided Design (ICCAD) (pp. 1-7). IEEE.

[319] Xu, D., Liu, C., Wang, Y., Tu, K., He, B., and Zhang, L. (2020). Accelerating generative neural networks on unmodified deep learning processors—A software approach. IEEE Transactions on Computers, 69(8), 1172-1184. .

[320] Cai, X., Wang, Y., Zhang, L. (2021) Optimus: Towards Optimal Layer-Fusion on Deep Learning Processors In The 22st ACM SIGPLAN/SIGBED Conference on Languages, Compilers, and Tools for Embedded Systems.

[321] Chen, W., Wang, Y., Lin, G., Gao, C., Liu, C., and Zhang, L.. CHaNAS: Coordinated Search for Network Architecture and Scheduling Policy. In The 22st ACM SIGPLAN/SIGBED Conference on Languages, Compilers, and Tools for Embedded Systems.

CHAPTER 8

Efficient Neural Network Architectures

Han Cai and Song Han

Massachusetts Institute of Technology

CONTENTS

Designing efficient neural network architectures is a widely adopted approach to improve efficiency, in addition to compressing an existing deep neural network. A CNN model typically consists of convolution layers, pooling layers, and fully-connected layers, where most of the computation comes from convolution layers. For example, in ResNet-50 [140], more than 99% multiply-accumulate operations (MACs) are from convolution layers. Therefore, designing efficient convolution layers is the core of building efficient CNN architectures.

This chapter first describes the standard convolution layer and then describes three efficient variants of the standard convolution layer. Next, we present three representative manually design efficient CNN architectures, including SqueezeNet [288], MobileNets [206, 190], and Shuf-fleNets [207, 322]. Finally, we describe automated methods for designing efficient CNN architectures.

DOI: 10.1201/9781003162810-8

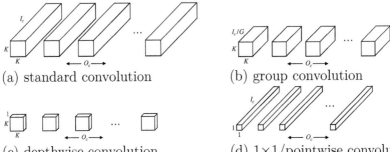

(a) standard convolution   (b) group convolution

(c) depthwise convolution   (d) 1×1/pointwise convolution

Figure 8.1 Standard convolution and three commonly used efficient variants. 1×1 (pointwise) convolution is a special kind of standard convolution, which uses a kernel size of 1 to reduce #Params and #MACs. Group convolution divides the input feature map into G group along the channel dimension and processes each group independently. Depthwise convolution is an extreme case of group convolution where G equals the number of input channels.

8.1   STANDARD CONVOLUTION LAYER

A standard convolution layer is parameterized by convolution kernel $\mathbf{K}$ of size $O_c \times I_c \times K \times K$ where O_c is the number of output channels, I_c is the number of input channels, K is the spatial dimension of the kernel (Figure 8.1a). Here, we assume the convolution kernel's width and height are the same for simplicity. It is also possible to have asymmetric convolution kernels [323].

Given input feature map $\mathbf{F_i}$ of size $I_c \times H \times W$, the output feature map $\mathbf{F_o}$ of size $O_c \times H \times W$ is computed as follows[1]:

$$\mathbf{F_o}[n, h, w] = \sum_{m,i,j} \mathbf{K}[n, m, i, j] \times \mathbf{F_i}[m, h+i-\lfloor K/2 \rfloor, w+j-\lfloor K/2 \rfloor]. \quad (8.1)$$

In the following discussions, we use $\mathbf{F_o} = Conv_{K \times K}(\mathbf{F_i}; \mathbf{K})$ to represent a standard convolution layer with kernel size K. According to equation (8.1), the computational cost of a standard convolution is

$$\#\text{MACs}(Conv_{K \times K}) = H \times W \times O_c \times I_c \times K \times K, \quad (8.2)$$

while the number of parameters is given as

$$\#\text{Params}(Conv_{K \times K}) = O_c \times I_c \times K \times K. \quad (8.3)$$

---

[1]Assuming the stride is 1 and zero-padding is applied to preserve the spatial size.

8.2 EFFICIENT CONVOLUTION LAYERS

1×1 (Pointwise) Convolution. 1×1 convolution (also called pointwise convolution) [324] is a special kind of standard convolution layer, where the kernel size K is 1 (Figure 8.1d). According to equations (8.2) and (8.3), replacing a $K \times K$ standard convolution layer with a 1×1 convolution layer will reduce #MACs and #Params by K^2 times. In practice, as the 1×1 convolution itself cannot aggregate spatial information, it is combined with other convolution layers to form CNN architectures. For example, 1×1 convolution is usually used to reduce/increase the channel dimension of the feature map in CNN.

Group Convolution. Different from 1×1 convolution that reduces the cost by decreasing the kernel size dimension, group convolution reduces the cost by decreasing the channel dimension. Specifically, the input feature map $\mathbf{F_i}$ is split into G groups along the channel dimension (Figure 8.1b):

$$\text{split}(\mathbf{F_i}) = (\mathbf{F_i}[0:c,:,:], \mathbf{F_i}[c:2c,:,:], \cdots, \mathbf{F_i}[I_c-c:I_c,:,:]), \quad \text{where } c = I_c/G.$$

Then each group is fed to a standard $K \times K$ convolution of size $\frac{O_c}{G} \times \frac{I_c}{G} \times K \times K$. Finally, the outputs are concatenated along the channel dimension. Compared to a standard $K \times K$ convolution, #MACs and #Params are reduced by $G\times$ in a group convolution.

Depthwise Convolution. The number of groups G is an adjustable hyper-parameter in group convolutions. A larger G leads to lower computational cost and fewer parameters. An extreme case is that G equals the number of input channels I_c. In that case, the group convolution layer is called a depthwise convolution (Figure 8.1c). #MACs and #Params of a depthwise convolution are

$$\#MACs(DWConv_{K \times K}) = H \times W \times O_c \times K \times K, \quad (8.4)$$

$$\#Params(DWConv_{K \times K}) = O_c \times K \times K, \quad (8.5)$$

$$\text{where } O_c = I_c.$$

8.3 MANUALLY DESIGNED EFFICIENT CNN MODELS

SqueezeNet (Figure 8.2). SqueezeNet [288] targets extremely compact model sizes for mobile applications. It has only 1.2 million parameters but achieved an accuracy similar to AlexNet (Table 8.1). SqueezeNet has 26 convolution layers and no fully-connected layer. The last feature

TABLE 8.1 Summarized Results of Manually Designed CNN Architectures on ImageNet

Network	#Params	#MACs	ImageNet	
			Top-1 Acc	Top-5 Acc
AlexNet [286]	60M	720M	57.2%	80.3%
GoogleNet [143]	6.8M	1550M	69.8%	89.5%
VGG-16 [287]	138M	15300M	71.5%	-
ResNet-50 [140]	25.5M	4100M	76.1%	92.9%
SqueezeNet [288]	1.2M	1700M	57.4%	80.5%
MobileNetV1 [206]	4.2M	569M	70.6%	89.5%
MobileNetV2 [190]	3.4M	300M	72.0%	-
MobileNetV2-1.4 [190]	6.9M	585M	74.7%	-
ShuffleNetV1-1.5x [322]	3.4M	292M	71.5%	-
ShuffleNetV2-1.5x [207]	3.5M	299M	72.6%	-
ShuffleNetV2-2x [207]	7.4M	591M	74.9%	-

map goes through a global average pooling and forms a 1000-dimension vector to feed the softmax layer. SqueezeNet has eight Fire modules. Each fire module contains a squeeze layer with 1×1 convolution and a pair of 1×1 and 3×3 convolutions. The SqueezeNet caffemodel achieved a top-1 accuracy of 57.4% and a top-5 accuracy of 80.5% on ImageNet 2012 [201]. SqueezeNet is widely used in mobile applications in which model size is a large constraint.

MobileNets (Figure 8.3). MobileNetV1 [206] is based on a building block called depthwise separable convolution (Figure 8.3a), which consists of a 3×3 depthwise convolution layer and a 1×1 convolution layer. The input image first goes through a 3×3 standard convolution layer with stride 2, then 13 depthwise separable convolution blocks. Finally, the

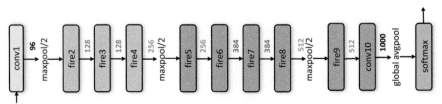

Figure 8.2 SqueezeNet Architecture [288].

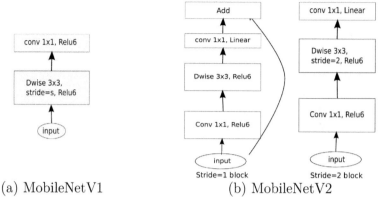

(a) MobileNetV1   (b) MobileNetV2

Figure 8.3 (a) Building block of MobileNetV1 [206]. It consists of a 3×3 depthwise convolution layer and a 1×1 convolution layer. (b) Building blocks of MobileNetV2 [190]. Each block consists of a 3×3 depthwise convolution layer and two 1×1 convolution layers. When the stride is 1, the block will have a skip connection. ReLU6 $(y = min(max(0, x), 6))$ is used for non-linearity, which is more friendly for quantization than ReLU $(y = max(0, x))$.

feature map goes through a global average pooling and forms a 1280-dimension vector fed to the final fully-connected layer with 1000 output units. With 569M MACs and 4.2M parameters, MobileNetV1 achieves 70.6% top-1 accuracy on ImageNet 2012 (Table 8.1).

MobileNetV2 [190], an improved version of MobileNetV1, also uses 3×3 depthwise convolution and 1×1 convolution to compose its building blocks. Unlike MobileNetV1, the building block in MobileNetV2 has three layers, including a 3×3 depthwise convolution layer and two 1×1 convolution layers (Figure 8.3b). The intuition is that the capacity of depthwise convolution is much lower than the standard convolution, and thus needs more channels to improve its capacity. From a cost perspective, the #MACs and #Params of a depthwise convolution only grow linearly (rather than quadratically as for a standard convolution) as the number of channels increases. Thus, even with a large channel number, the cost of a depthwise convolution layer is still moderate. Therefore, in MobileNetV2, the input feature map first goes through a 1×1 convolution to increase the channel dimension by a factor called expand ratio. Then the expanded feature map is fed to a 3×3 depthwise convolution, followed by another 1×1 convolution to reduce the channel dimension back to the original value. This structure is called inverted bottleneck and the

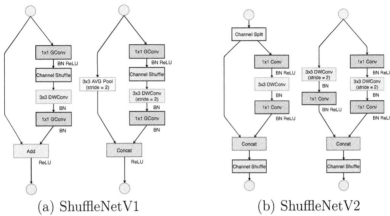

(a) ShuffleNetV1   (b) ShuffleNetV2

Figure 8.4 (a) Building blocks of ShuffleNetV1 [322]. 1×1 group convolutions are employed instead of 1×1 convolutions to improve efficiency.
Channel shuffle operation is introduced to exchange information between
different groups. (b) Building blocks of ShuffleNetV2 [207]. The input
feature map is divided into two groups at the beginning of the building
block, while the channel shuffle operation is moved to the end of the
building block.

block is called mobile inverted bottleneck block. Apart from the mobile
inverted bottleneck block, MobileNetV2 has another two improvements
over MobileNetV1. First, MobileNetV2 has skip connections for blocks
in which the stride is 1. Second, the activation function of the last
1×1 convolution in each block is removed. Empirical evidence [190, 325]
shows that this can improve the accuracy. One possible explanation is
that ReLU/ReLU6 does not fully preserve the information of the input
manifold. Removing unnecessary non-linearity prevents non-linearities
from destroying too much information. Combining these improvements,
MobileNetV2 achieves 72.0% top-1 accuracy on ImageNet 2012 with only
300M MACs and 3.4M parameters (Table 8.1).

ShuffleNets (Figure 8.4). The building blocks of ShuffleNetV1 [322]
are illustrated in Figure 8.4a. Similar to MobileNets, ShuffleNetV1 utilizes
3×3 depthwise convolution rather than standard convolution. Besides,
ShuffleNetV1 introduces two new operations, pointwise group convolution
and channel shuffle. The pointwise group convolution's motivation is to
reduce the computational cost of 1×1 convolution layers. However, it
has a side effect: a group cannot see information from other groups.

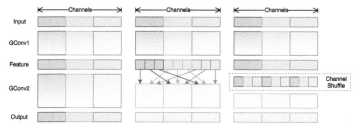

Figure 8.5 Illustration of the channel shuffle operation [322]. After shuffling, each group will contain information from all groups.

This will significantly hurt accuracy. The channel shuffle operation is thus introduced to address this side effect by exchanging feature maps between different groups. An illustration of the channel shuffle operation is provided in Figure 8.5. After shuffling, each group will contain information from all groups. On ImageNet 2012, ShuffleNetV1 achieves 71.5% top-1 accuracy with 292M MACs (Table 8.1).

The building blocks of ShuffleNetV2 [207] are shown in Figure 8.4b. In ShuffleNetV2, the input feature map is divided into two groups at the beginning of each building block. One group goes through the convolution branch that consists of a 3×3 depthwise convolution layer and two 1×1 convolution layers. The other group goes through a skip connection when the stride is 1 and goes through a 3×3 depthwise separable convolution when the stride is 2. In the end, the outputs are concatenated along the channel dimension, followed by a channel shuffle operation to exchange information between groups. With 299M MACs, ShuffleNetV2 achieves 72.6% top-1 accuracy on ImageNet 2012 (Table 8.1).

8.4   NEURAL ARCHITECTURE SEARCH

The success of the aforementioned efficient CNN models relies on hand-crafted neural network architectures that require domain experts to explore the large design space, trading off among model size, latency, energy, and accuracy. This is not only time-consuming but also suboptimal. Thus, there is a growing interest in developing automated methods to tackle this challenge.

Neural Architecture Search (NAS) refers to using machine learning techniques to automatically design neural network architectures. In the conventional NAS formulation [171], designing neural network architectures is modeled as a sequence generation problem, where an auto-regressive RNN controller is introduced to generate neural network architectures. This RNN controller is trained by repeatedly sampling neural network architectures, evaluating the sampled neural network architectures, and updating the controller based on the feedback. To find a good neural network architecture in the vast search space, this process typically requires to train and evaluate tens of thousands of neural networks (e.g., 12,800 in [326]) on the target task, leading to prohibitive computational cost (10^4 GPU hours). To address this challenge, many techniques are proposed that try to improve different components of NAS, including search space, search algorithm, and performance evaluation strategy.

Search Space. All the NAS methods need a pre-defined search space that contains basic network elements and how they connect with each other. For example, the typical basic elements of CNN models consist of (1) convolutions [326, 327]: standard convolution (1×1, 3×3, 5×5), asymmetric convolution (1×3 and 3×1, 1×7 and 7×1), depthwise-separable convolution (3×3, 5×5), dilated convolution (3×3); (2) poolings: average pooling (3×3), max pooling (3×3); (3) activation functions [203]. Then these basic elements are stacked sequentially [328] with identity connections [171]. The full network-level search space grows exponentially as the network deepens (Figure 8.6a). When the depth is 20, this search space contains more than 10^{36} different neural network architectures in [326].

Instead of directly searching on such an exponentially large space, restricting the search space is a very effective approach for improving the search speed. Specifically, [326, 329] propose to search for basic building cells (Figure 8.6b) that can be stacked to construct neural networks, rather than the entire neural network architecture. As such, the architecture complexity is independent of the network depth, and the learned cells (e.g., Figure 8.6c) are transferable across different datasets. It enables NAS to search on a small proxy dataset (e.g., CIFAR-10), and then transfer to another large-scale dataset (e.g., ImageNet) by adapting the number of cells. Within the cell, the complexity is further reduced by supporting hierarchical topologies [330], or increasing the number of elements (blocks) in a progressive (simple to complex) manner [168].

Search Algorithm. NAS methods usually have two stages at each search step: (1) the generator produces an architecture, and then (2) the evaluator trains the network and obtains the performance. As getting the performance of a sampled neural network architecture involves training a

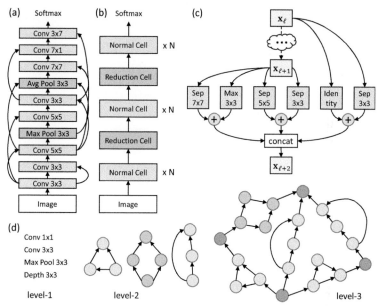

Figure 8.6 NAS search space [331]: (a) network-level search space [171]; (b) cell-level search space [326]; (c) an example of learned cell structure [168]; (d) three-level hierarchical search space [330].

neural network, which is very expensive, search algorithms that affect the sample efficiency play an important role in improving the search speed of NAS. Most of the search algorithms used in NAS fall into 5 categories: random search, reinforcement learning (RL), evolutionary algorithms, Bayesian optimization, and gradient-based methods. Among them, RL, evolutionary algorithms, and gradient-based methods provide the most competitive results.

RL-based methods model the architecture generation process as a Markov Decision Process, treat the validation accuracy of the sampled architecture as the reward and update the architecture generation model using RL algorithms, including Q-learning [328, 329], REINFORCE [171], PPO [326], etc. Instead of training an architecture generation model, evolutionary methods [327, 330] maintain a population of neural network architectures. This population is updated through mutation and recombination. While both RL-based methods and evolutionary methods optimize neural network architectures in the discrete space, DARTS [169] proposes continuous relaxation of the architecture representation:

$$y = \sum_i \alpha_i o_i(x), \qquad \text{where } \alpha_i \geq 0, \sum_i \alpha_i = 1. \qquad (8.6)$$

where $\{\alpha_i\}$ denotes architecture parameters, $\{o_i\}$ denotes candidate operations, x is the input, and y is the output. Such continuous relaxation allows optimizing neural network architectures in the continuous space using gradient descent, which greatly improves the search efficiency. Apart from the above techniques, the search efficiency of NAS can also be improved by exploring the architecture space with network transformation operations, starting from an existing network, and reusing the weights [332, 333, 334].

Performance Evaluation. To guide the search process, NAS methods need to get the performances (typically accuracy on the validation set) of sampled neural architectures. The trivial approach to get these performances is to train sampled neural network architectures on the training data and measure their accuracy on the validation set. However, it will result in excessive computational cost [171, 326, 327]. This motivates many techniques that aim at speeding up the performance evaluation step.

Alternatively, the evaluation step can also be accelerated using Hypernetwork [335], which can directly generate weights of a neural architecture without training it. As such, only a single Hypernetwork needs to be trained, which greatly saves the search cost. Similarly, One-shot NAS methods [336, 169, 183] focus on training a single super-net, from which small sub-networks directly inherit weights without training cost.

Auto-Designed vs. Human-Designed. Figure 8.7 reports the summarized results of auto-design CNN models and human-design CNN models on ImageNet. NAS not only saves engineer labor costs but also provides better CNN models over human-designed CNNs. Apart from ImageNet classification, auto-design CNN models have also outperformed manually designed CNN models on object detection [326, 173, 337, 338] and semantic segmentation [176, 339].

8.5 HARDWARE-AWARE NEURAL ARCHITECTURE SEARCH

While NAS has shown promising results, achieving significant MACs reduction without sacrificing accuracy, in real-world applications, we care about the real hardware efficiency (e.g., latency, energy) rather

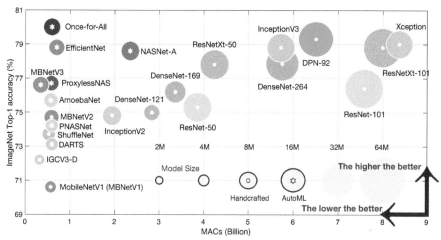

Figure 8.7 Summarized results of auto-design CNN models and human-design CNN models on ImageNet [340].

than #MACs. Unfortunately, MAC-efficiency does not directly translate to real hardware efficiency. Figure 8.9 shows the comparison between auto-designed CNN models (NASNet-A and AmoebaNet-A) and human-designed CNN model (MobileNetV2-1.4). Although NASNet-A and AmoebaNet-A have fewer MACs than MobileNetV2-1.4, they actually run slower than MobileNetV2-1.4 on hardware. It is because #MACs only reflects the computation complexity of convolution operations. Other factors like data access cost, parallelism, and cost of element-wise operations that significantly affect real hardware efficiency are not taken into consideration.

This problem motivates hardware-aware NAS techniques [341, 183, 184] that directly incorporate hardware feedback into the architecture search process. An example of the hardware-aware NAS framework is shown in Figure 8.8. Apart from accuracy, each sampled neural network

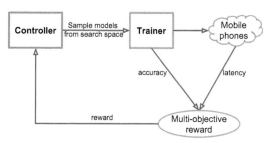

Figure 8.8 An example of the hardware-aware NAS framework [341].

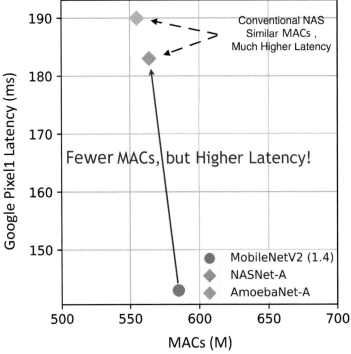

Figure 8.9 #MACs does not reflect real hardware efficiency.

architecture is measured on the target hardware to collect its latency information. A multi-objective reward is defined based on accuracy ACC and latency LAT:

$$\text{reward} = ACC \times (\frac{LAT}{T})^{\omega}, \tag{8.7}$$

where T is the target latency and ω is a hyper-parameter.

8.5.1 Latency Prediction

Measuring the latency on-device is accurate but not ideal for scalable neural architecture search. There are two reasons: (i) Slow speed. As suggested in TensorFlow-Lite, we need to average hundreds of runs to produce a precise measurement, approximately 20 seconds. This is far more slower than a single forward/backward execution. (ii) High cost. A lot of mobile devices and software engineering work are required to build an automatic pipeline to gather the latency from a mobile farm.

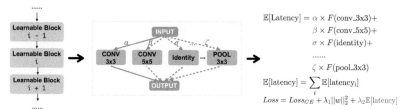

$$\mathbb{E}[\text{Latency}] = \alpha \times F(\text{conv\_3x3}) +$$
$$\beta \times F(\text{conv\_5x5}) +$$
$$\sigma \times F(\text{identity}) +$$
$$\dots$$
$$\zeta \times F(\text{pool\_3x3})$$
$$\mathbb{E}[\text{latency}] = \sum_i \mathbb{E}[\text{latency}_i]$$
$$Loss = Loss_{CE} + \lambda_1 ||w||_2^2 + \lambda_2 \mathbb{E}[\text{latency}]$$

Figure 8.10 Making latency differentiable by introducing latency loss [183].

Instead of direct measurement, an economical solution is to build a prediction model to estimate the latency [183]. In practice, this is implemented by sampling neural network architectures from the candidate space and profiling their latency on the target hardware platform. The collected data is then used to build the latency prediction model. For hardware platforms that sequentially execute operations, like mobile and FPGA, a simple latency lookup table that maps each operation to its estimated latency is sufficient to provide very accurate latency predictions (Figure 8.11). Another benefit of this approach is that it allows modeling the latency of a neural network as a regularization loss (Figure 8.10), enabling to optimize the trade-off between accuracy and latency in a differentiable manner.

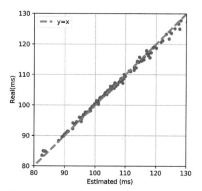

Figure 8.11 Predicted latency vs. real latency on Google Pixel 1 [183].

8.5.2 Specialized Models for Different Hardware

Given the high cost of building a new neural network model, it is common to deploy the same model for all hardware platforms. However, it is suboptimal, as different hardware platforms have different properties, such as the number of arithmetic units, memory bandwidth, cache size, etc. Using hardware-aware NAS techniques, it is possible to have a specialized neural network architecture for each target hardware.

Figure 8.12 demonstrates the detailed architectures of specialized CNN models on GPU and Mobile. We notice that the architecture

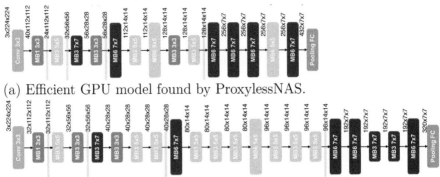

(a) Efficient GPU model found by ProxylessNAS.

(b) Efficient mobile model found by ProxylessNAS.

Figure 8.12 Efficient models optimized for different hardware. "MBConv3" and "MBConv6" denote mobile inverted bottleneck block with an expand ratio of 3 and 6, respectively. Insights: GPU prefers shallow and wide model with early pooling; Mobile prefers deep and narrow model with late pooling. Pooling layers prefer large and wide kernel. Early layers prefer small kernel. Late layers prefer large kernel [183].

shows different preferences when targeting different platforms: (i) The GPU model is shallower and wider, especially in early stages where the feature map has higher resolution; (ii) The GPU model prefers large MBConv operations (e.g., 7 × 7 MBConv6), while the Mobile model would go for smaller MBConv operations. This is because GPU has much higher parallelism than Mobile so it can take advantage of large MBConv operations. Another interesting observation is that the searched models on all platforms prefer larger MBConv operations in the first block within each stage where the feature map is downsampled. It might because larger MBConv operations are beneficial for the network to preserve more information when downsampling.

Table 8.2 shows the summarized results of specialized models on GPU and Mobile. An interesting observation is that models optimized for GPU do not run fast on Mobile and, vice versa. Therefore, it is essential to learn specialized neural networks for different hardware architectures to achieve the best efficiency on different hardware.

8.5.3 Handling Many Platforms and Constraints

Although specialized CNN models are superior over non-specialized counterparts, designing specialized CNNs for every scenario is still difficult, either with human-based methods or hardware-aware NAS. Since such

TABLE 8.2 Hardware Prefers Specialized Models [183]. With a similar accuracy, the specialized model (ProxylessNAS-Mobile) reduces the latency by 1.8× compared to the non-specialized CNN model (MobileNetV2-1.4). Besides, models optimized for GPU does not run fast on Mobile, and vice versa.

Network Network	ImageNet Top-1 (%)	GPU latency	Mobile latency
MobileNetV2-1.4 [190]	74.7	-	143ms
ProxylessNAS-GPU [183]	75.1	5.1ms	124ms
ProxylessNAS-Mobile [183]	74.6	7.2ms	78ms

methods need to repeat the network design process and retrain the designed network from scratch for each case. Their total cost grows linearly as the number of deployment scenarios increases, which will result in excessive energy consumption and CO_2 emission [342]. It makes them unable to handle the vast amount of hardware devices (23.14 billion IoT devices till 2018[2]) and highly dynamic deployment environments (different battery conditions, different latency requirements, etc.).

To tackle this challenge, one promising solution is to build a once-for-all (OFA) network [340, 343] that can be directly deployed under diverse architectural configurations, amortizing the training cost. The inference is performed by selecting only part of the OFA network. It flexibly supports different depths, widths, kernel sizes, and resolutions without retraining. An example of OFA is illustrated in Figure 8.13 (left). Specifically, the model training stage is decoupled from the neural architecture search stage. In the model training stage, the focus is to improve the accuracy of all sub-networks derived by selecting different parts of the OFA network. A subset of sub-networks is sampled in the model specialization stage to train an accuracy predictor and latency predictors. Given the target hardware and constraint, a predictor-guided architecture search [168] is conducted to get a specialized sub-network, and the cost is negligible[3]. As such, the total cost of specialized neural network design is reduced from O(N) to O(1) (Figure 8.13 middle).

---

[2]https://www.statista.com/statistics/471264/iot-number-of-connected-devices-worldwide/

[3]https://github.com/mit-han-lab/once-for-all/blob/master/tutorial/ofa.ipynb

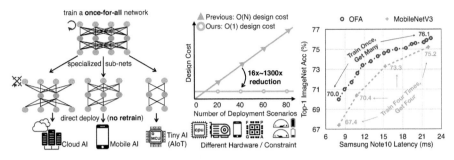

Figure 8.13 Left: a single once-for-all network is trained to support versatile architectural configurations including depth, width, kernel size, and resolution. Given a deployment scenario, a specialized sub-network is directly selected from the once-for-all network without training. Middle: this approach reduces the cost of specialized deep learning deployment from O(N) to O(1). Right: once-for-all network followed by model selection can derive many accuracy-latency trade-offs by training only once, compared to conventional methods that require repeated training [340].

Table 8.3 reports the comparison between OFA and state-of-the-art hardware-aware NAS methods on the mobile phone (Pixel1). The cost of OFA is *constant* while others are *linear* to the number of deployment scenarios (N). With $N = 40$, the total CO_2 emissions of OFA is 16×

TABLE 8.3 Summarized Results on Pixel1 Phone [340]. The first group corresponds to human-designed CNN models. The second group corresponds to conventional NAS. The third group corresponds to hardware-aware NAS. The final group corresponds to OFA. "#75" denotes the specialized sub-networks are fine-tuned for 75 epochs after grabbing weights from the OFA network. "CO_2e" denotes CO_2 emission which is calculated based on [342]. AWS cost is calculated based on the price of on-demand P3.16xlarge instances.

Network	ImageNet Top1 (%)	MACs	Mobile latency	Search cost (GPU hours)	Training cost (GPU hours)	Total cost ($N = 40$)		
						GPU hours	CO_2e (lbs)	AWS cost
MobileNetV2 [190]	72.0	300M	66ms	0	150N	6k	1.7k	$18.4k
MobileNetV2 #1200	73.5	300M	66ms	0	1200N	48k	13.6k	$146.9k
NASNet-A [326]	74.0	564M	-	48,000N	-	1,920k	544.5k	$5875.2k
DARTS [169]	73.1	595M	-	96N	250N	14k	4.0k	$42.8k
MnasNet [341]	74.0	317M	70ms	40,000N	-	1,600k	453.8k	$4896.0k
FBNet-C [184]	74.9	375M	-	216N	360N	23k	6.5k	$70.4k
ProxylessNAS [183]	74.6	320M	71ms	200N	300N	20k	5.7k	$61.2k
SinglePathNAS [167]	74.7	328M	-	288 + 24N	384N	17k	4.8k	$52.0k
AutoSlim [344]	74.2	305M	63ms	180	300N	12k	3.4k	$36.7k
MobileNetV3-Large [212]	75.2	219M	58ms	-	180N	7.2k	1.8k	$22.2k
OFA	76.0	230M	58ms	40	1200	1.2k	0.34k	$3.7k
OFA #75	76.9	230M	58ms	40	1200 + 75N	4.2k	1.2k	$13.0k
OFA$_{Large}$ #75	80.0	595M	-	40	1200 + 75N	4.2k	1.2k	$13.0k

less than ProxylessNAS, 19× less than FBNet, and 1,300× less than MnasNet.

8.6 CONCLUSION

Over the past few years, deep neural networks have achieved unprecedented success in vision tasks; however, their superior performance comes at the cost of high computational complexity. This limits their applications on many edge devices, where the hardware resources are tightly constrained by the form factor, battery, and heat dissipation.

This chapter offers a systematic overview of efficient CNN architectures to enable both researchers and practitioners to get started in this field quickly. We first introduce various efficient convolution layers that have been widely used in efficient CNN architectures, including 1×1 convolution, group convolution, and depthwise convolution. We then describe three representative manually-designed efficient CNN architectures (SqueezeNet, MobileNets, and ShuffleNets) built upon these efficient convolution layers. To reduce the design cost of these handcrafted solutions, we then describe recent efforts on neural architecture search and hardware-aware neural architecture search, which can outperform the manual design with minimal human efforts. Finally, we describe the once-for-all technique to efficiently handle many hardware platforms and efficiency constraints without repeating the costly search and re-training phases.

CHAPTER 9

Design Methodology for Low-Power Image Recognition Systems

Soonhoi Ha, EunJin Jeong, Duseok Kang, Jangryul Kim, and Donghyun Kang

Seoul National University

CONTENTS

DOI: 10.1201/9781003162810-9

In the development of an embedded image recognition system, there are many issues to consider, such as which hardware platform and algorithm to use, how to optimize the software with resource constraints and how to optimize multiple design objectives, and so on. This chapter presents a systematic design methodology that could be applied to the design of embedded systems with a concrete example of image recognition systems. Based on the proposed methodology, we won the first prize in LPIRC (Low-Power Image Recognition Challenge) 2017. LPIRC seeks the best system-level solutions for detecting objects in images while using as little energy as possible. To win the challenge, we needed to jointly optimize speed, accuracy, and energy since the scoring function is defined as the ratio of accuracy (measured by mean average precision, mAP) and the energy consumption (measured by WattHour) in the processing of 20,000 images within 10 minutes. If the processing speed is lower than 33 FPS, the score decreases proportionally.

Since there is no restriction on hardware and software in LPIRC, we needed to select a hardware platform and an image recognition algorithm. We chose NVIDIA Jetson TX2 as the hardware platform and Tiny YOLO as the algorithm. By applying the well-known software optimization techniques in a systematic way, we could build the award-winning solution in only three months, without developing any new technique. We have refined the methodology to choose a different algorithm on the same hardware platform and could build another winning solution in track 2 of LPIRC 2018.

In the first two sections, the proposed methodology we used for LPIRC is explained in detail. Recently new hardware platforms have been developed that contain CNN hardware accelerators as well as GPU (Graphics Processing Units). A representative example of this writing is NVIDIA Jetson AGX Xavier. Since it is a heterogeneous system that contains multiple hardware accelerators, how to exploit the computing power of those accelerators maximally becomes an important issue to consider in the proposed design methodology. We have developed a novel technique to maximally utilize multiple accelerators to achieve 21.7 times better score than our previous solution in LPIRC 2018, which is presented in the third section. In the conclusion section, we will introduce a NAS(Neural Architecture Search) technique as a future research topic.

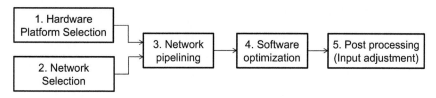

Figure 9.1 Overall flow of the proposed optimization methodology.

9.1 DESIGN METHODOLOGY USED IN LPIRC 2017

The overall flow of the proposed design methodology is shown in Figure 9.1. The first step is to select a hardware platform. NVIDIA Jetson TX2 was known as the embedded AI computing device with better performance per watt than the other candidate devices available to us. It has a Tegra GPU with 256 Pascal CUDA cores running at 1.3 GHz, in addition to a 2 GHz quad-core ARM A57 and a dual-core Denver 2. Thanks to CUDA and cuDNN (CUDA deep neural network library) provided by NVIDIA, it is easy to program deep learning networks on the GPU. cuDNN provides highly tuned implementations for standard operations used in CNNs such as convolution, pooling, normalization, and so on.

After selecting Jetson TX2 as the hardware platform, we needed to select a neural network for image recognition, also known as object detection, which is to find all objects and the associated bounding boxes that contain the objects in each image.

The third step is to pipeline the network to increase the throughput performance of an application by overlapping the processing of processing elements, CPU, and GPU. Since the throughput performance is determined by the slower processing element, it is important to make the execution latency of pipeline stages well balanced. After pipeline decision is made, we apply software optimization techniques on each processing element, CPU, and GPU. We detect a bottleneck processing element and apply appropriate software optimization techniques to reduce the execution time. The software optimization techniques used in our solution for LPIRC 2017 are presented below.

After applying the software optimization techniques for each processing element, we adjust the assumed input parameters to make the best compromise among design objectives in the post-optimization step. For LPIRC 2017, we adjusted the clock frequencies of processing elements to explore the trade-off between speed and energy consumption. In the following, we explain each step in detail.

9.1.1 Object Detection Networks

The object detection technique has been developed remarkably since Girshick et al. (2014) [255] first proposed a CNN-based detector, called R-CNN, which is a pioneer of two-stage convolutional object detectors. It consists of a region proposal stage based on a selective search algorithm and a set of CNNs to classify the image contained in each proposed region. Since proposed regions are usually overlapped, and a separate CNN network is applied for each region, it incurs excessive computation redundancy. To improve the computation efficiency of R-CNN, a series of extensions have been made. Fast R-CNN [258] uses a single CNN to extract intermediate features of the input image first and the *ROIPooling* layer to generate a set of region proposals. By sharing the feature extraction CNN, it could reduce the time complexity significantly. Faster R-CNN [345] could reduce more by replacing an external proposal generator with a CNN, called region proposal network (RPN). In this approach, region proposals are generated from a regular grid of "anchors" that can be shifted and scaled by convolution operations. The RPN is followed by the second stage, where region proposals are used to crop features from the same intermediate feature map, which are fed to the remainder of the feature extractor. R-FCN [144] achieves additional speed-up over Faster R-CNN by exchanging the feature extractors and feature cropping in the second stage to minimize the amount of per-region computation.

Despite a series of extensions, the speed of two-stage object detection networks is far below the required throughput or latency performance for real-time object detection. To overcome this difficulty, several neural networks have been proposed based on a one-stage convolutional network that directly predicts the bounding boxes and associated class probabilities without proposal generation. YOLO [346, 347, 348], SSD [349], and RetinaNet [350] are representative single stage detectors. According to these studies, the one-stage object detection networks without the region proposal network could achieve reasonable accuracy while execution speed is much faster than the aforementioned two-stage networks. Since speed gain prevails the accuracy loss in the scoring of the challenge, we chose to use a single stage detector.

We compared two popular single stage detectors, YOLOv2 and SSD, in terms of FPS (Frame Per Second) performance and mAP accuracy. We ran two networks on Jetson TX2 after retraining them with the ImageNet dataset and evaluated them with the ImageNet test dataset [201]. As shown in the first two rows in Table 9.1, YOLOv2 is proven to dominate

TABLE 9.1 Performance of Single state Object Setectors

Model (Framework)	FPS	mAP	FPS × mAP
SSD (Caffe)	3.5	0.43	1.51
YOLOv2 (Darknet)	5.81	0.51	2.96
Tiny YOLO (Darknet)	17.4	0.32	**5.57**

SSD in terms of performance and accuracy. But the FPS performance is still far below the required performance which is 33.3 FPS. Since it was not clear how much speed-up could be obtained by software optimization, we decided to compare YOLOv2 and Tiny YOLO. As shown in the table, there is no dominance relation between YOLOv2 and Tiny YOLO. Since the score is proportional to both FPS and mAP, we chose Tiny YOLO as the object detection network, ignoring the effect of software optimization at the early stage of design.

9.1.2 Throughput Maximization by Pipelining

While deep neural networks usually have one or more fully connected layers, Tiny YOLO is a fully convolutional network that consists of 9 convolution layers and 1 detection layer. Figure 9.2 shows the profiling result of those layers as well as input and output layers in Tiny YOLO. The execution time of each layer is measured by elapsed time in the processing of 20,121 images and the time unit is second. The input image size is 416×416. The figure also displays the kernel size and the channel size in each convolution layer. The first six convolution layers are followed by a max pooling layer to reduce the image size by half so that the final image size becomes 13×13.

Note that the input layer and the output layer should be processed on the CPU side. A sequence of input images is read from the disk or the camera and fed to the first convolution layer after format conversion. The output tensor from the last convolution layer includes all predictions of bounding boxes and the associated objects. To reduce the workload by filtering out region proposals, the output layer performs a post-processing step, called non-maximum suppression (NMS) [351]. As shown in the figure, the output layer takes a significant portion of the processing time if the network is executed sequentially. We pipelined the network by adding a pipeline buffer between the detection layer and the output layer. By overlapping the NMS computation in the CPU with the network

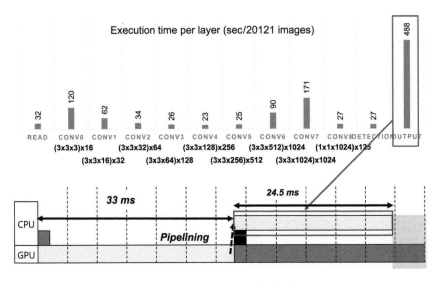

Figure 9.2 CPU-GPU pipelining of Tiny YOLO network.

processing in the GPU, we could achieve a 42.6% reduction of the inference time.

9.1.3 Software Optimization Techniques

After partitioning the layers onto processing elements, we apply software optimization techniques for the bottleneck processing element since the throughput performance is determined by the longest pipeline stage. A popular method to optimize the network is *pruning* which makes the weight values, which have a negligible effect on the accuracy, to zero [156] so that the number of MAC (multiply and accumulate) operations can be reduced. Unfortunately, pruning is not beneficial for our target hardware, NVIDIA Jetson TX2, since the architecture is not designed to utilize

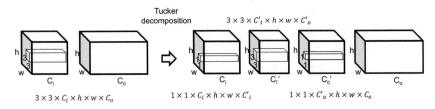

Figure 9.3 Tucker decomposition of a convolution layer.

zero values in the convolution operation. In this section, we explain the software optimization techniques in the application order in our LPIRC 2017 solution.

9.1.3.1 Tucker Decomposition

As illustrated in Figure 9.2, GPU becomes the bottleneck after pipelining. Since convolution layers are the most time consuming and require large memory space, several approximate computing methods have been developed to reduce the computation time and memory requirements. Tucker decomposition is one of such methods we adopted in the proposed solution [352].

Consider a convolution layer that convolves a feature map of size $h \times w \times C_i$ and a filter of size $3 \times 3 \times C_i$ to produce an output feature map of size $h \times w \times C_o$. The number of multiplications involved in the convolution operation is $3 \times 3 \times C_i \times h \times w \times C_o$. A single convolution layer is decomposed into three small convolution layers as displayed in Figure 9.3. The key idea is to reduce the number of channels involved in the main convolution layer with 3×3 kernel. On the input side, the number of channels is reduced from C_i to C_i' by 1×1 convolution. The number of output channels of the main convolution layer is also reduced from C_o to C_o'. Finally, 1×1 convolution is used to expand the number of output channels to C_o. If the number of input channels is small in the original convolution, we can omit the first 1×1 convolution in Tucker decomposition.

The reduction ratio of channels affects the accuracy as well as the computation time. So it is necessary to determine the reduction ratio carefully. Since we need to train the network again for each selection of C_i' and C_o', we gave up finding the best reduction ratio since we could not bear long training time. As a rule of thumb, we set C_i' and C_o' to the half of C_i and C_o, respectively.

Figure 9.4 displays the profiled execution time of layers before and after Tucker decomposition. Note that Tucker decomposition is not always beneficial: the second and third convolution layers become slower if Tucker decomposition is applied since the number of channels is not large enough to make the computation gain prevail the overhead of decomposition. Thus, we applied Tucker decomposition from the fourth convolution layer. The speed-up gain by Tucker decomposition is about 40% while the accuracy loss after retraining is less than 2%.

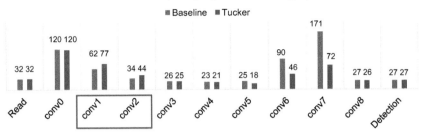

Figure 9.4 Profiled execution time of layers before and after Tucker decomposition.

9.1.3.2 CPU Parallelization

After Tucker decomposition is applied to the convolution layers, the CPU became the performance bottleneck. Since NMS is the most time-consuming task, we reduced the NMS execution time by exploiting the parallelism of the NMS algorithm on a multi-core CPU. The NMS algorithm used in Tiny YOLO is applied to each class. Among predictions made from the CNN, it first invalidates predictions whose confidence value is smaller than a given threshold (0.1 in our implementation). For each class, we select a region proposal with the highest confidence and compare this proposal with the other proposals. If the intersection between the selected proposal with another proposal is greater than a threshold, it is removed due to similarity with the selected proposal. Hence we could parallelize the NMS algorithm at the class level by using the OpenMP library [353]. As a result, the CPU time was greatly reduced, and the GPU became the performance bottleneck again.

9.1.3.3 16-bit Quantization

Quantization is a popular optimization method of DNNs, which changes the data representation from a 32-bit float to a lower precision data type [354]. Recent researches show that bit width can be reduced by 8 to 10 bits without noticeable accuracy loss in neural networks [196]. While quantization is a very effective method to reduce the hardware complexity, its effectiveness is rather limited in software implementations. Jetson TX2 does not support a smaller data size but a 16-bit float type, called *fp16* only, and the cuDNN library supports 16-bit precision in Jetson TX2.

Execution time per layer (sec / 20121 images)

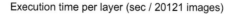

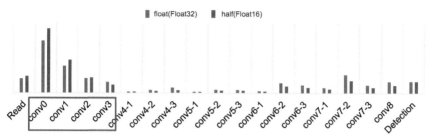

Figure 9.5 Profiled execution time of layers before and after 16-bit quantization: actual time values are omitted for simple illustration

To reduce the GPU computation time further, we reduced the precision of data representation from a 32-bit floating point to a 16-bit floating point. Since the cuDNN library supports 16-bit precision in Jetson TX2, we re-implemented the GPU kernels with fp16 APIs. In addition, we converted the input feature maps and the filter weights for fp16 quantization.

By reducing the data width, it is expected that the required memory size and memory access overhead are also reduced. Using the *nvprof* profiler provided by NVIDIA, we observed that the number of L2 cache accesses is actually reduced to half by using 16-bit quantization. If simultaneous multiplication of two 16-bit values is supported, it is expected that the computation time can be halved by reducing the bit width of all variables to 16 bits. But it is not the case in reality.

We applied fp16 operations to all convolution layers and profiled the execution time of layers. Surprisingly, as shown in Figure 9.5, the profiling results showed that 16-bit quantization is not beneficial in the first three convolution layers. We conducted two preliminary experiments with two different input sizes: 13 × 13 (Figure 9.6a) and 208 × 208 (Figure 9.6b). Figure 9.6 shows the execution time of the cuDNN convolution operation, which depends on the number of input channels and filters. It is observed that layers with 16-bit precision data are slower than the same size layers with 32-bit precision data when the input size is small and the number of channels is not large. Since the performance gain from 16-bit quantization depends on the configuration of the convolution layer, it is necessary to profile each layer to check if 16-bit quantization is beneficial and apply 16-bit quantization selectively.

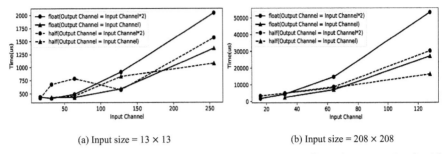

(a) Input size = 13 × 13   (b) Input size = 208 × 208

Figure 9.6 Performance comparison between 32-bit(float) and 16-bit(half) precision convolution with two different input sizes.

9.1.3.4 Post Processing

In the post processing step, we consider the trade-off among three objectives and explore other optimization possibilities. After a sequence of software optimizations, the estimated total inference time for 20,000 images was reduced to 460 seconds, which is 140 seconds earlier than the time limit, 10 minutes. It means that we could use the spare time to minimize the energy consumption further by adjusting the operating frequencies of processing elements. Jetson TX2 allows us to set up the clock frequency manually.

Table 9.2 shows the measurement results on the execution time and energy consumption consumed to process 2,000 images with various pairs of CPU and GPU frequencies. Energy consumption was measured with the WT310E powermeter. With the maximum clock frequencies for both CPU (2,035MHz) and GPU (1,301MHz), it took 48 seconds to process

TABLE 9.2 CPU-GPU Frequency Exploration Results

CPU) Freq.(MHz)	GPU) Freq.(MHz)	Exec. Time(s)	Energy Consumption(Wh)
2,035	1,301	48	0.205224
1,728	1,301	49	0.195594
1,114	1,301	55	0.192547
1,728	1,122	52	0.184183
1,114	1,122	58	0.180682
1,728	944	59	0.184728

TABLE 9.3 Step-by-step Performance Improvement Results

Step	Running Time	FPS	Improvement ratio
Baseline	1150s	17.4	1.0x
Pipelining	660s	30.3	1.74x
Tucker	530s	37.7	1.25x
CPU Parallelization	502s	39.8	1.06x
16-bit Quantization	460s	43.5	1.09x

2,000 images, as displayed in the first row in the table. Since GPU is the bottleneck processor, we could reduce the CPU frequency to 1,728 MHz without noticeable FPS degradation (second row in the table). Further reduction of CPU or GPU clock frequencies increases the execution time but reduces the energy consumption. We explored some frequency pairs until the execution time reaches the time limit (60 seconds). From this exploration of frequencies, we finally set the frequencies for CPU and GPU to 1,114MHz and 1,122MHz, respectively, to achieve the minimum energy consumption within the time limit.

Table 9.3 summarizes the performance improvement after each optimization technique is applied to the baseline network. The largest gain comes from pipelining, followed by Tucker decomposition. The overall FPS improvement is 2.5 times, which is sufficient to meet the throughput constraint. Our proposed solution was 2.7 times better than the winner of LPIRC 2016.

9.2   IMAGE RECOGNITION NETWORK EXPLORATION

Unlike LPIRC 2017, LPIRC 2018 offered three Tracks for competition, and we won the first prize in Track 2 that fixes the hardware platform to NVIDIA Jetson TX2 and the software platform to Caffe2. Since our solution for LPIRC 2017 used the same hardware platform, we could skip the first step in the proposed methodology. Since there were several rooms for further performance improvement over our LPIRC 2017 solution, we refined the methodology to achieve a 2.1 times better score in LPIRC 2018. The key refinement is made to select the image recognition algorithm, considering the estimated performance improvement *after* software optimization.

Since two-stage detectors are too slow, we compared one-stage object detectors that can satisfy the real-time constraint with reasonable

TABLE 9.4 Speed and Accuracy Comparison among Object Detection Networks for the COCO Dataset [356]

Network	mAP(IoU = 0.5)	FPS
YOLOv2 608x608 [347]	48.1	40
Tiny YOLO [347]	23.7	244
YOLOv3-320 [348]	51.5	45
SSD300 [349]	41.2	46
RetinaNet-50-500 [350]	50.9	14
R-FCN [144]	51.9	12
Faster RCNN w/ FPN [357]	59.1	6

accuracy: YOLO, SSD, RetinaNet, and DenseNet. Table 9.4 shows the speed in terms of FPS (frame per second) and mAP accuracy of each network on the server with TITAN X for COCO object detection dataset [355].

As shown in the table, there is a trade-off between accuracy and performance among networks. In addition, each network has a different potential for software optimization. Hence we refined the methodology with the systematic network exploration technique. In this section, we first give a brief overview of the selected networks.

9.2.1 Single Stage Detectors

YOLO is a representative one-stage detector structure that is displayed in Figure 9.7a. Redmon et al. [346] proposed a grid-based algorithm that obtains the prediction candidates through a 1x1 convolution after obtaining features through a sequence of convolutional layers. The final output tensor is in the form of S x S x (B x 5 + C), where each grid of S x S final spatial resolution has B anchors and C class-probability pairs, and each anchor is associated with 5-tuple information on its position, size, and confidence value. Then, the final box proposals are obtained through the non-maximum suppression (NMS) process. Redmon et al. [347] have improved the network and introduced YOLOv2 and its light-weighted version, Tiny YOLO. Since the box proposal is performed at the last stage, the detection accuracy of YOLO networks becomes worse as the object size decreases. To remedy this drawback, YOLOv3 [348] uses an upsampling layer to propose boxes from a larger feature map and consists

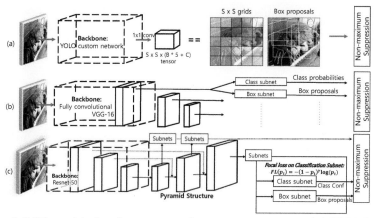

Figure 9.7 Three kinds of one-stage detector structures. (a) YOLO [346, 347], (b) SSD [349], and (c) RetinaNet [350]

of more convolution layers. By doing so, it could improve the accuracy with a smaller input image as illustrated in Table 9.4.

DenseNet [358] was originally proposed as a fast deep learning network for image classification. It typically consists of four dense blocks that consist of a cascade of convolution layers inside. The input of each convolution layer in a dense block is a concatenation of all feature maps of previous layers in the dense block. We make a new detection network that replaces the YOLO backbone with a DenseNet for feature extraction, keeping the remaining layers of YOLOv2. Thus it is categorized as a YOLO-based network. DenseNet has two key parameters to define the number of convolution layers in each dense block and the number of filters in each convolution layer denoted k. Considering the trade-off between speed and accuracy, we include a DenseNet with 85 convolution layers and 48 filters in the comparison.

In the case of one-stage detection without a separate region proposal network, detection performance for small objects may deteriorate. To solve this weakness, SSD [349] adds extra layers to the backbone and performs prediction not only at the grids of the final layer but also at the grids of middle layers as illustrated in Figure 9.7b. It differs from YOLO in that each of the multiple target layers has corresponding subnets for obtaining class probabilities and box proposals from its grids. In addition, SSD utilizes its NMS operation to handle a greater number of proposals than YOLO—it first keeps a certain number of box proposals with high probability for each class and selects final predictions among them. We use SSD300 downloaded from the author's github [349]. It uses the VGG network as a feature extractor. While it runs on the Caffe framework, its Caffe model is translated into a Caffe2 model that is assumed in the challenge.

Observing that the performance loss of one-stage detectors is mainly caused by the dominant contribution of easy-to-classify samples in the training phase, Lin et al. [350] propose a new form of the loss function, called *Focal Loss*. It scales the loss function dynamically by making the scaling factor decay to zero as confidence in the correct class increases. It automatically down-weights the contribution of easy examples and focuses the model on hard examples during training. Based on this idea, they proposed a new one-stage detector, called RetinaNet. As shown in Figure 9.7c, RetinaNet generates its prediction candidates from multiple target layers that compose the Feature Pyramid Network(FPN) [357, 349] through the subnets like SSD, with applying focal loss on the classification subnetwork. This effort helps RetinaNet achieve comparable accuracy as two-stage detectors on COCO detection while maintaining the speed advantage. The officially released version of RetinaNet on the Caffe2 framework is the 800-scale version that aims to achieve high accuracy. For real-time applications on the Jetson TX2 platform, we modified the configuration of the network. We have selected the options to use ResNet-50 body and reduce the FPN level and the number of subnet layers. By using various scales from 224 to 480 in training, the modified RetinaNet can perform well with a wide range of input sizes. In addition, Caffe2 Detectron APIs [359] are extended to apply the proposed software optimization techniques.

9.2.2 Software Optimization Techniques

We applied three software optimization techniques that are explained in Section 9.1.3: Tucker decomposition, CPU parallelization, and 16-bit quantization. Since Tucker decomposition is an approximation technique, we cannot avoid accuracy loss. To minimize the accuracy loss, all networks need to be retrained after the technique is applied. Since the degree of accuracy loss depends on the network structure, we apply Tucker decomposition differently, observing the accuracy degradation percentile. In the case of YOLOv2 [347] and Tiny YOLO [360], the degradation of accuracy (mAP) is less than 1% when we apply Tucker decomposition to all layers at once and then retrain the network. But accuracy is dropped by more than 3% in SSD when applying Tucker decomposition at once. Hence Tucker decomposition is applied incrementally, selecting two heavy

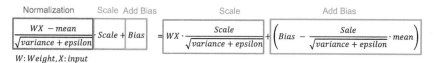

Figure 9.8 How to merge batch normalization to the previous convolution layer.

layers first to transform and fine-tune iteratively. In this manner, we could attain several optimal points of the network with an acceptable accuracy drop of less than 2%.

In the case of RetinaNet [350] that uses ResNet-50 as the feature extractor, Tucker decomposition is not applied since ResNet-50 has the bottleneck architecture where two cascaded 3x3 convolution layers are replaced by three cascaded layers of 1x1, 3x3, another 1x1 convolution. The bottleneck architecture serves the same purpose of computation reduction, resembling the layer structure produced by Tucker decomposition. Since DenseNet [358] and YOLOv3 has a similar bottleneck architecture, Tucker decomposition is not applied to those networks.

In addition, we applied another optimization to merge batch normalization to convolution operation. Batch normalization is a technique to normalize the input data to have unit variance and zero mean. Since batch normalization is a linear transformation, it can be merged with the previous convolution layer by adjusting the filter weights at compile-time, as displayed in Figure 9.8. We perform this optimization to all networks, but SSD does not use batch normalization in order to reduce the overall network execution time.

9.2.3 Post Processing

As shown in Table 9.4, the baseline performance of networks except Tiny YOLO is far below the required throughput performance, 33.3 FPS. Since the performance improvement by software optimization is not large enough. We reduce the image size in the post processing step. If we reduce the input image size, we can reduce the computation complexity proportionally. How input size reduction affects the accuracy depends on the network.

The last layer of YOLOv2 and Tiny YOLO, called the region layer, produces grid outputs whose size is 1/32 of the input size. One grid contains the position, width, height, and class information for the objects

TABLE 9.5 Exploration of input size and batch size for YOLOv2

Input size	416x416				320x320				256x256			
Batch size	1	4	8	16	1	4	8	16	1	4	8	16
mAP	50.4				46.5				42.2			
FPS	13.3	23.1	25.6	25.9	17.9	34.2	37.5	38.3	43.8	50.1	51.7	50.4

in the corresponding part of the image. If we reduce the input size, one grid represents a larger portion of the image, degrading the accuracy for small objects. Compared with the other networks, however, the accuracy drop is graceful. For a network that has a multi-layer-prediction structure like SSD, reducing the input size is not a simple job. Because SSD is characterized by non-scalable and static anchor generation, unaligned convolutions, and the last extra layer whose output dimension is 1x1, it is necessary to change the network structure and anchor boxes and retrain the network if we want to reduce the input size. Even though RetinaNet also has the multi-layer-prediction scheme, it allows various input image sizes without network restructuring since it uses aligned convolutions and stride-aware anchor generation.

Since multiple images can be downloaded from the server at once, we could use batch processing that processes multiple images concurrently to increase the utilization of the processing element. Batch processing may not be applicable for real-time object detection. Table 9.5 shows the exploration result of YOLOv2 before Tucker decomposition is applied. By reducing the input size to 256×256 and increasing the batch size to 16, we could obtain a 3.9x FPS increase, sacrificing mAP by 16.3% (50.4 vs 42.2).

9.2.4 Network Exploration

We compare six one-stage networks in terms of speed and accuracy for two different datasets, considering the performance improvements from optimizations as shown in Figure 9.9. Five networks are compared for ImageNet, and six networks for the COCO dataset, respectively. At first, it should be noted that significant speed improvement can be obtained by software optimizations in all networks. The baseline implementation without optimization is marked with a star(*) mark in the figure. For each network, the observed FPS improvement between the baseline implementation and the optimized one with less than 1% accuracy drop indicates the effect of software optimization. Tiny YOLO and YOLOv2 show the largest and the second larger gain from software optimization

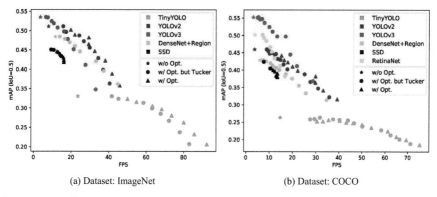

(a) Dataset: ImageNet   (b) Dataset: COCO

Figure 9.9 Comparison between optimized networks.

in both datasets. The gain from software optimization is minimal for SSD. The effect of Tucker decomposition corresponds to the difference between two data points: circle mark and triangle mark. While the FPS performance is improved, slight accuracy loss can be noticed.

Second, from Figure 9.9, we can observe that YOLO-based network forms the Pareto-optimal front for both datasets: For a given FPS performance, a YOLO-based network shows the best accuracy. Before software optimization, no dominance relation can be found since there is a trade-off between FPS performance and mAP accuracy between networks. From the figure, it is observed that YOLOv3 experiences a steeper accuracy drop as the input size decreases. As a result, YOLOv2 shows the best accuracy when the FPS performance lies between 20 FPS and 40 FPS in both datasets, which is usually required for real-time streaming applications. The DenseNet with YOLO region layer shows comparable accuracy when FPS performance is about 40. For the range of low FPS performance requirements, YOLOv3 shows the highest accuracy followed by RetinaNet, particularly for the COCO dataset.

9.2.5   LPIRC 2018 Solution

By comparing the aforementioned networks, we found that YOLOv2 outperforms the other networks on Jetson TX2 under the given throughput constraint. Table 9.6 shows how much the YOLOv2 network is improved after each step of the proposed design methodology. Since the total energy consumption is inversely proportional to the network speed, the score (mAP/Wh) is estimated as mAP × speed. This optimized YOLOv2 network won the first prize in the LPIRC Track 2 competition.

TABLE 9.6 Performance Improvement after Each Step of the Design Methodology(YOLOv2, the Input Size Is 416x416)

Description	mAP(A)	FPS(B)	A x B	Normalized score
Baseline	51.1	7.97	407	1.00
Pipelining	51.1	8.85	452	1.11
Tucker decomposition	50.2	15.1	758	1.86
16-bit Quantization	50.2	19.9	999	2.45
input size reduction (256 x 256)	43.0	38.1	1640	4.03
Batch size = 16	43.0	90.3	3880	9.54

YOLOv2 is tested on the Darknet framework in the experiments, and it needs to be translated to Caffe2 framework for Track 2. We implemented custom Caffe2 operators to support Darknet-specific operators and optimization techniques such as pipelining and 16-bit quantization. Additionally, various batch sizes for the network are tested to find the best batch size for the submission. Through the steps illustrated above, the estimated score for the YOLOv2 has increased about 9.54 times compared with the baseline, and this result surpassed the other object detection networks on the Jetson TX2. The largest gain in score comes from the post processing step, image size reduction, and batch processing. Before post processing, SW optimization techniques boost the performance by 2.45 times, which is similar to the case of LPIRC 2017 solution.

9.3 NETWORK PIPELINING FOR HETEROGENEOUS PROCESSOR SYSTEMS

Recently, an embedded device tends to equip hardware accelerators for deep learning applications in addition to a multi-core CPU and a GPU as deep learning (DL) inference applications are increasingly popular. As an example, NVIDIA Jetson AGX Xavier board contains an octa-core CPU, one GPU, and two Deep learning accelerators (DLAs) [361]. In this section, we use Jetson AGX Xavier as the hardware platform in the proposed methodology. Even though DLA is a power-efficient accelerator, its computational power is weaker than GPU. The Xavier board has a unified memory of 16GB shared by all processing elements. It means that communication between processing elements can be easily conducted by memory operation with the potential risk of access contention.

Figure 9.10 TensorRT workflow.

For efficient software development, TensorRT is provided as the SDK for high-performance inference on an NVIDIA device. Built on CUDA, it enables a user to optimize an inference network leveraging cuDNN libraries and development tools. Like most deep learning SDKs, TensorRT assumes that the inference is executed on a single processing element (PE), GPU or NPU, not both. Figure 9.10 shows the workflow of TensorRT that creates an optimized inference engine by the *Builder* module with optimization configuration parameters and network definition. While some optimizations are performed internally, a user may set some parameters to guide optimization methods such as data precision. The *Runtime* module loads the optimized engine and generates the execution context which performs the inference. Note that it is possible to build multiple execution contexts from the same engine and map them to distinct streams to perform inference concurrently. Figure 9.11a illustrates the schedule diagram of the baseline TensorRT implementation, where a single execution context is mapped to a GPU stream. Since TensorRT does not support CPU, pre/post-processing steps are executed on CPU as shown in the figure. In the figure, we use colors to distinguish iterations of the network execution. By using multiple threads on the CPU side, we could pipeline the execution as shown in Figure 9.11b.

9.3.1 Network Pipelining Problem

To utilize all available accelerator, GPU and two DLAs, maximally, a main extension is made in the third step in Figure 9.1, Network pipelining.

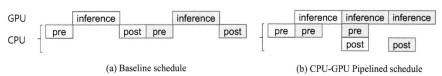

Figure 9.11 Schedule diagram for baseline tensorRT program (a) and CPU-GPU pipelining (b).

Figure 9.12a shows a simple schedule after the inference network is split into two stages and assign the first stage, *Infer1*, to a DLA and the second stage, *Infer2*, to the GPU. By utilizing both GPU and DLA, we may achieve better throughput performance as depicted. For pipelining on heterogeneous processing elements, we need to determine how many stages to be made, how to split the network, and how to assign the stages to the PEs. Since the design space of pipelining is huge, how to explore the design space is a challenging problem. To tackle this challenge, we devise a heuristic to decide the cut-points for network splitting for a given number of pipeline stages.

In addition, NVIDIA GPU and DLA can parallelize multiple instances of the assigned computation kernel using multiple streams, which increases the utilization further. Independently of the number of threads in the pre- or post-processing step, we can create more than one streams in GPU and DLA. In Figure 9.12b, it is assumed that the number of streams in GPU is two, the same as the number of threads in the pre-processing step. We may use more streams in the GPU by creating as many buffers as the number of streams at the pipeline-interface. This technique is called *Intra-PE parallelization*. For easy buffer management, it is good to set the number of streams as a multiple of the number of threads in the pre- and post-processing steps. It is observed that increasing the number of streams is not always beneficial. As changing the number of pre/post-processing threads can increase the performance, the throughput can be increased by changing the number of streams in the same processing elements. If the number of buffers exceeds a certain level, the performance is saturated and/or the program fails to run due to the limitation of the hardware platform. Through preliminary experiments, we found the upper bound on the number of buffers first. With the maximum number of buffers, we explore the design space of pipelining with the proposed

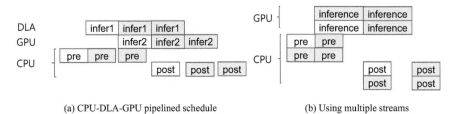

(a) CPU-DLA-GPU pipelined schedule   (b) Using multiple streams

Figure 9.12 Simple examples of CPU-DLA-GPU pipelining (a) and using multiple streams (b).

heuristic. After pipelining decision is made, we reduce the number of buffers as long as the performance is not degraded in order to minimize the resource requirement.

9.3.2 Network Pipelining Heuristic

For a given network with N layers, the size of possible combinations of cut-points is as large as $_{N-1}C_c$ to select c cut-points. To avoid the exponential complexity of exhaustive search, a 2-phase heuristic is devised to find a sub-optimal set of cut-points for a given number of pipeline stages. The first phase is a *global* search over a sampled set of cut-points: we prune the search space by sampling. After we define a set of cut-points in the first phase, we perform the second phase, called *local* search, to examine the other cut-points around the sampled cut-points. While the main objective is to maximize the FPS performance, we use the GPU utilization metric to determine how to explore the design space; moving the cut-points to the direction of increasing the GPU utilization.

Algorithm 5 shows the first phase of the proposed heuristic. First, we define a set of candidate cut-points, C, by sampling. It includes concatenation/summation layers, which combine multiple inputs, and regularly sampled layers in each single path segment of the network. Candidate cut-points are usually located at the backbone of the object detection network because some layers in the neck or head of the object detection network are not mappable to a DLA. Among candidate cut-points, we randomly select a set of initial cut-points, denoted by cut_{init}. We define two threshold values for GPU utilization, $thres_{coarse}$ and $thres_{local}$. Before defining the threshold values, we obtain the maximum GPU utilization by running the network on the GPU only. Note that since we use the maximum number of streams to maximize the GPU utilization, the GPU utilization after pipelining will be lower than the maximum GPU utilization. We set the former threshold, $thres_{coarse}$, to a slightly lower than the maximum GPU utilization. The latter thresold, $thres_{local}$, is set to a smaller value than $thres_{coarse}$, using it as the lower bound of the final GPU utilization.

The first phase starts with the set of initial cut-points, cut_{init} (line 1). We obtain the initial value for FPS by executing the program with this initial set (line 2). If the GPU utilization is lower than $thres_{coarse}$ (line 6), we need to move the cut-points to the direction of increasing the GPU utilization. For fast search, we use a binary search (lines 8–9). For example, if a two-stage pipeline with one DLA and GPU has a cut-point located on the 24th candidate cut-points, MOVE_BINARY policy moves the cut-point to the 12th candidate cut-point. If the GPU utilization is higher than $thres_{coarse}$, we move the cut-point to neighbor candidate cut-points (lines 6–7). For each movement of cut-points, we execute the program and update the best FPS and the best cut-points set (line 13). If the FPS performance with cut_{cur} is improved over that of cut_{prev}, $MovePolicy$ is maintained on the next search (lines 4–5).

If the current GPU utilization, $gpuUtil_{cur}$, becomes no smaller than $thres_{local}$, the second phase of the heuristic, local search, is performed (lines 14–15). The local search heuristic is described in Algorithm 6. During local search, each cut-point of cut_{cur} is moved one by one to search over the cut-points that do not belong to the set of candidate cut-points and checks the FPS improvement.

Algorithm 5 Overall algorithm of the network pipelining heuristic

---

Require: C : A set of candidate cut-points
Require: $cut_{init/cur/prev/best} = \{c_0, c_1, ...\}$: a set of initial/current/previous/best cut-points
Require: $thres_{coarse}$: GPU Threshold value to determine if the binary search will be used
Require: $thres_{local}$: Lower bound of the GPU utilization
Require: $fps_{cur/prev/best}$: FPS of current/previous/best program run.
Require: $gpuUtil_{cur}$: GPU utilization of current program run.
1: $cut_{cur} = cut_{init}$, $cut_{prev} =$None, $fps_{prev} = MAX$
2: Execute the program with cut_{cur} and set fps_{cur}
3: **while** $cut_{cur}! = cut_{prev}$ **do**
4:     **if** $fps_{prev} < fps_{cur}$ and $MovePolicy$ is set **then**
5:         Use previous $MovePolicy$
6:     **else if** $gpuUtil_{cur} \geq thres_{coarse}$ **then**
7:         $MovePolicy =$\{MOVE_ONE, SHIFT_ONE\}
8:     **else**
9:         $MovePolicy =$\{MOVE_BINARY\}
10:     **end if**
11:     $cut_{prev} = cut_{cur}, fps_{prev} = fps_{cur}$
12:     $cut_{cur} =$ Find a new un-searched cut-point set obtained by the $MovePolicy$ set above
13:     Execute the program with cut_{cur} and update cut_{cur}, cut_{best} and fps_{best}
14:     **if** $gpuUtil_{cur} \geq thres_{local}$ **then**
15:         Perform Local search with cut_{cur}
16:     **end if**
17: **end while**

Algorithm 6 Local search heuristic of finding pipeline cut-points

---

Require: $numCut$: the number of cut-points
Require: $cut_{init/cur/prev} = \{c_0, c_1, ...\}$: a set of initial/current/previous cut-points
Require: $fps_{cur/prev}$: FPS of current/previous program run.
 1: $cut_{cur} = cut_{init}$, $fps_{prev} = \text{MAX}$
 2: **for** $cut\_curser = 0$; $cut\_curser < numCut$; $cut\_curser = cut\_curser + 1$
　　do
 3:　　$MovePolicy = \{\text{MOVE\_ONE\_LEFT, MOVE\_ONE\_RIGHT}\}$, $i = 0$
 4:　　**while** $i < length(MovePolicy)$ **do**
 5:　　　　$cut_{prev} = cut_{cur}, fps_{prev} = fps_{cur}$
 6:　　　　$cut_{cur} = $ Move cut_cursor point of cut_{cur} based on $MovePolicy[i]$
 7:　　　　Execute the program with cut_{cur} and update cut_{cur}, cut_{best} and
　　fps_{best}
 8:　　　　**if** $fps_{cur} > fps_{prev}$ **then**
 9:　　　　　　continue
10:　　　　**else**
11:　　　　　　$cut_{cur} = cut_{prev}, fps_{cur} = fps_{prev}$
12:　　　　**end if**
13:　　　　$i = i + 1$
14:　　**end while**
15: **end for**

9.3.3　Software Framework for Network Pipelining

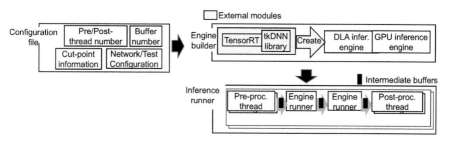

Figure 9.13 TensorRT-based Software framework for network pipelining.

Since there is no public framework that enables us to use for network pipelining on the Jetson hardware platform, we have developed a TensorRT-based framework called JEDI (Jetson-aware Embedded Deep learning Inference acceleration) [1]. Figure 9.13 shows the overall flow of JEDI starting from the configuration file which contains the network

---

[1]The framework is available at https://github.com/cap-lab/jedi

information and the pipelining decision made from the proposed heuristic. It includes cut-point information as well as configurable parameters such as the number of buffers/streams, the number of pre/post-processing threads, and mapping information to processing elements.

Based on the configuration information, the engine builder creates a separate inference engine for each pipeline stage. To this end, we modified the tkDNN [362] library that makes TensorRT easy to use. A pair of buffers are created for each pipeline edge, one for the output of the sending engine and the other for the input of the receiving engine. Since there may be more than one pipeline edge, careful buffer management is needed for correct operation.

After inference engines are built, the inference runner module creates threads based on the configuration parameters, including pre/post-processing threads and engine runner threads. An engine runner thread is created for each inference engine. It maps the execution contexts to the streams and takes charge of synchronization with the neighbor pipelining stages: mapping to the stream is suspended until the input buffers and the output buffers are all available. Since an engine runner thread may have multiple execution contexts, synchronization delay can be hidden by interleaved execution of streams. As shown in Figure 9.13, the inference runner module creates a set of threads for each network instance.

9.3.4 Experimental Results

In the proposed methodology, we need to select the network algorithm after the hardware platform is determined. Since YOLO-based network outperforms the other single-stage detectors, we take the following three networks for comparison: YOLOv2, YOLOv4, and Tiny YOLOv4. Table 9.7 shows the baseline inference results of those networks using Jetpack 4.3 and TensorRT 6. The table shows two FPS results, one for GPU and one for DLA since the baseline TensorRT allows us to use a single accelerator for inference. Note that the GPU performance is higher than DLA. Since the DLA does not support *leaky relu* or *mish* activation, we replace those with *relu* activation and retrain the networks. If there is any layer that the DLA does not support among the layers mapped to the DLA, the layer is actually executed on the GPU, which is called *GPU fallback*. As for the dataset, we use the *COCO2014 trainval* for training and the *COCO2017 val* for inference with image size 416x416 and batch is set to one. While the board allows us to set the performance and energy parameters with the *nvpmodel* command, we use the maximum frequency

TABLE 9.7 Baseline inference results of the benchmarks

Network	# of layers	GPU FPS	GPU Util(%)	DLA FPS
Yolov2 [347]	54	82	45.49	47
Yolov4 [363]	269	49	65.73	21
Yolov4-tiny [363]	57	115	25.01	76

on the *MAXN* power mode which does not limit the power budget. The *tegrastats* command is used for measuring power consumption and GPU utilization. On the other hand, the DLA utilization cannot be obtained. In all experiments, we perform each experiment five times and get the average value unless stated differently.

Several observations can be made from the baseline inference results as shown in Figure 9.7. First, since Jetson AGX Xavier is much faster than Jetson TX2, all networks can satisfy the throughput requirement of LPIRC. Second, the GPU utilization is not high even though the entire inference is run on the GPU. Lastly, the performance gap between GPU and DLA is reduced as the network size is small because the pre-processing step takes a more significant effect on the performance. It guides us to use multiple threads for the pre-processing step.

There is no single best configuration of intra-network pipelining for all benchmarks: how many stages to use and how to map the stages to PEs. Thus we compare four pipelining options that are listed in Table 9.8. Since object detection networks have some layers that cannot be performed on a DLA at the head, we assign the last pipeline stage to GPU in all options. In options C and D, we distinguish two DLAs with *DLA0* and *DLA1*. In options B and D, the GPU is assigned two pipeline stages since the GPU has more computation power than DLA.

For each network, we vary the pipelining option manually and apply the proposed pipelining heuristic to make the best pipelining decision for each option. Comparing the performance among four pipelining options, we select the best option for each network. Table 9.9 shows

TABLE 9.8 Four options for Network Pipelining

Option	# of pipeline stages	Composition of PEs
A	2	DLA0-GPU
B	3	GPU-DLA0-GPU
C	3	DLA0-DLA1-GPU
D	4	GPU-DLA0-DLA1-GPU

TABLE 9.9 Comparison of Object Detection networks after Optimization (image size $= 256 \times 256$): configuration $=$ (# of pre-processing threads, # of post-processing threads, # of streams or buffers)

Network	HW	Mapping	FPS	mAP@ 0.5(%)	Power(W)	Score	Configuration
YOLOv2	TX2	GPU	30	33.2	25.7	0.07	(2,1,1)
YOLOv2	Xavier	GPU	63	33.2	12.7	0.29	(2,1,1)
YOLOv2	Xavier	C	298	32.6	18.3	0.94	(4,2,8)
YOLOv4	Xavier	D	106	53.3	18.7	0.53	(2,2,8)
Tiny YOLOv4	Xavier	GPU	327	34.0	13.0	1.52	(4,2,8)

the comparison result between three networks, varying the hardware platform within 30W power budget. The first row corresponds to our LPIRC 2018 solution that uses the following optimizations: CPU-GPU pipelining, tucker decomposition, batch normalization, and image size reduction. Note that we used COCO dataset in this experiment while LPIRC used ImageNet dataset since the COCO dataset is commonly used for objection detection network comparison and the available YOLO networks only are trained with the COCO dataset. We use the *COCO2014 trainval* for training and the *COCO2017 test* for the inference with the image size of 256x256 and batch is set to one. So the score reported in this table for the LPIRC 2018 solution is different from what we obtained in LPIRC 2018, 0.25.

Here is the summary of the comparison results.

1. When we ran the LPIRC solution which does not use the TensorRT on Jetson AGX Xavier, we could obtain 2.1 times FPS gain and about 50% power saving since the Xavier board is much more efficient than Jetson TX2. By changing the hardware only, we achieved 4.1 times better score than our LPIRC 2018 as shown in the second row.

2. With the same YOLOv2 network but using TensorRT on the Xavier board, we could improve the score by 3.24 times by the proposed pipelining heuristic as shown in the third row. The best pipelining option is option C that uses all three accelerators, GPU and two DLAs. Note that the FPS performance is significantly increased by using more CPU threads and eight streams as well as three-stage pipelining. Since DLA does not support *leaky relu* activation

function, we replaced it with *relu* activation and retrained the network, which results in a slight decrease in mAP.

3. YOLOv4 is known as the state-of-the-art object detector in terms of mAP, which is also confirmed in our experiments as shown in the fourth row. Since it is significantly larger than YOLOv2, it is more difficult to find an optimal pipelining configuration. The best pipelining option found by the proposed heuristic is option D that consists of four pipeline stages. Since its FPS is about 3 times lower than YOLOv2, the estimated score is much smaller than YOLOv2. Even though it has a low estimated score, it could be the best solution since it satisfies the throughput constraint and it gives much higher accuracy than the other detectors.

4. The best estimated score is obtained with Tiny YOLOv4. While the mAP performance is similar to YOLOv2, it runs much faster than YOLOv2 when it is run on the GPU. Since the network is not long and the GPU fallback may occur in the middle of the network, pipelining does not give any gain. The entire speed-up comes from multithreading and multiple streams.

5. Experimental results confirm the value of the proposed pipelining technique that explores the wide design space of parallelism for a given network and the importance of comparing networks after optimization.

9.4 CONCLUSION AND FUTURE WORK

In this chapter, we present a design methodology for the design of low-power image recognition systems. The overall design flow consists of five steps as shown in Figure 9.1. After selecting a hardware platform and a neural network in the first and the second step, we map the network onto the hardware platform in the third step. The goal of mapping is to maximize the throughput by pipelining. After the pipelining decision is made, various software optimization techniques are applied to each processing element, aiming to minimize the execution time of the associated pipeline stage. Lastly, we perform post-processing at the system level to jointly optimize multiple objectives such as speed, accuracy, and energy consumption.

In the selection of the hardware platform, it is necessary to consider the trade-off between performance and power consumption. For LPIRC 2017 and 2018, it was easy to choose the hardware platform since NVIDIA Jetson TX2 board outperforms other available hardware platforms and NVIDIA provides the cuDNN library to alleviate the programmer of the burden of optimizing various operations. The board has a single DNN accelerator which is GPU. Recently, hardware platforms that have specialized hardware accelerators, called NPU(neural processing unit), have been introduced and NVIDIA AGX Xavier is an example. Since NPUs are actively researched and developed, we will have more diverse hardware platforms in the future. Then, it will be more challenging to select the best hardware platform since it affects the subsequent steps in the design methodology. For instance, an NPU may run a normal convolution efficiently while it performs poorly for depth-wise convolution. Then any network based on depth-wise convolution will be ignored in the second step. It is envisioned that we need to revise the overall flow by adding an early performance estimation step after the pipelining decision is made and adding a feedback arc from this step to the first step.

It is also a future work to make an NPU that is friendly for software optimization. While low-bit quantization is a very effective technique to reduce the memory size and memory access delay, we could use only 16-bit quantization due to hardware limitations. If an NPU supports 4-bit or 8-bit arithmetic efficiently, more aggressive software optimization could be performed. If the NPU can skip zero value computation during convolution, pruning will be another effective optimization method. It means that the software optimization techniques are heavily dependent upon the hardware platform.

In the second step of the proposed design methodology, we explore the existing networks to find the best network. Recently, automated neural architecture search (NAS) emerges as the default technique to find a CNN architecture with higher accuracy than manually-designed architectures. Since the performance of a network depends on the hardware platform, the NAS technique needs to be customized to a given hardware platform. While numerous NAS techniques have been proposed with various search strategies recently, their assumed hardware platforms are mostly GPUs. As NPUs are getting popular, NPU-aware NAS attracts research attention recently [364]. If a hardware platform has NPUs inside, it might be better to design the network specialized to the platform than using a network that is designed for GPU execution. Thus, it will be a future research topic to find the customized object detector for the chosen hardware platform by a NAS technique.

In the future, we may want to run multiple networks concurrently, serving different purposes. We will extend the proposed methodology to co-optimize multiple networks together.

CHAPTER 10

Guided Design for Efficient On-device Object Detection Model

Tao Sheng and Yang Liu

Amazon

CONTENTS

The low-power computer vision (LPCV) challenge is an annual competition for the best technologies in image classification and object detection measured by both efficiency (execution time and energy consumption) and accuracy (precision/recall). Our Amazon team has won three awards from LPCV challenges: 1st prize for interactive object detection challenge in 2018 and 2019 and 2nd prize for interactive image classification

DOI: 10.1201/9781003162810-10

challenge in 2018. This paper is to share our award-winning methods, which can be summarized as four major steps. First, 8-bit quantization friendly model is one of the key winning points to achieve the short execution time while maintaining the high accuracy on edge devices. Second, network architecture optimization is another winning keypoint. We optimized the network architecture to meet the 100ms latency requirement on Pixel2 phone. The third one is dataset filtering. We removed the images with small objects from the training dataset after deeply analyzing the training curves, which significantly improved the overall accuracy. And the forth one is non-maximum suppression optimization. By combining all the above steps together with the other training techniques, for example, cosine learning function and transfer learning, our final solutions were able to win the top prizes out of large number of submitted solutions across worldwide.

10.1 INTRODUCTION

Competitions encourage diligent development of technologies. Historical examples include Ansari XPRIZE competitions for suborbital spaceflight, numerous Kaggle competitions such as identifying salt deposits beneath the Earth's surface from seismic images, and the PASCAL VOC, ILSVRC (a.k.a ImageNet), and COCO [365] competitions for accurate computer vision techniques. The Low-Power Image Recognition Challenge (LPIRC) aims to accelerate the development of computer vision solutions that are fast, accurate, and low-power for edge devices.

Started in 2015, LPIRC [366], [367] is an annual competition identifying the best computer vision solutions for classifying and detecting objects in images, while using as little energy as possible. Although many competitions are held every year, LPIRC is the only one integrating both state-of-the-art computer vision techniques and low power requirement. In June 2018, the competition was held very successfully co-located with CVPR 2018 in Salt Lake City. Due to the large number of contestants and strong interests of sponsors, the LPIRC was held again in November 2018 (called LPIRC-II), and the winners of which were announced at NeurIPS 2018. In June 2019, the competition was held co-located with CVPR 2019 in Long Beach, California. LPIRC offered two tracks in 2018, Track 1 is sponsored by Google focusing on the evaluation of accuracy and execution time using TensorFlow and TFLite running on Pixel2 phone. Track 2 is sponsored by Facebook focusing on the evaluation of energy consumption using the Caffe2 framework running on NVIDIA

Jetson TX2. We as Amazon team participated the Track 1 in LPIRC-II in 2018 and LPIRC-2019 due to the main interest on the edge phone devices. Here is the description of LPIRC Track 1.

10.1.1 LPIRC Track 1 in 2018 and 2019

The goal of LPIRC is to achieve the best accuracy within the wall-time constraint by evaluating the accuracy and the execution time using TensorFlow models. Although no power or energy constraint is explicitly measured for this track, latency correlates reasonably with energy consumption. This track is further divided into three categories (with separate awards):

Real-time image classification: This is the original task where focusing on ImageNet classification models operating at 30 ms/image. Submissions are evaluated based on classification accuracy/time while focusing on the real-time regime running on Google's Pixel2 phone.

Interactive image classification: Similar to the above category but the latency budget is extended to 100 ms/image.

Interactive object detection: The newly introduced category in 2018 focusing on COCO object detection models at 100 ms/image. Submissions are evaluated based on detection mAP and latency running on Google's Pixel2 phone.

10.1.2 Three Awards for Amazon team

After diligent hard-work, our Amazon team won the two awards out of all submitted solutions at 2018: the first place for interactive object detection challenge and the second place for interactive image classification challenge. Then in LPIRC-2019, our team won the first place again for interactive object detection challenge. We would like to describe the details of our winning methods in this paper. The rest of the paper will be organized as follows: Section 10.2 will talk about some technical details about getting an efficient deep learning model for edge devices. In Section 10.3, we illustrate the winning methods and show the experimental results. And Section 10.4 concludes this paper.

10.2 BACKGROUND

Quantization is crucial for deep learning inference on edge devices, which have very limited budget for power and memory consumption. Such platforms often rely on fixed-point computational hardware blocks, such as Digital Signal Processor (DSP), to achieve higher power efficiency over floating-point processor, such as CPU and GPU. Although quantization may not impact inference accuracy for many over-parameterized deep leanring models, such as VGGNet [368], InceptionNet [369], ResNet [370], etc., deploying those models on edge devices is not quite feasible due to large computational complexity and memory footprint.

Recently, there has been tramendous progress on the innovation of many lightweight deep learning networks. These models can trade off accuracy with efficiency by replacing conventional convolution with depthwise separable convolution. For example, the MobileNets [371] [372] drastically shrink the parameter size and memory footprint, thus are getting increasingly popular on edge devices. The downside is that the separable convolution core layer in MobileNets causes large quantization loss, thus results in significant feature representation degradation in the 8-bit fixed-point inference pipeline.

To demonstrate the quantization issue, we selected the TensorFlow implementation of MobileNets [371] [372] and InceptionV3 [369], and compared their accuracies on floating-point pipeline against 8-bit fixed-point pipeline. The results are summarized in Table 10.1. The top-1 accuracy of InceptionV3 drops slightly after applying the 8-bit quantization, while the accuracy loss is significant for MobileNetV1, MobileNetV2, and MobileNetV1-SSD [349]. Although the accuracy of InceptionV3 is much less impacted by 8-bit quantization, its computational complexity is eight times more than MobileNetV1, which means it is not well suited to use InceptionV3 on edge devices. Therefore, how to make the lightweight deep learning models quantization friendly is essential to achieve a good tradeoff between accuracy and latency. There are two approaches: one is the quantization-aware training proposed by [373], and the other one is the re-architectured MobileNet network proposed by [374]. Due to the availablity of TFLite quantization toolkit, we utilize the quantization-aware training approach in both interactive classification challenge and interactive detection challenge.

TABLE 10.1 Float vs. 8-bit Accuracy/Latency Comparison on ImageNet2012 and COCO validatation Dataset

Networks	Float Pipeline	8-bit Pipeline	Comments
InceptionV3	78.00%/1433ms	76.92%/637ms	Only standard convolution
MobileNetV1	70.50%/160ms	1.8%/70.2ms	Mainly separable convolution
MobileNetV2	71.9%/117ms	1.23%/80.3ms	Mainly separable convolution
MobileNetV1-SSD	21%/331ms	6.7%/145ms	Mainly separable convolution

10.3   AWARD-WINNING METHODS

In this section, we will describe the steps to achieve the award-winning models: First, we built the proper deep learning networks and trained the networks by 8-bit quantization-aware framework. Second, we optimized the network architecture to meet the latency requirement while maintaining the high accuracy. Third, we applied dataset filtering to training images. Forth, we optimized the non-maximum suppression in detection model by choosing the best threshold. Besides the above steps, we also applied the other training techniques to further improve the accuracy, for example, cosine learning function and transfer learning.

10.3.1   Quantization Friendly Model

As mentioned in Section 10.2, we built our deep learning models for both classification and detection based on MobileNets [371] [372]. For the classification task, we chose the MobileNetV2 [372] architecture with best configuration of the input resolution and depth multiplier. For the detection task, we used MobileNetV1 [371] as the base network in 2018, and MobileNetV2 [372] as the base network in 2019, and add SSD [349] layers after the base network for object detection as illustrated in Figure 10.1. We trained the models by the quantization-aware framework proposed by [373] to achieve high efficiency for both tasks on Pixel2 phone.

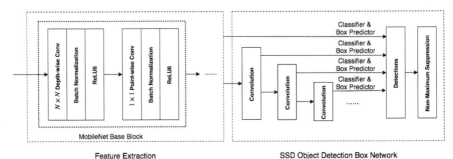

Figure 10.1 MobileNet SSD Object Detection Network Architecture where MobileNet V1/V2 is used as backbone, and SSD is used as detection head. We chose MobileNet-SSD to meet the latency requirement while maintaining the high accuracy on Pixel2 phone.

10.3.2   Network Architecture Optimization

To speed up the training process, we adopted the idea of transfer learning by starting the training with a pre-trained model in tensorflow [268]. The pre-trained weights was obtained on the 300×300 image resolution. However, the quantized MobileNetV1 SSD model with 300×300 input resolution took 145 ms on Pixel2 phone, which exceeded the speed budget by around 45 ms. Therefore, we need to optimize the detection network to meet the latency requirement. We also found that it's important to make the layer geometry as multiple of eight due to the underneath 8-bit runtime optimazation in the convolution kernels. In the following sections, we will introduce the training approaches and the optimal model architecture to achieve the best results on the LPIRC competition in both 2018 and 2019.

10.3.3   Training Hyper-parameters

For detection task, when we started to train the quantized MobileNetV1 SSD model using COCO dataset [365] and Tensorflow [268] framework with GPUs, it is very critical to choose the proper training parameters so that the network can converge to an optimal state in the fastest speed. Based on recent literature [375], stochastic gradient descent with warm restart has gained a lot of attention and been proven to be fast and stable in many empirical studies. In particular, we chose to use a variant of this technique—cosine decay learning rate for our fine-tune network training.

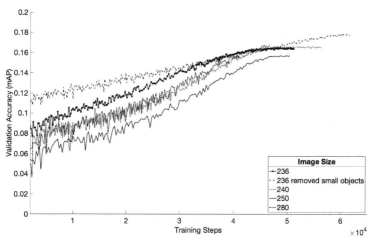

Figure 10.2 Object detection accuracy (mAP) vs. input image size. The 236×236 model reached almost the same accuracy as the 280×280 model, which means we can significantly reduce the computation complexity, while maintaining the similar accuracy. The other interesting observation is that 250×250 detection model has even lower accuracy than 240×240 detection model and 236×236 detection model. Our understanding is that the layer geometry of the network needs to be multiple of eight to reduce the possible negative impact of padding or striding. The accuracy of 236×236 model got further improved after removing the training images with small objects.

10.3.4   Optimal Model Architecture

The fastest way to improve the latency of a deep learning network is to reduce the input image resolution. However, changing the input image resolution without re-training the network will significantly degrade the model accuracy. Hypothetically, if the change of resolution is not drastic, fine-tuning the network to adapt to the new image resolution should be viable. To validate, we have run the experiments to fine-tune 300×300 detecion model to smaller image sizes. The mAP accuracy of pre-trained 300×300 detection model is 18.0%. From Figure 10.2, the object detection mAP accuracy has decreased significantly for smaller image sizes before fine-tuning. As expected, fine-tuning the networks eventually improved the mAP accuracy over the training steps. Figure 10.2 and Table 10.2 also show the trade offs between speed and accuracy. As shown in Figure 10.2, changing input image resolution with fine-tuning does not have

TABLE 10.2 Input Image Resolution vs. Latency on Pixel2 Phone

Input Image Resolution (pixels)	300	280	270	240	236
Latency (ms)	145	131	120	95	93

significant impact to its accuracy. The 236×236 detection model reached almost the same accuracy as the 280×280 detection model with only 0.02 accuracy loss. It suggests that we can significantly reduce the computatial complexity, while maintaining the similar accuracy. The other interesting observation is that 250×250 detection model has even lower accuracy than 240×240 detection model and 236×236 detection model. Our understanding is that the layer geometry of the network needs to be multiple of eight to reduce the possible negative impact of padding or striding. By fine-tuning the network to 236×236 pixels, we achieved the speed of 93 ms on Pixel2 phone while maintaining a good accuracy in 2018.

10.3.5 Neural Architecture Search

Neural architecture search (NAS) [209] have been shown to be able to automatically acquire a neural architecture that achieves competitive performance. Despite its great success, most of the NAS approaches are focused on searching for a universal model for all computer vision applications using a proxy task (e.g., searching a model on ImageNet classification but applying it on semantic segmentation task). Moreover, most NAS approaches adopt proxy latency such as FLOPS instead of real on-device latency during neural architecture searching to tradeoff the model accuracy and overhead. Unfortunately, it has been demonstrated that using proxy task and proxy latency is inaccurate and the searched model may achieve poor performance when being applied on different tasks and platforms. In LPIRC-2019, we performed neural architecture search directly on object detection task and used on-device (Pixel2 Phone) latency for searching to ensure that the searched model meets the latency requirement while achieving the optimal accuracy.

10.3.6 Dataset Filtering

COCO [365] dataset is a very challenging dataset for object detection, because it contains objects in a wide range of scales. It can be divided

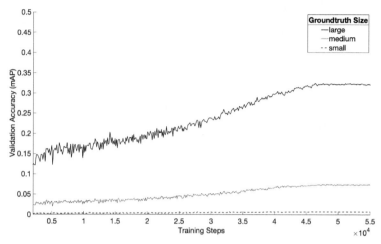

Figure 10.3 Object detection accuracy of objects with different sizes: Entire training steps.

into three categories of training images with regard to the object size: large, medium, and small. Let A be the area size of the object in each category, the categories are defined as following:

$$Large : A > 96 \times 96$$
$$Medium : 32 \times 32 < A < 96 \times 96 \qquad (10.1)$$
$$Small : A < 32 \times 32$$

The size 96/32 is determined by the COCO [365] dataset. As seen in Figure 10.3, the impact to objects in each category is very different. The mAP accuracy of large objects is significantly better than medium and small objects. The mAP accuracy improvement in small objects is too small and it has almost no contribution to the overall mAP. However, the error back-propagation from small objects is weighted equally as other objects. In the extreme case, the small objects are likely to be chosen as the hard examples and largely contribute to the back-propagated error. Due to the noise generated by image down-sampling, small objects would have a very low signal-to-noise ratio and the quality of objective error measured from small objects are significantly degraded. Figure 10.4 shows an example of the increase in mAP of small objects would negatively impact the mAP of medium and large objects. Therefore, we removed the images with small objects from our training dataset to focus more on large and medium objects. As shown in Figure 10.2, this dataset filtering

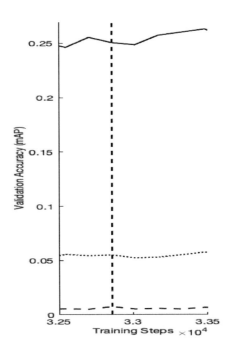

Figure 10.4 Object detection accuracy of objects with different sizes: A zoom-in section

optimization improved the object detection accuracy (mAP) by more than 7.8% relatively, and increased the overall mAP from 16.5% to 17.8%.

10.3.7   Non-maximum Suppression Threshold

Lastly, the non-maximum suppression threshold [349] can be tuned to further trade off the speed versus accuracy. The 236×236 detection model has left us a few minisecond room to reach the latency budget 100ms, which is an ideal range for fine-tuning non-max suppression threshold. Table 10.3 summarizes the experimental results of applying different non-max suppression thresholds to the optimized 236×236 MobileNetV1 SSD model. Finally, we chose to use threshold 0.15 which has the best trade-off between accuracy and latency.

TABLE 10.3 Non-maximum Suppression threshold vs. Accuracy on 236×236 MobileNetV1 SSD

Non-max suppression threshold	0.30	0.25	0.20	0.15	0.10
Accuracy (mAP %)	17.85	18.11	18.29	18.40	18.43

10.3.8 Combination

By combing all the above methods, we can achieve the desired detection model with latency within 100 ms and the highest accuracy. As shown in Table 10.4, our optimized 236×236 detection model can achieve even higher accuracy than original 300×300 detection model with much faster model execution on Pixel2 phone. This detection model won the first place for interactive detection challenge on LPIRC COCO hold-out test dataset in 2018, and further optimized model won the first prize again in 2019. With similar methods applied to ImageNet classification except the dataset filtering and non-maximum suppression optimization, we also trained a very good 8-bit quantization friendly classification model, which won the second place for interactive classification challenge LPIRC-II ImageNet hold-out test dataset in 2018.

TABLE 10.4 Comparison between Our Optimized 236×236 Detection Model vs. Original 300×300 Detection Model

Model	Input	Accuracy	Latency
Model	Resolution	(mAP %)	(ms)
Original detection model	300×300	18.0	145
Our optimized detection model	236×236	18.4	93

10.4   CONCLUSION

This paper explained our award-winning methods from the highly participated public low power image recognition challenge 2018 and 2019. As
computer vision is widely used in many battery-powered edge devices, the
needs for low-power computer vision will become increasingly important.
Our Amazon team will continue to focus on low-power computer vision
area and engage with LPIRC community.

III

Invited Articles

CHAPTER 11

Quantizing Neural Networks for Low-Power Computer Vision

Marios Fournarakis, Markus Nagel, Rana Ali Amjad, Yelysei Bondarenko, Mart van Baalen, and Tijmen Blankevoort

Qualcomm

CONTENTS

DOI: 10.1201/9781003162810-11

In recent years, deep learning has been widely adopted in computer vision applications. While for many tasks neural networks outperform traditional computer vision algorithms they often come at a high computational cost. Reducing the power and latency of neural network inference is key if we want to integrate state-of-the-art networks into edge devices with strict power and compute requirements. Neural network quantization is one of the most effective ways of achieving these savings but the additional noise it induces can lead to accuracy degradation.

In this chapter, we introduce state-of-the-art algorithms for mitigating the impact of quantization noise on the network's performance while maintaining low-bit weights and activations. We start with a hardware motivated introduction to quantization and then consider two main classes of algorithms: Post-Training Quantization (PTQ) and Quantization-Aware-Training (QAT). PTQ requires no re-training or labelled data and is thus a lightweight push-button approach to quantization. In most cases, PTQ is sufficient for achieving 8-bit quantization with close to floating-point accuracy. QAT requires fine-tuning and access to training data but enables lower bit quantization with competitive results. For both solutions, we provide tested pipelines based on existing literature and extensive experimentation that lead to state-of-the-art performance for common computer vision models and tasks.

11.1   INTRODUCTION

With the rise in popularity of deep learning as a general-purpose tool to inject intelligence into electronic devices, the necessity for small, low-latency and energy efficient neural networks solutions has increased. Today neural networks can be found in many electronic devices and services, from smartphones, smart glasses, and home appliances, to drones, robots, and self-driving cars. All these devices have either strict time restrictions on the execution of neural networks or stringent power requirements for long-duration performance.

One of the most impactful ways to decrease the computational time and energy consumption of neural networks is quantization. In neural network quantization, the weights and activation tensors are stored in lower bit-precision than the 16 or 32-bit precision they are usually trained in. When moving from 32 to 8 bits, the memory overhead of storing tensors decreases by a factor of 4 while the computational cost for matrix multiplication reduces quadratically by a factor of 16. We also observe that networks are generally robust to quantization, meaning they can be quantized to lower bit-widths with a relatively small impact on the network's accuracy. Besides, neural network quantization can often be applied along with other common methods for neural network optimization, such as neural architecture search, compression and pruning. It is an essential step in the model efficiency pipeline for any practical use-case of deep learning. However, neural network quantization is not entirely free. Low bit-width quantization introduces noise to the network that can lead to a drop in accuracy. While some networks are robust to this extra noise, other networks require extra work to exploit the full potential of quantization.

In this chapter, we introduce the state-of-the-art in neural network quantization. We start with a general introduction to quantization and discuss hardware and practical considerations. We then consider two different regimes of quantizing neural networks: Post-Training Quantization (PTQ) and Quantization-Aware Training (QAT). PTQ methods, discussed in Section 11.3, take an already-trained network and quantize it with little or no data requiring minimal hyperparameter tuning and no end-to-end training. This makes them a lightweight push-button approach to quantizing neural networks with low engineering effort and computational cost. In contrast, QAT discussed in Section 11.4, relies on retraining the neural networks with quantization simulation in the training pipeline. While this requires more effort in training and hyperparameter tuning, it generally further closes the gap to the full-precision precision accuracy compared to PTQ. For both regimes, we introduce a standard pipeline based on existing literature and extensive experimentation that leads to state-of-the-art performance for common computer vision models.

11.2   QUANTIZATION FUNDAMENTALS

In this section, we introduce the basic principles of neural network quantization. We start with a hardware motivation and then introduce standard quantizers and their properties. Later, we discuss practical considerations related to layers commonly found in computer vision models and implications for fixed-point accelerators.

11.2.1   Hardware Background

Before diving into the technical details, we first explore the hardware background of quantization and how it enables efficient inference on device. Figure 11.1 provides a schematic overview of how a matrix-vector multiplication, $\mathbf{y} = \mathbf{W}\mathbf{x} + \mathbf{b}$, is calculated in a neural network (NN) accelerator. This is the building block of larger matrix-matrix multiplications and convolutions found in neural networks. Such hardware blocks aim at improving the efficiency of NN inference by performing as many calculations as possible in parallel.

The two fundamental components of this NN accelerator are the *processing elements* $C_{n,m}$ and the *accumulators* A_n. Our toy example in Figure 11.1 has 16 processing elements arranged in a square grid and 4 accumulators. The calculation starts by loading the accumulators with the bias value $\mathbf{b}_n$. We then load the weight values $\mathbf{W}_{n,m}$ and the input values $\mathbf{x}_m$ into the array and compute their product in the respective processing elements $C_{n,m} = \mathbf{W}_{n,m}\,\mathbf{x}_m$ in a single cycle. Their results are

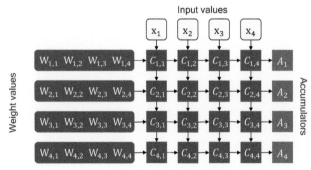

Figure 11.1 A schematic overview of matrix-multiply logic in neural network accelerator hardware.

then added in the accumulators:

$$A_n = \mathbf{b}_n + \sum_m C_{n,m} \tag{11.1}$$

The above operation is also referred to as *Multiply-Accumulate* (MAC). This step is repeated many times for larger matrix-vector multiplications. Once all cycles are completed, the values in the accumulators are then moved back to memory to be used in the next neural network layer. Neural networks are commonly trained using FP32 weights and activations. If were to also perform inference in FP32, the processing elements and the accumulator would have to support floating-point logic. Besides, we need to transfer the data from memory to the compute units. The MAC operations and data transfer consume the bulk of the energy spent during neural network inference. Hence, significant benefits can be achieved by using a lower bit fixed-point or *quantized* representation for these quantities. Low-bit fixed-point representations, such as INT8, not only reduces the amount of data transfer but also the size and energy consumption of the MAC operation [376]. This is because the cost of digital arithmetic typically scales linearly to quadratically with the number of bits used and because fixed-point addition is more efficient than its floating-point counterpart [376].

To move from floating point operations to the efficient fixed-point alternative, we need a scheme for converting floating vectors to an integer presentation. A real-valued vector, $\mathbf{x}$, can be expressed approximately as real number multiplied by vector of integer values:

$$\widehat{\mathbf{x}} = s_{\mathbf{x}} \cdot \mathbf{x}_{\text{int}} \approx \mathbf{x}, \tag{11.2}$$

where $s_{\mathbf{x}}$ is called a *scale factor* and $\mathbf{x}_{\text{int}}$ is an integer vector, e.g., INT8. We denote this *quantized* version of the vector as $\widehat{\mathbf{x}}$. By quantizing the weights and activations we can write the quantized version of the accumulation equation:

$$\hat{A}_n = \widehat{\mathbf{b}}_n + \sum_m \widehat{\mathbf{W}}_{n,m} \widehat{\mathbf{x}}_m \tag{11.3}$$

$$= \widehat{\mathbf{b}}_n + \sum_m \left(s_{\mathbf{w}} \mathbf{W}_{n,m}^{\text{int}} \right) \left(s_{\mathbf{x}} \mathbf{x}_m^{\text{int}} \right) \tag{11.4}$$

$$= \widehat{\mathbf{b}}_n + s_{\mathbf{w}} s_{\mathbf{x}} \sum_m \mathbf{W}_{n,m}^{\text{int}} \mathbf{x}_m^{\text{int}} \tag{11.5}$$

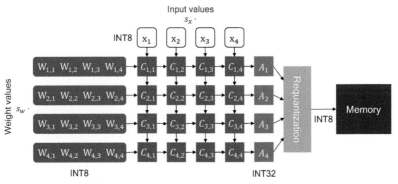

Figure 11.2 A schematic of matrix-multiply logic in an neural network accelerator for quantized inference.

Note that we used a separate scale factor for weights, $s_{\mathbf{w}}$, and activations, $s_{\mathbf{x}}$. This provides flexibility and reduces the quantization error (more in Section 11.2.2). But each scale factor is applied to the whole tensor. This scheme allows us to factor the scale factors out of the summation in equation 11.5 and perform MAC operations in fixed-point format. We intentionally ignore bias quantization for now, because the bias is normally stored in higher bit-width (32-bits) and its scale factor depends on that of the weights and activations [377].

Figure 11.2 shows how the neural network accelerator changes when we introduce quantization. In our example, we use INT8 arithmetic, but this could be any quantization format for the sake of this discussion. It is important to maintain a higher bit-width for the accumulators. Otherwise, we risk incurring loss due to overflow as more products are accumulated during the computation. A typical accumulator for INT8 MAC operations is 32-bits wide, which allows a minimum of 16 8×8-bit MAC calculations.

The activations stored in the 32-bit accumulators need to be written to memory before they can be used by the next layer. To reduce data transfer and the complexity of the next layer's operations, we want to quantize these activations back to INT8. This requires a *requantization* step which is shown in Figure 11.2.

11.2.2 Uniform Affine Quantization

In this section, we define the quantization scheme that we will use in this chapter. This scheme is called *uniform quantization* and is the most commonly used quantization scheme because it permits efficient implementation of fixed-point arithmetic.

Uniform affine quantization, also known as *asymmetric quantization*, is defined by three quantization parameters: the *scale factor s*, the *zero-point z*, and the *bit-width b*. The scale factor and the zero-point are used to map a floating point value to the integer grid, whose size depends on the bit-width. The scale factor is commonly represented as a floating-point number and specifies the *step-size* of the quantizer. The zero-point is an integer that ensures that real zero is quantized without an error. This is important to ensure that common operations like zero padding or ReLU do not cause quantization error.

Once the three quantization parameters are defined we can proceed with the quantization operation. Starting from a real-valued vector $\mathbf{x}$ we first map it to the *unsigned* integer grid $\{0, \dots, 2^b - 1\}$:

$$\mathbf{x}_{\text{int}} = \text{clamp}\left(\left\lfloor \frac{\mathbf{x}}{s} \right\rceil + z; 0, 2^b - 1\right), \tag{11.6}$$

where $\lfloor \cdot \rceil$ is the round-to-nearest operator and clamping is defined as:

$$\text{clamp}\,(x; a, c) = \begin{cases} a, & x < a, \\ x, & a \leq x \leq c, \\ c, & x > c. \end{cases} \tag{11.7}$$

To approximate the real-valued input $\mathbf{x}$ we perfrom a *de-quantization* step:

$$\mathbf{x} \approx \hat{\mathbf{x}} = s\,(\mathbf{x}_{\text{int}} - z) \tag{11.8}$$

Combining the two steps above we can provide a general definition for the *quantization function*, $q(\cdot)$, as:

$$\hat{\mathbf{x}} = q(\mathbf{x}; s, z, b) = s\left[\text{clamp}\left(\left\lfloor \frac{\mathbf{x}}{s} \right\rceil + z; 0, 2^b - 1\right) - z\right] \tag{11.9}$$

Through the de-quantization step, we can also define the quantization grid limits $(q_{\text{min}}, q_{\text{max}})$ where $q_{\text{min}} = sz$ and $q_{\text{max}} = s(2^b - 1 - z)$. Any values of $\mathbf{x}$ that lie outside this range will be clipped to its limits, incurring a *clipping error*. If we want to reduce the clipping error we can expand the quantization range by increasing the scale factor. However, increasing the scale factor leads to increased *rounding error* as the rounding error lies in the range $\left[-\frac{1}{2}s, \frac{1}{2}s\right]$. In Section 11.3.1, we explore in more detail how to choose the quantization parameters to achieve the right trade-off between these two errors.

11.2.2.1 Symmetric Uniform Quantization

Symmetric quantization is a simplified version of the general asymmetric case. The symmetric quantizer restricts the zero-point to 0. This reduces the computational overhead of dealing with zero-point offset during the accumulation operation in equation 11.5. But the lack of offset restricts the mapping between integer and floating-point domain. As a result, the choice of either signed or unsigned integer grid.

$$\widehat{\mathbf{x}} = s\,\mathbf{x}_{\text{int}} \tag{11.10a}$$

$$\mathbf{x}_{\text{int}} = \text{clamp}\left(\left\lfloor \frac{\mathbf{x}}{s} \right\rceil; 0, 2^b - 1 \right) \qquad \text{for unsigned integers} \tag{11.10b}$$

$$\mathbf{x}_{\text{int}} = \text{clamp}\left(\left\lfloor \frac{\mathbf{x}}{s} \right\rceil; -2^{b-1}, 2^{b-1} - 1 \right) \qquad \text{for signed integers} \tag{11.10c}$$

Unsigned symmetric quantization is well-suited for skewed distributions, such as ReLU activations (see Figure 11.3). On the other hand, signed symmetric quantization can be chosen for distributions that are roughly symmetric about zero. 11.2.4.1).

11.2.2.2 Power-of-two Quantizer

Power-of-two quantization is a special case of symmetric quantization, in which the scale factor is restricted to a power of two, $s = 2^{-k}$. This choice brings hardware efficiencies because scaling with s corresponds to simple bit-shifting. However, the restricted expressiveness of the scale factor can complicate the trade-off between rounding and clipping error.

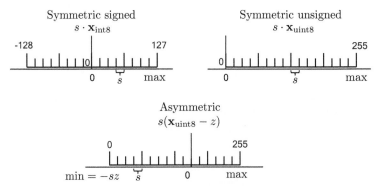

Figure 11.3 A visual explanation of the different available uniform quantization grids. s is the scaling factor, z the zero-point. The floating-point grid is in black, the integer quantized grid in blue.

11.2.2.3 Quantization Granularity

So far, we have defined a single set of quantization parameters (quantizer) per tensor, one for the weights and one for activations, as seen in equation 11.5. This is called *per-tensor quantization* . We can also define a separate quantizer for individual segments of a tensor (e.g., output channels of a weight tensor), thus increasing the *quantization granularity*. In neural network quantization, per-tensor quantization is the the most common choice of granularity due to its simpler hardware implementation: all accumulators in equation 11.5 use the same scale factor, $s_{\mathbf{w}} s_{\mathbf{x}}$. Although increased granularity can be used to improve performance. For example, for weight tensors, we can specify a different quantizer per output channel. This is known as *per-channel* quantization and its implications are discussed in more detailed in Section 11.2.4.2.

Other works go beyond per-channel quantization parameters and apply separate quantizers per group of weights or activations [378, 379, 380]. Increasing the granularity of the groups generally improves accuracy at the cost of some extra overhead. The overhead is associated with accumulators handling sums of values with varying scale factors. Most existing fixed-point accelerators do not currently support such logic and for this reason, we will not consider them in this work. However, as research in this area grows, more hardware support for these methods can be expected in the future.

11.2.3 Quantization Simulation

To test how well a neural network would run on a quantized device, we often simulate the quantized behavior on the same general purpose hardware we use for training neural networks. This is called *quantization simulation.* We aim to approximate fixed-point operations using floating-point hardware. Such simulations are significantly easier to implement compared to running experiments on actual quantized hardware or using quantized kernels. They allow the user to efficiently test various quantization options and it enables GPU acceleration for quantization-aware training as described in Section 11.4. In this section, we first explain the fundamentals of this simulation process and then discuss techniques on how to better match simulated performance with actual on-device performance.

Previously, we saw how matrix-vector multiplication is calculated in dedicated fixed-point hardware. In Figure 11.4a we generalize this process for a convolutional layer but we also include an activation function to

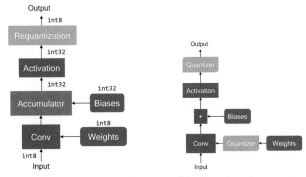

(a) Diagram for quantized on-device inference with fixed-point operations.

(b) Simulated quantization using floating-point operations.

Figure 11.4 Schematic overview of quantized forward pass for convolutional layer: (a) Compute graph of actual on-device quantized inference. (b) Simulation of quantized inference for general-purpose floating-point hardware.

make it more realistic. During on-device inference, all the inputs (biases, weight, and input activations) to the hardware are in a fixed-point format. However, when we simulate quantization using common deep learning frameworks and general-purpose hardware these quantities are in floating-point. This is why we introduce quantizer blocks in the compute graph to induce quantization effects.

Figure 11.4b shows how the same convolutional layer is modeled in a deep-learning framework. Quantizer blocks are added in between the weights and the convolution to simulate weight quantization, and after the activation function to simulate activation quantization. The bias is often not quantized because it is stored in higher precision. In Section 11.2.3.2 we discuss in more detail when it is appropriate to position the quantizer after the non-linearity. The quantizer block implements the quantization function of equation 11.9 and each quantizer is defined by a set of quantization parameters (scale factor, zero-point, bit-width). Both the input and output of the quantizer are in floating-point format but the output lies on the quantization grid.

11.2.3.1 *Batch Normalization Folding*

The batch normalization layer [381] is a standard component of most modern neural networks used in Computer Vision. Batch normalization normalizes the output of a linear layer before adding a scale and an offset (see equation 11.11). For on-device inference, these operations are folded into the previous or next linear layers in a step called *batch normalization folding* [382, 377]. This removes the batch normalization operations entirely from the network, as the calculations are absorbed into a linear layer. Besides reducing the computational overhead of the additional scaling and offset, this prevents extra data movement and the quantization of the layer's output. More formally, during inference batch normalization defined as an affine map of an output x:

$$\text{BatchNorm}(x) = \gamma \left(\frac{x - \mu}{\sqrt{\sigma^2 + \epsilon}} \right) + \beta, \tag{11.11}$$

where μ and σ are the running mean and variance computed as exponential running mean of batch-statistics, and γ and β are learned affine hyper-parameters per-channel. If batch normalization is applied right after a linear layer $\mathbf{y} = \text{BatchNorm}(\mathbf{Wx})$, we can rewrite the terms such that the batch normalization operation is fused with the linear layer itself. Assuming a weight matrix $\mathbf{W} \in \mathbb{R}^{n \times m}$ we apply batch normalization to each output $\mathbf{y}_k$ for $k = \{1, \ldots, n\}$:

$$\begin{aligned}
\mathbf{y}_k &= \text{BatchNorm}(\mathbf{W}_{k,:}\mathbf{x}) \\
&= \gamma_k \left(\frac{\mathbf{W}_{k,:}\mathbf{x} - \mu_k}{\sqrt{\sigma_k^2 + \epsilon}} \right) + \beta_k \\
&= \frac{\gamma_k \mathbf{W}_{k,:}}{\sqrt{\sigma_k^2 + \epsilon}}\mathbf{x} + \left(\beta_k - \frac{\gamma_k \mu_k}{\sqrt{\sigma_k^2 + \epsilon}} \right) \\
&= \widetilde{\mathbf{W}}_{k,:}\mathbf{x} + \widetilde{\mathbf{b}}_k, \tag{11.12}
\end{aligned}$$

where:

$$\widetilde{\mathbf{W}}_{k,:} = \frac{\gamma_k \mathbf{W}_{k,:}}{\sqrt{\sigma_k^2 + \epsilon}} \tag{11.13}$$

$$\widetilde{\mathbf{b}}_k = \beta_k - \frac{\gamma_k \mu_k}{\sqrt{\sigma_k^2 + \epsilon}} \tag{11.14}$$

11.2.3.2 Activation Function Fusing

In our naive quantized accelerator introduced in Section 11.2.1, we saw that the requantization of activations happens after the matrix multiplication or convolutional output values are calculated. However, in practice, we often have a non-linearity directly following the linear operation. It would be wasteful to write the linear layer's activations to memory, and then load them back into a calculation core to apply a non-linearity. For this reason, many hardware solutions come with a hardware unit that applies the non-linearity before the requantization step. If this is the case, we only have to simulate the requantization that happens after the non-linearity. For example, ReLU non-linearities are readily modeled by the requantization block, as you can just set the minimum representable value of that activation quantization to 0.

Other more complex activation functions, such as sigmoid or Swish [383], require more dedicated support. If this support is not available, we need to add a quantization step before and after the non-linearity in the graph. This can have a big impact on the accuracy of quantization simulation. Although newer activations like Swish functions [383] provide accuracy improvement in floating-point, they may be difficult and inefficient to deploy on fixed-point hardware .

11.2.3.3 Other Layers and Quantization

There are many other types of layers being used in neural networks. How these are modeled depends greatly on the specific hardware implementation. Sometimes the mismatch between simulated quantization and on-target performance is down to layers not being properly quantized. Here, we provide some guidance on how to simulate quantization for a few commonly used layers:

Max pooling Activation quantization is not required because the input and output of values are on the same quantization grid.

Average pooling The average of integers is not necessarily an integer. For this reason, a quantization step is required after average-pooling. However, we use the same quantizer for the inputs and outputs as the quantization range does not change.

Element-wise addition Despite its simple nature, this op is difficult to simulate accurately. During addition, the quantization ranges of both inputs have to match exactly. If these ranges do not match, extra care is needed to make addition work as intended. There is no single accepted solution for this but adding a requantization step can simulate the added noise coarsely. Another approach is to optimize the network by tying the quantization grids of the inputs. This would prevent the requantization step but may require fine-tuning.

Concatenation The two branches that are being concatenated generally do not share the same quantization parameters. This means that their quantization grids may not overlap making a requantization step necessary. As with element-wise addition, it is possible to optimize your network to have shared quantization parameters for the branches being concatenated.

11.2.4 Practical Considerations

When quantizing neural networks with multiple layers we are confronted with a large space of quantization choices including the quantization scheme, granularity, bit-width, etc. In this section, we explore some of the practical considerations that help reduce the search space.

Note that in this chapter we only consider *homogeneous* bit-width. This means that the bit-width chosen for either weights or activations remains constant across all layers. Homogeneous bit-width is more universally supported by hardware but some recent works also explore the implementation of heterogeneous bit-width or *mixed-precision* [384, 385, 386].

11.2.4.1 Symmetric vs. Asymmetric Quantization

For each weight and activation quantization step, we have to choose a quantization scheme. On one hand, asymmetric quantization is more expressive because there is an extra offset parameter, but on the other hand there is a possible computational overhead. To see why this is the case, consider what happens when asymmetric weights, $\widehat{\mathbf{W}} = s_{\mathbf{w}}(\mathbf{W}_{\text{int}} - z_{\mathbf{w}})$, are multiplied with asymmetric activations $\widehat{\mathbf{x}} = s_{\mathbf{x}}(\mathbf{x}_{\text{int}} - z_{\mathbf{x}})$.

$$\widehat{\mathbf{W}}\widehat{\mathbf{x}} = s_{\mathbf{w}}(\mathbf{W}_{\text{int}} - z_{\mathbf{w}})s_{\mathbf{x}}(\mathbf{x}_{\text{int}} - z_{\mathbf{x}})$$
$$= s_{\mathbf{w}}s_{\mathbf{x}}\mathbf{W}_{\text{int}}\mathbf{x}_{\text{int}} - s_{\mathbf{w}}s_{\mathbf{x}}z_{\mathbf{w}}\mathbf{x}_{\text{int}} - s_{\mathbf{w}}s_{\mathbf{x}}\mathbf{W}_{\text{int}}z_{\mathbf{x}} + s_{\mathbf{w}}s_{\mathbf{x}}z_{\mathbf{w}}z_{\mathbf{x}} \quad (11.15)$$

The first term is exactly what we would have if both operations were in symmetric format. The third and fourth terms depend only on the scale, offset and weight values, which are known in advance. Thus these two terms can be pre-computed and added to the bias term of a layer at virtually no cost. The second term, however, depends on the input data $\mathbf{x}$. This means that for each batch of data we need to compute an additional term during inference. This can lead to significant overhead in both latency and power, as it is equivalent to adding an extra channel.

For this reason, it is a common approach to use *asymmetric activation quantization* and *symmetric weight quantization.*

11.2.4.2 Per-tensor and Per-channel Quantization

In Section 11.2.2.3 we spoke about different levels of quantization granularity. Per-tensor quantization of weights and activations has been a standard for a while because it is supported by all quantized inference accelerators. However, per-channel quantization of the weights can improve accuracy, especially when the distribution of weights varies significantly from channel to channel. Looking back at the quantized MAC operation of equation 11.5, we can see how per-channel weight quantization can be achieved. Each accumulator could apply a separate weight scale factor. Per-channel quantization of activations is much harder to implement because we cannot factor the scale factor out of the summation, see equation 11.5.

For these reasons, in recent times *per-channel quantization* for the weights has become a common practice. However, not all hardware supports per-channel quantization, so it is important to check what is possible in your intended target device.

11.3 POST-TRAINING QUANTIZATION

Post-training quantization (PTQ) algorithms take a pre-trained FP32 network and convert it directly into a fixed-point network without the need for the original training pipeline. These methods can be data-free or may require a small calibration set, which is often easily accessible. Additionally, having almost no hyperparameter tuning makes them usable via a single API call as a black-box method to quantize a pretrained NN in a computationally efficient manner. This frees the neural network designer from having to be an expert in NN quantization and thus allows for a much wider application of NN quantization.

A fundamental step in the PTQ process is finding good quantization ranges for each quantizer. In Section 11.2.2 we briefly discussed how the choice of quantization range affects the quantization error. In this chapter, we discuss various common methods used in practice to find good quantization parameters. In the following subsections, we will discuss several common issues observed during PTQ and introduce the most successful techniques to overcome them. Finally, in Section 11.3.5, we introduce a standard post-training quantization pipeline which we found to work best in most common scenarios.

11.3.1 Quantization Range Setting

Quantization range setting refers to the method of determining clipping thresholds of the quantization grid, q_{min} and q_{max} (see equation 11.9). The key trade-off in range setting is between clipping error and rounding error, described in Section 11.2.2, and their impact on the final task loss for each quantizer being configured. Each of the methods described here provides a different trade-off between the two quantities. These methods typically optimize local cost functions instead of task loss. This is because in PTQ we do not want to employ end-to-end network computations or backpropagation, and we also want to have computationally fast methods.

Weights can usually be quantized without any need for calibration data. However, determining parameters for activation quantization often requires a few batches of calibration data.

Min-max To cover the whole dynamic range of the tensor, we can define the quantization parameters as follows

$$q_{min} = \min \mathbf{V}, \tag{11.16}$$

$$q_{max} = \max \mathbf{V}, \tag{11.17}$$

where $\mathbf{v}$ denotes the tensor to be quantized. This leads to no clipping error. However, this approach is sensitive to outliers; large outliers may cause excessive rounding errors.

Mean squared error (MSE) One way to alleviate the issue of being strongly impacted by the outliers is to use MSE-based range setting. In this range setting method we find q_{min} and q_{max} that minimize the MSE between the original and the quantized tensor:

$$\underset{q_{min}, q_{max}}{\arg\min} \ \left\| \mathbf{v} - \hat{\mathbf{v}}(q_{min}, q_{max}) \right\|_F^2 , \tag{11.18}$$

where $\hat{\mathbf{v}}(q_{min}, q_{max})$ denotes the quantized version of $\mathbf{v}$ and $\|\cdot\|_F$ is the Frobenius norm. This alleviates the impact of outliers on the quantization grid. The optimization problem is commonly solved using grid search, golden section method or analytical approximations with closed-form solution [387]. Several variants of this range setting method exist in literature but they are all very similar in terms of objective function and optimization.

Cross entropy   For certain layers, all values in the tensor being quantized may not be equally important. One such scenario is the quantization of logits of a NN that performs classification when we are only interested in the prediction accuracy. In this case, it is important to preserve the order of the largest value after quantization. MSE may not be a suitable metric for this, as it weighs all the values in a tensor equally regardless of the numerical values and order. For a large number of classes, this may lead to a range setting that tries to avoid quantization error in the large number of small or negative logits that are in fact unimportant for prediction accuracy. In this specific case, it is beneficial to minimize the following cross-entropy loss function

$$\underset{q_{\min}, q_{\max}}{\arg\min} \quad C\left(\psi\left(\mathbf{v}\right) \| \psi\left(\hat{\mathbf{v}}(q_{\min}, q_{\max})\right)\right), \tag{11.19}$$

where $C(\cdot\|\cdot)$ denotes the cross-entropy function, ψ is the softmax function, and $\mathbf{v}$ is the logits vector.

BN-based range setting   Range setting for activation quantizers often requires some calibration data. If a layer has batch-normalized activations, the per-channel mean and standard deviation of the activations are equal to the learned batch normalization shift and scale parameters, respectively. These can then be used to find suitable parameters for activation quantizer as follows [388]:

$$q_{\min} = \min\left(\boldsymbol{\beta} - \alpha\boldsymbol{\gamma}\right), \tag{11.20}$$

$$q_{\max} = \max\left(\boldsymbol{\beta} + \alpha\boldsymbol{\gamma}\right), \tag{11.21}$$

where $\boldsymbol{\beta}$ and $\boldsymbol{\gamma}$ are vectors of per-channel learned shift and scale parameters, and $\alpha > 0$. [388] uses $\alpha = 6$ so that only large outliers are clipped.

Comparison   In Table 11.1 we compare range setting methods for weight quantization. For high bit-widths, the MSE and min-max approaches are mostly on par. However, on lower bit-widths the MSE approach clearly outperforms the min-max. In Table 11.2 we present a similar comparison for activation quantization. We note that MSE

TABLE 11.1 Ablation Study for Different Methods of Range Setting of (symmetric uniform) Weight Quantizers While Keeping the Activations in FP32. Average ImageNet Validation Accuracy (%) over 5 runs.

Model (FP32 acc.)	ResNet18 (69.68)			MobileNetV2 (71.72)		
Bit-width	W8	W6	W4	W8	W6	W4
Min-Max	**69.57**	63.90	0.12	**71.16**	64.48	0.59
MSE	69.45	**64.64**	**18.82**	71.15	**65.43**	**13.77**
Min-Max (Per-channel)	69.60	69.08	44.49	71.21	68.52	18.40
MSE (Per-channel)	**69.66**	**69.24**	**54.67**	**71.46**	**68.89**	**27.17**

TABLE 11.2 Ablation Study for Different Methods of Range Setting of (asymmetric uniform) Activation Quantizers While Keeping the Weights in FP32. Average ImageNet Validation Accuracy (%) over 5 Runs.

Model (FP32 acc.)	ResNet18 (69.68)			MobileNetV2 (71.72)		
Bit-width	A8	A6	A4	A8	A6	A4
Min-Max	69.60	68.19	18.82	70.96	64.58	0.53
MSE	69.59	67.84	31.40	71.35	67.55	13.57
MSE + Xent	**69.60**	**68.91**	**59.07**	**71.36**	**68.85**	**30.94**
BN ($\alpha = 6$)	69.54	68.73	23.83	71.32	65.20	0.66

combined with cross-entropy for the last layer, denoted as MSE + Xent, outperforms other methods, especially at lower bit-widths. The table also clearly demonstrates the benefit of using cross-entropy for the last layer instead of the MSE objective.

11.3.2 Cross-Layer Equalization

A common issue for quantization error is that elements in the same tensor can have significantly different magnitudes. As discussed in the previous section, range setting for the quantization grid tries to find a good trade-off between clipping and rounding error. Unfortunately, in some cases, the difference in magnitude is so large that even for moderate quantization (e.g., INT8) we cannot find a suitable trade-off. [388] showed that this is especially prevalent in depth-wise separable layers since only a few weights are responsible for each output feature and this might result in higher variability of the weights. Further, they noted that batch

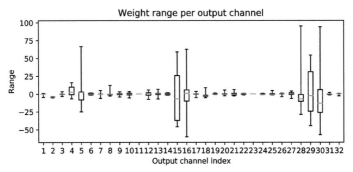

Figure 11.5 Per (output) channel weight ranges of the first depthwise-separable layer in MobileNetV2 after BN folding. In the boxplot the min and max value, the 2nd and 3rd quartile and the median are plotted for each channel.

norm folding adds to this effect and can result in a strong imbalance between weights connected to various output channels (see Figure 11.5). While the latter is less of an issue for a more fine-grained quantization granularity (e.g. per-channel quantization), this remains a big issue for the predominant per-tensor quantization. Several papers [382, 388, 389] noted that efficient models with depth-wise separable convolutions, such as MobileNetV1 [206] and MobileNetV2 [390], show a significant drop for PTQ or even result in random performance.

A solution to overcome such imbalances without the need to use per-channel quantization is introduced by [388]. A similar approach was introduced in concurrent work by [391]. In both papers the authors observe that for many common activation functions (e.g. ReLU, PreLU) a positive scaling equivariance holds:

$$f(sx) = sf(x) \tag{11.22}$$

for any non-negative real number s. This equivariance holds for any homogeneous function of degree one and can be extended to also hold for any piece-wise linear function by scaling its parameterization (e.g. ReLU6). We can exploit this positive scaling equivariance in consecutive layers in neural networks. Given two layers, $\mathbf{h} = f(\mathbf{W}^{(1)}\mathbf{x} + \mathbf{b}^{(1)})$ and

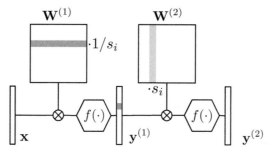

Figure 11.6 Illustration of the rescaling for a single channel. Scaling a channel in the first layer by a factor s_i leads a reparametrization of the equivalent channel in the second layer by $1/s_i$.

$\mathbf{y} = f(\mathbf{W}^{(2)}\mathbf{h} + \mathbf{b}^{(2)})$, through scaling equivariance we have that:

$$\begin{aligned}\mathbf{y} &= f(\mathbf{W}^{(2)} f(\mathbf{W}^{(1)}\mathbf{x} + \mathbf{b}^{(1)}) + \mathbf{b}^{(2)}) \\ &= f(\mathbf{W}^{(2)}\mathbf{S}\hat{f}(\mathbf{S}^{-1}\mathbf{W}^{(1)}\mathbf{x} + \mathbf{S}^{-1}\mathbf{b}^{(1)}) + \mathbf{b}^{(2)}) \\ &= f(\widetilde{\mathbf{W}}^{(2)}\hat{f}(\widetilde{\mathbf{W}}^{(1)}\mathbf{x} + \widetilde{\mathbf{b}}^{(1)}) + \mathbf{b}^{(2)}), \end{aligned} \tag{11.23}$$

where $\mathbf{S} = \mathrm{diag}(\mathbf{s})$ is a diagonal matrix with value $\mathbf{S}_{ii}$ denoting the scaling factor $\mathbf{s}_i$ for neuron i. This allows us to reparameterize our model with $\widetilde{\mathbf{W}}^{(2)} = \mathbf{W}^{(2)}\mathbf{S}$, $\widetilde{\mathbf{W}}^{(1)} = \mathbf{S}^{-1}\mathbf{W}^{(1)}$ and $\widetilde{\mathbf{b}}^{(1)} = \mathbf{S}^{-1}\mathbf{b}^{(1)}$. In case of CNNs the scaling will be per-channel and broadcast accordingly over the spatial dimensions. We illustrate this rescaling procedure in Figure 11.6.

To make the model more robust to quantization, we can find a scaling factor $\mathbf{s}_i$ such that the quantization noise in the layers rescaled is minimal. The *cross-layer equalization* (CLE) procedure introduced in [388] achieves this by equalizing dynamic ranges across consecutive layers. They prove that an optimal weight equalization is achieved by setting $\mathbf{S}$ such that:

$$\mathbf{s}_i = \frac{1}{\mathbf{r}_i^{(2)}}\sqrt{\mathbf{r}_i^{(1)}\mathbf{r}_i^{(2)}}, \tag{11.24}$$

where $\mathbf{r}_i^{(j)}$ is the dynamic range of channel i of weight tensor j. The algorithm of [391] introduces a similar scaling factor that also takes the intermediate activation tensor into account, however they do not have a proof of optimality for it.

Absorbing high biases [388] further noticed that in some cases, especially after CLE, high biases can lead to differences in the dynamic ranges of the activations. Therefore they propose a procedure to, if possible, absorb high biases into the next layer. In order to absorb $\mathbf{c}$ from layer one (followed by a ReLU activation function f) into layer two, we can do the following reparameterization:

$$
\begin{aligned}
\mathbf{y} &= \mathbf{W}^{(2)}\mathbf{h} + \mathbf{b}^{(2)} \\
&= \mathbf{W}^{(2)}(f(\mathbf{W}^{(1)}\mathbf{x} + \mathbf{b}^{(1)}) + \mathbf{c} - \mathbf{c}) + \mathbf{b}^{(2)} \\
&= \mathbf{W}^{(2)}(f(\mathbf{W}^{(1)}\mathbf{x} + \widetilde{\mathbf{b}}^{(1)}) + \mathbf{c}) + \mathbf{b}^{(2)} \\
&= \mathbf{W}^{(2)}\widetilde{\mathbf{h}} + \widetilde{\mathbf{b}}^{(2)},
\end{aligned}
\tag{11.25}
$$

where $\widetilde{\mathbf{b}}^{(2)} = \mathbf{W}^{(2)}\mathbf{c} + \mathbf{b}^{(2)}$, $\widetilde{\mathbf{h}} = \mathbf{h} - \mathbf{c}$, and $\widetilde{\mathbf{b}}^{(1)} = \mathbf{b}^{(1)} - \mathbf{c}$. In step two, we use the fact that for a layer with ReLU function f, there is a non-negative vector $\mathbf{c}$ such that $r(\mathbf{Wx} + \mathbf{b} - \mathbf{c}) = r(\mathbf{Wx} + \mathbf{b}) - \mathbf{c}$. The trivial solution $\mathbf{c} = \mathbf{0}$ holds for all $\mathbf{x}$. However, depending on the distribution of $\mathbf{x}$ and the values of $\mathbf{W}$ and $\mathbf{b}$, there can be some values $\mathbf{c}_i > 0$ for which this equality holds for (almost) all $\mathbf{x}$ in the empirical distribution. This value is equal to

$$
\mathbf{c}_i = \max\left(0, \min_{\mathbf{x}} \left(\mathbf{W}_i^{(1)}\mathbf{x} + \mathbf{b}_i^{(1)}\right)\right),
\tag{11.26}
$$

where $\min_{\mathbf{x}}$ is evaluated on a small calibration dataset. In order to remove dependence on data, [388] proposed to estimate the right hand side of equation 11.26 by the shift and scale parameters of the batch normalization layer which results in $\mathbf{c}_i = \max(0, \boldsymbol{\beta}_i - 3\boldsymbol{\gamma}_i)$[1].

Experiments   In Table 11.3 we demonstrate the effect of CLE and bias absorption for quantizing MobileNetV2 to 8-bit. As skip connections break the equivariance between layers, we apply cross-layer equalization only to the layers within each residual block. Similar to [382], we observe that model performance is close to random when quantizing MobileNetV2 to INT8. Applying CLE brings us back within 2% of FP32 performance, close to the performance of per-channel quantization. We note that absorbing high biases results in a small drop in FP32 performance (as it is an approximation), but it boosts quantized performance by 1% due to more precise activation quantization. Together, CLE and bias absorption followed by per-tensor quantization yield better results than per-channel quantization.

---

[1] Assuming $\mathbf{x}$ is normally distributed, the equality will hold for approximately 99.865% of the inputs.

TABLE 11.3 Impact of Cross-layer Equalization (CLE) for MobileNetV2. ImageNet Validation Accuracy (%), Evaluated at Full Precision and 8-bit Quantization

Model	FP32	INT8
Original model	71.72	0.12
+ CLE	71.70	69.91
+ absorbing bias	71.57	**70.92**
Per-channel quantization	71.72	70.65

11.3.3 Bias Correction

Another common issue is that quantization error is often biased. This means that the expected output of the original and quantized layer (or network) is shifted ($\mathbb{E}\left[\mathbf{W}\mathbf{x}\right] \neq \mathbb{E}\left[\widehat{\mathbf{W}}\mathbf{x}\right]$). Such biased error is common in depth-wise separable layers as only few weights (usually 9 for a 3×3 kernel) are connected to each output feature. The main contributor to this error is often the clipping error, as a few strongly clipped outliers will likely lead to a shift in the expected distribution.

Several papers [388, 391, 392] noted this issue and introduce methods to correct for the expected shift in distribution. For a quantized layer $\widehat{\mathbf{W}}$ with quantization error $\Delta\mathbf{W} = \widehat{\mathbf{W}} - \mathbf{W}$, the expected output distribution is

$$
\begin{aligned}
\mathbb{E}\left[\widehat{\mathbf{y}}\right] &= \mathbb{E}\left[\widehat{\mathbf{W}}\mathbf{x}\right] \\
&= \mathbb{E}\left[(\mathbf{W} + \Delta\mathbf{W})\mathbf{x}\right] \\
&= \mathbb{E}\left[\mathbf{W}\mathbf{x}\right] + \mathbb{E}\left[\Delta\mathbf{W}\mathbf{x}\right]
\end{aligned}
\tag{11.27}
$$

Thus the biased error is given by $\mathbb{E}\left[\Delta\mathbf{W}\mathbf{x}\right]$. Since $\Delta\mathbf{W}$ is constant, we have that $\mathbb{E}\left[\Delta\mathbf{W}\mathbf{x}\right] = \Delta\mathbf{W}\,\mathbb{E}\left[\mathbf{x}\right]$. In case $\Delta\mathbf{W}\,\mathbb{E}\left[\mathbf{x}\right]$ is nonzero, the output distribution is shifted. To counteract this shift we can substract it from the output:

$$
\mathbb{E}\left[\mathbf{y}_{\text{corr}}\right] = \mathbb{E}\left[\widehat{\mathbf{W}}\mathbf{x}\right] - \Delta\mathbf{W}\,\mathbb{E}\left[\mathbf{x}\right] = \mathbb{E}\left[\mathbf{y}\right]
\tag{11.28}
$$

Note, this correction term is a vector with the same shape as the bias and can thus be absorbed into it without any additional overhead at inference time. There are several ways of calculating the bias correction term, the two most common of which are *empirical bias correction* and *analytic bias correction*.

Empirical bias correction If we have access to a calibration dataset the bias correction term can simply be calculated by comparing the activations of the quantized and full precision model. In practice, this can be done layer-wise by computing

$$\Delta \mathbf{W} \, \mathbb{E}\,[\mathbf{x}] = \mathbb{E}\left[\widehat{\mathbf{W}}\mathbf{x}\right] - \mathbb{E}\,[\mathbf{W}\mathbf{x}]. \tag{11.29}$$

Analytic bias correction [388] introduce a method to analytically calculate the biased error, without the need for data. For common networks with batch normalization and ReLU functions they use the BN statistics of the preceding layer in order to compute the expected input distribution $\mathbb{E}\,[\mathbf{x}]$. The BN parameters γ and β correspond to the mean and standard deviation of the BN layers output. Assuming input values are normally distributied, the effect of ReLU on the distribution can be modeled using the clipped normal distribution. They show that

$$\mathbb{E}\,[\mathbf{x}] = \mathbb{E}\,[\mathrm{ReLU}\,(\mathbf{x}^{\mathrm{pre}})]$$
$$= \gamma \mathcal{N}\left(\frac{-\beta}{\gamma}\right) + \beta\left[1 - \Phi\left(\frac{-\beta}{\gamma}\right)\right], \tag{11.30}$$

where $\mathbf{x}^{\mathrm{pre}}$ is the pre-activation output, which is assumed to be normally distributed with the per-channel means $\boldsymbol{\beta}$ and per-channel standard deviations $\boldsymbol{\gamma}$, $\Phi(\cdot)$ is the standard normal cumulative distribution function (CDF), and the notation $\mathcal{N}(x)$ is used to denote the standard normal probability density function (PDF). Note, all vector operations are element-wise (per-channel) operations. After calculating the input distribution $\mathbb{E}\,[\mathbf{x}]$ the correction term can be simply derived by multiplying it with the weight quantization error $\Delta \mathbf{W}$.

Experiments In Table 11.4 we demonstrate the effect of bias correction for quantizing MobileNetV2 to 8-bit. Applying analytical bias correction improves quantized model performance from random to over 50%, indicating that the biased error introduced due to quantization significantly harms model performance. When combining bias correction with CLE we see that both techniques are complementary. Together, they achieve near FP32 performance without using any data.

11.3.4 AdaRound

Neural network weights are usually being quantized by projecting each FP32 value to the *nearest* quantization grid point, as indicated by $\lfloor\cdot\rceil$ in

TABLE 11.4 Impact of Bias correction for MobileNetV2. ImageNet Validation Accuracy (%) Evaluated at Full precision and 8-bit Quantization

Model	FP32	INT8
Original Model	71.72	0.12
+ bias correction	71.72	**52.02**
CLE + bias absorption	71.57	70.92
+ bias correction	71.57	**71.19**

equation 11.6 for a uniform quantization grid. We refer to this quantization strategy as rounding-to-nearest. The rounding-to-nearest strategy is motivated by the fact that, for a fixed quantization grid, it yields the lowest MSE between the floating-point and quantized weighs. However, recent work [393] showed that rounding-to-nearest is not optimal in terms of the task loss when quantizing weights in the post-training regime. To illustrate this the authors quantized only the weights of the first layer of ResNet18 to 4 bits using 100 different stochastic rounding samples [394] and evaluated the performance of the network for each rounding choice. The best rounding choice among these outperformed rounding-to-nearest by more than 10%. Figure 11.7 illustrates this by plotting the performance of these rounding choices on the y-axis. In this section, we introduce AdaRound [393], a systematic approach to finding good weight rounding choices for PTQ. AdaRound is a theoretically well-founded and computationally efficient method that shows significant performance improvement in practice.

As the main aim is to minimize the impact of quantization on the final task loss, we start by formulating the optimization problem in terms of this loss

$$\underset{\Delta w}{\arg\min} \quad \mathbb{E}\left[\mathcal{L}\left(\mathbf{x}, \mathbf{y}, \mathbf{w} + \Delta \mathbf{w}\right) - \mathcal{L}\left(\mathbf{x}, \mathbf{y}, \mathbf{w}\right)\right], \tag{11.31}$$

where $\Delta \mathbf{w}$ denotes the perturbation due to quantization and can take two possible values for each weight, one by rounding the weight up and the other by rounding the weight down. We want to solve this binary optimization problem efficiently. As a first step, we approximate the cost function using a second-order Taylor series expansion. This alleviates the need for performance evaluation for each new rounding choice during the optimization. We further assume that the model has converged, implying that the contribution of the gradient term in the approximation

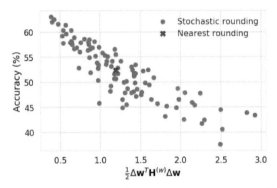

Figure 11.7 Correlation between the cost in equation 11.32 vs ImageNet validation accuracy (%) of 100 stochastic rounding vectors $\hat{\mathbf{w}}$ for 4-bit quantization of only the first layer of ResNet18.

can be ignored, and the Hessian is block-diagonal, which ignores cross-layer correlations. This leads to the following Hessian-based quadratic unconstrained binary optimization (QUBO) problem:

$$\underset{\Delta\mathbf{w}^{(\ell)}}{\arg\min} \quad \mathbb{E}\left[\Delta\mathbf{w}^{(\ell)^T}\mathbf{H}^{(\mathbf{w}^{(\ell)})}\Delta\mathbf{w}^{(\ell)}\right] \tag{11.32}$$

The clear correlation in Figure 11.7 between the validation accuracy and objective of equation 11.32 indicates that the latter serves as a good proxy for the task loss (equation 11.31), even for 4-bit weight quantization. Despite the performance gains (see Table 11.5), equation 11.32 cannot be widely applied for NN weight rounding for main two reasons:

- The memory and computational complexity of calculating the Hessian is impractical for general use-cases.

- The QUBO problem of equation 11.32 is NP-Hard.

TABLE 11.5 Impact of Various Approximations and Assumptions Made in Section 11.3.4 on the ImageNet Validation Accuracy (%) for ResNet18 Averaged over 5 Runs. N/A Implies that the Corresponding Experiment Was Computationally Infeasible.

Rounding	First layer	All layers
Nearest	52.29	23.99
$\mathbf{H}^{(\mathbf{w})}$ task loss (Eq. 11.32)	68.62	N/A
Cont. Relaxation MSE (Eq. 11.34)	69.58	66.56
AdaRound (Eq. 11.37)	**69.58**	**68.60**

To tackle the first problem, the authors introduced additional suitable assumptions that allow simplifying the objective of equation 11.32 to the following local optimization problem that minimizes the MSE of the output activations for a layer.

$$\underset{\Delta \mathbf{W}_{k,:}^{(\ell)}}{\arg\min} \quad \mathbb{E}\left[\left(\Delta \mathbf{W}_{k,:}^{(\ell)} \mathbf{x}^{(\ell-1)}\right)^2\right] \tag{11.33}$$

Equation 11.33 requires neither the computation of Hessian nor any other backward or forward propagation information from the subsequent layers. Note that the approximations and the analysis that have been used to link QUBO problem of equation 11.32 with the local optimization problem of equation 11.33 is independent of the rounding problem. Hence this analysis also benefits the design of algorithms for other problems, including model compression and NAS [395].

The optimization of equation 11.33 is still an NP-hard optimization problem. To find a good approximate solution with reasonable computational complexity, the authors relax the optimization problem to the following continuous optimization problem

$$\underset{\mathbf{V}}{\arg\min} \quad \left\|\mathbf{W}\mathbf{x} - \widetilde{\mathbf{W}}\mathbf{x}\right\|_F^2 + \lambda f_{reg}\left(\mathbf{V}\right), \tag{11.34}$$

where $\|\cdot\|_F^2$ denotes the Frobenius norm and $\widetilde{\mathbf{W}}$ are the soft-quantized weights defined as

$$\widetilde{\mathbf{W}} = s \cdot \text{clamp}\left(\left\lfloor \frac{\mathbf{W}}{s} \right\rfloor + h\left(\mathbf{V}\right); n, p\right). \tag{11.35}$$

$\mathbf{V}_{i,j}$ is the continuous variable that we optimize over and h can be any monotonic function with values between 0 and 1, i.e., $h\left(\mathbf{V}_{i,j}\right) \in [0,1]$. In [393], the authors use a rectified sigmoid as h. The objective of equation 11.34 introduces an additional term in the cost function which is a regularizer to encourage the continuous optimization variables $h\left(\mathbf{V}_{i,j}\right)$ to converge to either 0 or 1 to be valid solution of the discrete optimization in equation 11.33. The regularizer used in [393] is

$$f_{\text{reg}}\left(\mathbf{V}\right) = \sum_{i,j} 1 - \left|2h\left(\mathbf{V}_{i,j}\right) - 1\right|^\beta, \tag{11.36}$$

where β is annealed during the course of optimization to initially allow free movement of $h\left(\mathbf{V}_{i,j}\right)$ and later to force them to converge to 0 or

1. In addition, to avoid error accumulation across layers of a NN and to account for the non-linearity, the authors propose the following final optimization problem

$$\underset{\mathbf{V}}{\arg\min} \left\| f_a\left(\mathbf{W}\mathbf{x}\right) - f_a\left(\widetilde{\mathbf{W}}\hat{\mathbf{x}}\right) \right\|_F^2 + \lambda f_{reg}\left(\mathbf{V}\right), \qquad (11.37)$$

where $\hat{\mathbf{x}}$ is the layer's input with all preceding layers quantized and f_a is the activation function. The objective of equation 11.37 can be effectively and efficiently optimized using stochastic gradient descent. This approach of optimizing weight rounding is known as AdaRound.

To summarize, the way we round weights during the quantization operation has a significant impact on the performance of the network. AdaRound provides a theoretically sound, computationally fast push-button weight rounding method. It requires only a small amount of unlabeled data samples, no hyperparameter tuning or end-to-end fine-tuning, and can be applied to fully connected and convolutional layers of any neural network.

11.3.5 Standard PTQ Pipeline

In this section, we present a best-practice pipeline for PTQ based on relevant literature and our own experience. We illustrate the recommended pipeline in Figure 11.8. This pipeline shows good PTQ results for many computer vision models and tasks. Depending on the model, some steps might not be required, or other choices could lead to equal or better performance.

Cross-layer equalization First we apply cross-layer equalization (CLE), which is a pre-processing step for the full precision model to make it more quantization friendly. CLE is particularly important

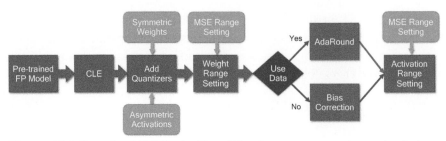

Figure 11.8 Standard PTQ pipeline. Blue boxes represent required steps and the turquoise boxes recommended choices.

for models with depth-wise separable layers and for per-tensor quantization, but it often also shows improvements for other layers and quantization choices.

Add quantizers Next we choose our quantizers and add quantization operations in our network as described in Section 11.2.3. The choice of quantizer might depend on the specific target HW; for common AI accelerators we recommend using symmetric quantizers for the weights and asymmetric quantizers for the activations. If supported by the HW/SW stack then it is favorable to use per-channel quantization for weights.

Weight range setting To set the quantization parameters of all weight tensors we recommend using the layer-wise MSE-based criteria. In the specific case of per-channel quantization, using the min-max method can be favorable in some cases.

AdaRound In case we have a small calibration dataset[2] available we next apply AdaRound in order to optimize the rounding of the weights. This step is crucial to enable low-bit weight quantization (e.g., 4 bits) in the PTQ.

Bias correction In case we do not have such a calibration dataset we can use analytical bias correction instead.

Activation range setting As the final step we determine the quantization ranges of all data-dependent tensors in the network (i.e., activations). We use the MSE-based criteria for most of the layers, which requires a small calibration set to find the minimum MSE loss. Alternatively, we can use the BN-based range setting to have a full data-free pipeline.

11.3.6 Experiments

We now evaluate the performance of the aforementioned PTQ pipeline for various computer vision models and tasks. Our results are summarized

---

[2]Usually between 500 and 1000 unlabeled images are sufficient as a calibration set.

TABLE 11.6 Performance (average over 5 runs) of Our Standard PTQ Pipeline for Various Computer Vision Models and Tasks. DeeplabV3 (MobileNetV2 backbone) Is Evaluated on Pascal VOC (Mean Intersection over Union), Others on ImageNet (Accuracy).

Bit-width	FP32	Per-tensor		Per-channel	
		W8A8	W4A8	W8A8	W4A8
ResNet18	69.68	69.60	68.62	69.56	68.91
ResNet50	76.07	75.87	75.15	75.88	75.43
InceptionV3	77.40	77.68	76.48	77.71	76.82
MobileNetV2	71.72	70.99	69.21	71.16	69.79
EfficientNet lite	75.42	75.25	71.24	75.39	74.01
DeeplabV3	72.94	72.44	70.80	72.27	71.67

in Table 11.6. DeepLabV3 (with a MobileNetV2 backbone) is evaluated on Pascal VOC and all other networks on ImageNet.

In all cases, we observe that 8-bit quantization of weights and activation (W8A8) leads to only marginal loss of accuracy compared to floating-point (within 0.7%) for all models. For W8A8 quantization we also see no significant gains from using per-channel quantization. However, the picture change when weights are more aggressively quantized to 4-bits (W4A8). For ResNet18/50 and InceptionV3 the accuracy drop is still within 1% of floating-point for both per-tensor and per-channel quantization. However, for more efficient networks, such as MobileNetV2 and EfficientNet lite, the drop increases to 2.5% and 4.2%, respectively for per-tensor quantization. This is likely due to the quantization of the depth-wise separable convolutions. Here, per-channel quantization can show a significant benefit, for example in Efficient-Net lite per-channel quantization increases the accuracy by 2.8% compared to per-tensor quantization bringing it within 1.5% of full-precision accuracy.

11.4 QUANTIZATION-AWARE TRAINING

The post-training quantization techniques described in the previous section are the first go-to tool in our quantization toolkit. They are very effective and fast to implement because they do not require retraining of the network with labeled data. However, they have limitations, especially when aiming for low-bit quantization, e.g., 4-bit and below. Post-training techniques may not be enough to mitigate the large quantization error incurred by low-bit quantization. In these cases, we resort to *quantization-aware training* (QAT). QAT models the quantization noise source (see Section 11.2.3) during training. This allows the model to find more optimal solutions than post-training quantization. However, the higher accuracy comes with the usual costs of neural network training, i.e, longer training times, need for labeled data and hyper-parameter search.

In this section, we explore how back-propagation works in networks with simulated quantization and provide a standard pipeline for training models with QAT effectively. We will also discuss the implications of batch normalization folding and per-channel quantization in QAT and provide results for common computer vision models.

11.4.1 Simulating Quantization for Backward Path

In Section 11.2.3, we saw how to simulate quantization using floating-point in deep learning frameworks. However, if we look at the computation graph of Figure 11.4, to train such a network we need to back-propagate through the simulated quantizer block. This poses an issue because the gradient of the round-to-nearest operation in equation 11.6 is zero or undefined everywhere , which makes gradient-based training impossible. A way around this would be to approximate the gradient using the *straight-through estimator* (STE) [312], which approximates the gradient of the rounding operator as 1:

$$\frac{\partial \lfloor y \rceil}{\partial y} = 1 \tag{11.38}$$

Using this approximation we can now calculate the gradient of the quantization operation from equation 11.9. For clarity we assume symmetric quantization, namely $z = 0$, but the same result applies to asymmetric quantization since the zero-point is a constant. We use n and p to define the integer grid limits, such that $n = q_{min}/s$ and $p = q_{max}/s$. The gradient of equation 11.9 *w.r.t* its input, $\mathbf{x}_i$, is given by:

$$\frac{\partial \hat{\mathbf{x}}_i}{\partial \mathbf{x}_i} = \frac{\partial q(\mathbf{x}_i)}{\partial \mathbf{x}_i}$$

$$= s \cdot \frac{\partial}{\partial \mathbf{x}_i} \text{clamp}\left(\left\lfloor \frac{\mathbf{x}_i}{s} \right\rceil; n, p\right) + 0$$

$$= \begin{cases} s \cdot \dfrac{\partial \lfloor \mathbf{x}_i/s \rceil}{\partial (\mathbf{x}_i/s)} \dfrac{\partial (\mathbf{x}_i/s)}{\partial \mathbf{x}_i} & \text{if } q_{\min} \leq \mathbf{x}_i \leq q_{\max}, \\[2ex] s \cdot \dfrac{\partial n}{\partial \mathbf{x}_i} & \text{if } \mathbf{x}_i < q_{\min}, \\[2ex] s \cdot \dfrac{\partial p}{\partial \mathbf{x}_i} & \text{if } \mathbf{x}_i > q_{\max}. \end{cases}$$

$$= \begin{cases} 1 & \text{if } q_{\min} \leq \mathbf{x}_i \leq q_{\max}, \\ 0 & \text{otherwise.} \end{cases} \tag{11.39}$$

Using this gradient definition we can now back-propagate through the quantization blocks. Figure 11.9 shows a simple computational graph for the forward and backward pass used in quantization-aware training. The forward pass is identical to that of Figure 11.4 but in the backward pass we effectively skip the quantizer block due to the STE assumption. In earlier QAT work the quantization ranges for weights and activations were updated at each iteration most commonly using the min-max range [382]. In later work [396, 397, 398], the STE is used to calculate the gradient *w.r.t.* the quantization parameters, z and s. Using the chain

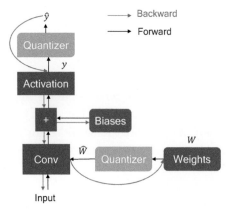

Figure 11.9 Forward and backward computation graph for quantization aware training with STE assumption.

rule and the STE, we first calculate the gradient *w.r.t.* the scale-factor:

$$\frac{\partial \widehat{\mathbf{x}}_i}{\partial s} = \frac{\partial}{\partial s}\left[s \cdot \text{clamp}\left(\left\lfloor\frac{\mathbf{x}_i}{s}\right\rceil; n, p\right)\right]$$

$$= \begin{cases} -\mathbf{x}_i/s + \lfloor\mathbf{x}_i/s\rceil & \text{if } q_{\min} \leq \mathbf{x}_i \leq q_{\max}, \\ n & \text{if } \mathbf{x}_i < q_{\min}, \\ p & \text{if } \mathbf{x}_i > q_{\max}. \end{cases} \quad (11.40)$$

Originally, we restricted the zero-point to be an integer. To make zero-point learnable we convert into a real number and apply the rounding operator. The modified quantization function is defined as:

$$\widehat{\mathbf{x}} = q(\mathbf{x}; s, z) = s \cdot \left[\text{clamp}\left(\left\lfloor\frac{\mathbf{x}}{s}\right\rceil + \lfloor z\rceil; n, p\right) - \lfloor z\rceil\right] \quad (11.41)$$

The gradient *w.r.t.* to z is calculated by applying the STE once again to the rounding operator:

$$\frac{\partial \widehat{\mathbf{x}}_i}{\partial z} = \begin{cases} 0 & q_{\min} \leq \mathbf{x}_i \leq q_{\max}, \\ -s & \text{otherwise.} \end{cases} \quad (11.42)$$

11.4.2   Batch Normalization Folding and QAT

In Section 11.2.3.1, we introduced batch norm folding that absorbs the batch norm scaling and addition into a linear layer to allow for more efficient inference. During quantization-aware training we want to simulate closely inference behavior, which is why we have to account for BN-folding during training. Note that in some QAT literature, the BN-folding effect is ignored. While this is fine when we employ *per-channel* quantization (more below in this section), keeping BN unfolded for per-tensor quantization will results in one of the two: (1) we have a per-channel rescaling during inference due to the BN layer, in this case we might as well use per-channel quantization in the first place or (2) we fold BN during deployment into the weight tensor in which we likely suffer from a significant accuracy drop as we trained the network to adapt to a different quantization noise.

A simple but effective approach to modeling BN-folding in QAT is to *statically fold* the BN scale and offset into the linear layer's weights and bias, as we saw in equations 11.13 and 11.14. This corresponds to re-parametrization of the weights and effectively removes the batch normalization operation from the network entirely. When starting from

TABLE 11.7 Ablation Study with Various Ways to Include BN into QAT. The learning rate Is Individually Optimized for Each Configuration. Average ImageNet Validation Accuracy (%) over 3 Runs.

| Model (FP32 acc.) | ResNet18 (69.68) | | MobileNetV2 (71.72) | |
Bit-width	W4A8	W4A4	W4A8	W4A4
Static folding BN	**69.76**	**68.32**	**70.17**	**66.43**
Double forward [382]	69.42	68.20	66.87	63.54
Static folding per-channel)	69.58	68.15	**70.52**	66.32
Keep original BN BN (per-channel)	**70.01**	**68.83**	70.48	**66.89**

a converged pre-trained model, static folding is very effective as we can see from the result of Table 11.7.

An alternative approach by [377] both updates the running statistics during QAT and applies BN-folding using a correction. This approach is more cumbersome and computationally costly because it involves a *double forward pass*: one for the batch-statistics and one for the quantized linear operation. However, based on our experiments (see Table 11.7) static-folding performs on par or better despite its simplicity.

Per-channel quantization In Section 11.2.4.2, we mentioned that per-channel quantization of the weights can improve accuracy when it is supported by our hardware. The static folding re-parametrization is also valid for per-channel quantization. However, per-channel quantization provides additional flexibility as it allows us to absorb the batch norm scaling operation into the per-channel scale-factor. Let us see how this is possible by revisiting the BN folding equation from Section 11.2.3.1 but this time introduce per-channel quantization of the weights such that $\widehat{\mathbf{W}}_{k,:} = q(\mathbf{W}_{k,:}; s_{\mathbf{w},k}) = s_{\mathbf{w},k}\mathbf{W}_{k,:}^{\text{int}}$. By applying batch norm the output of a linear layer similar to equation 11.12 we get:

$$
\begin{aligned}
\widehat{\mathbf{y}}_k &= \text{BatchNorm}\left(\widehat{\mathbf{W}}_{k,:}\mathbf{x}\right) \\
&= \frac{\boldsymbol{\gamma}_k \widehat{\mathbf{W}}_{k,:}}{\sqrt{\boldsymbol{\sigma}_k^2 + \epsilon}}\mathbf{x} + \left(\boldsymbol{\beta}_k - \frac{\boldsymbol{\gamma}_k \boldsymbol{\mu}_k}{\sqrt{\boldsymbol{\sigma}_k^2 + \epsilon}}\right) \\
&= \frac{\boldsymbol{\gamma}_k s_{\mathbf{w},k}}{\sqrt{\boldsymbol{\sigma}^2 + \epsilon}}\mathbf{W}_{k,:}^{\text{int}}\,\mathbf{x} + \widetilde{\mathbf{b}}_k \\
&= \tilde{s}_{\mathbf{w},k}\left(\mathbf{W}_{k,:}^{\text{int}}\mathbf{x}\right) + \widetilde{\mathbf{b}}_k
\end{aligned}
\tag{11.43}
$$

We can see that it is now possible to absorb the batch normalization scaling parameters into the per-channel scale-factor. For QAT this means that we can keep the BN layer during training and merge after QAT the BN scaling factor into the per-channel quantization parameters. In practice, this modeling approach is on par or better for per-channel quantization compared to static folding as we can see from the last two rows of Table 11.7.

11.4.3 Initialization for QAT

In this section, we will explore the effect of initialization for QAT. It is common practice in literature to start from a pre-trained FP32 model [396, 382, 377, 397]. While it is clear that starting from an FP32 model is beneficial, the effect of the quantization initialization on the final QAT result is less studied. Here we explore the effect of using several of our PTQ techniques as an initial step before doing QAT.

Effect of range estimation   To assess the effect of the initial range setting (see Section 11.3.1) for weights and activations we perform two sets of experiments,which are summarized in Table 11.8. In the first experiment, we quantize the weights to 4-bits and keep the activations in 8-bits. We compare the min-max initialization with the MSE-based initialization for the weights quantization range. While the MSE initialized model has a significantly higher starting accuracy, the gap closes after training for 20 epochs.

To explore the same effect for activation quantization, we perform a similar experiment, where we now quantize the activation to 4-bits and compare min-max initialization with MSE-based initialization. The

TABLE 11.8 Ablation Study for Various Ways to Initialize the Quantization Grid. The Learning Rate Is Individually Optimized for Each Configuration. ImageNet Validation Accuracy (%) Averaged Over 3 Runs.

Model (FP32 acc.)	ResNet18 (69.68)		MobileNetV2 (71.72)	
	PTQ	QAT	PTQ	QAT
W4A8 w/ min-max weight init	0.12	69.61	0.56	69.96
W4A8 w/ MSE weight init	18.58	**69.76**	12.99	**70.13**
W4A4 w/ min-max act init	7.51	68.23	0.22	**66.55**
W4A4 w/ MSE act init	9.62	**68.41**	0.71	66.29

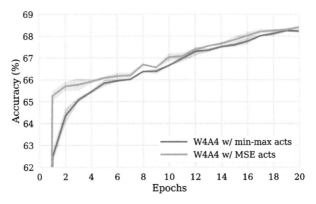

Figure 11.10 Influence of the initial activation range setting on the QAT training behavior of ResNet18. ImageNet validation accuracy (%) vs. training epoch.

observations from weight range initialization hold here as well. In Figure 11.10 we show the full training curve of this experiment. In the first few epochs, there is a significant advantage for using MSE initialization, which almost vanishes in the later stage of training. In conclusion, a better initialization can lead to better QAT results, but the gain is usually small and vanishes the longer the training lasts.

Effect of CLE   In Table 11.9 we compare the effect of other PTQ improvements such as CLE and bias correction. While for ResNet18 we do not see a significant difference in the final QAT performance, for MobileNetV2 we observe that it cannot be trained without CLE. This is likely due to the catastrophic performance drop caused by per-tensor quantization, which we discussed in Section 11.3.2.

In conclusion, for models that have severe issues with plain PTQ we may need advanced PTQ techniques such as CLE to initialize QAT. In most other cases, an improved PTQ initialization leads only to a minor improvement in the final QAT performance.

11.4.4   Standard QAT Pipeline

In this section, we present a best-practice pipeline for QAT based on relevant literature and our own experience. We illustrate the recommended pipeline in Figure 11.11. This pipeline shows good QAT results over a variety of computer vision models and tasks, and can be seen as the go-to

TABLE 11.9 Ablation Study with Various PTQ Initialization. The Learning Rate Is Individually Optimized for Each Configuration. ImageNet Validation Accuracy (%) Averaged Over 3 Runs.

Model (FP32 acc.)	ResNet18 (69.68)		MobileNetV2 (71.72)	
	PTQ	QAT	PTQ	QAT
W4A8 baseline	18.58	69.74	0.10	0.10
W4A8 w/CLE	16.29	**69.76**	12.99	**70.13**
W4A8 w/CLE + BC	38.58	69.72	46.90	70.07

tool for getting good low-bit quantization performance. As discussed in previous sections, we always start from a pre-trained model and follow some PTQ steps in order to have faster convergence and higher accuracy.

Cross-layer equalization Similar to PTQ we first apply CLE to the full precision model. As we saw in Table 11.9 this step is necessary for models that suffer from imbalanced weight distributions such as MobileNet architectures. For other networks and per-channel quantization, this step can be optional.

Add quantizers Next we choose our quantizers and add quantization operations in our network as described in Section 11.2.3. The choice for quantizer might depend on the specific target HW, for common AI accelerators we recommend using symmetric quantizers for the weights and asymmetric quantizers for the activations. If supported by the HW/SW stack then it is favorable to use per-channel quantization for weights. At this stage, we will also take care

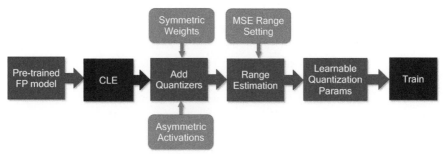

Figure 11.11 Standard quantization-aware training pipeline. The blue boxes represent the steps and the turquoise boxes recommended choices.

that our simulation of batch normalization is correct as discussed in Section 11.4.2.

Range estimation Before training we have to initialize all quantization parameters. A better initialization will help faster training and might improve the final accuracy, though often the improvement is small (see Table 11.8). In general, we recommend setting all quantization parameters using the layer-wise MSE-based criteria. In the specific case of per-channel quantization, using the min-max setting can be favorable in some cases.

Learnable Quantization Parameters Unless there is a good reason, we recommend making the quantizer parameters learnable, as discussed in Section 11.4.1. Learning the quatization parameters directly, rather than updating them at every epoch leads to higher performance especially when dealing with low-bit quantization. However, using learnable quantizers requires special care when setting up the optimizer for the task. When using SGD-type optimizers, the learning rate for the quantization parameters needs to be reduced compared to the rest of the network parameters. The learning rate adjustment can be avoided if we use optimizers with adaptive learning rates such as Adam or RMSProp.

11.4.5 Experiments

Using our QAT pipeline, we quantize common computer vision networks for image classification and semantic segmentation, similar to Section 11.3.6 of PTQ. Our results are presented in Table 11.10 for different bit-widths and quantization granularities. DeepLabV3 is trained for 80 epochs on Pascal VOC while all other models are trained for 20 epochs on ImageNet. We use Adam optimizer for all models and evaluate them on the validation set of the benchmark they have been trained on. We present the results with the best learning rate per quantization configuration and perform no further hyper-parameter tuning.

We observe that for networks without depth-wise separable convolutions (first 3 rows of Table 11.10), W8A8 and W4A8 quantization perform on par with and even outperform the floating-point model in certain cases. This could be due to regularizing effect of training with quantization noise or due to the additional fine-tuning during QAT. For the more aggressive W4A4 case we notice a small drop but still within 1% of the floating-point accuracy.

TABLE 11.10 Performance (Average Over 3 Runs) of Our Standard QAT Pipeline for Various Computer Vision Models and Tasks. DeeplabV3 (MobileNetV2 backbone) Is Evaluated on Pascal VOC (Mean Intersection Over Union), Others on ImageNet (Accuracy)

Bit-width	FP32	Per-tensor			Per-channel		
		W8A8	W4A8	W4A4	W8A8	W4A8	W4A4
ResNet18	69.68	70.38	69.76	68.32	70.43	70.01	68.83
ResNet50	76.07	76.21	75.89	75.10	76.58	76.52	75.53
InceptionV3	77.40	78.33	77.84	77.49	78.45	78.12	77.74
MobileNetV2	71.72	71.76	70.17	66.43	71.82	70.48	66.89
EfficientNet lite	75.42	75.17	71.55	70.22	74.75	73.92	71.55
DeeplabV3	72.94	73.99	70.90	66.78	72.87	73.01	68.90

Quantizing networks with depth-wise separable layers (last 3 rows of Table 11.10) is more challenging; a trend we also observed from the PTQ results in Section 11.3.6 and discussed in the literature [399]. Whereas 8-bit quantization incurs close to no accuracy drop, quantizing weights to 4 bits leads to a larger drop, e.g., approximately 4% drop for EfficientNet lite with per-tensor quantization. Per-channel quantization can improve performance significantly bringing DeepLabV3 to floating-point accuracy and reducing the gap of MobileNetV2 and EfficientNet lite to less than 1.5%. Quantizing both weights and activations to 4-bits remains a challenging for such networks, even with per-channel quantization it can lead to a drop of up to 5%.

11.5   SUMMARY AND CONCLUSIONS

Deep learning has become an integral part of most computer vision algorithms and can now be found in many electronic devices and services, from smartphones and home appliances to drones, robots, and self-driving cars. As the popularity and reach of deep learning in our everyday life increases, so does the need for fast and power-efficient neural network inference. Neural network quantization is one of the most effective ways of reducing the energy and latency of neural networks.

Quantization allows us to move from floating-point representations to a fixed-point format and in combination with dedicated hardware utilizing efficient fixed-point operations has the potential to achieve significant power gains and accelerate inference. However, to exploit these savings, we require robust quantization methods that can maintain high accuracy, while reducing the bit-width of weights and activations. To this end, we considered two main classes of quantization algorithms: Post-Training Quantization (PTQ) and Quantization-Aware-Training (QAT).

Post-training quantization techniques take a pre-trained FP32 network and converts it into a fixed-point network without the need for the original training pipeline. This makes them a lightweight, push-button approach to quantization with low engineering effort and computational cost. We describe a series of recent advances in PTQ and introduce a PTQ pipeline that leads to near floating-point accuracy results for a range of modern computer vision networks in image classification and semantic segmentation. In particular, using the proposed pipeline we can achieve 8-bit quantization of weight and activations (W8A8) within only 1% of the floating-point accuracy for all networks. We further show that many networks can be quantized even to 4-bit weights with only a small additional drop in performance.

Quantization-aware training models the quantization noise during training through simulated quantization operations. This training procedure allows for better solutions to be found compared to PTQ while enabling more effective and aggressive activation quantization. Similar to PTQ, we introduce a standard training pipeline utilizing the latest algorithms in the area. We also pay special attention to how to treat batch normalization folding during QAT and show that simple static-folding outperforms other more computationally expensive approaches. We demonstrate that with our QAT pipeline we can achieve 4-bit quantization of weights, and for some models even 4-bit activations, with only a small drop of accuracy compared to floating-point.

CHAPTER 12

Building Efficient Mobile Architectures

Mark Sandler and Andrew Howard

Google

CONTENTS

In this chapter we overview a set of basic techniques for designing and tuning neural network architectures. Our goal is to help both practitioners and researchers to develop broad intuition of how to adapt existing architectures to specific applications and hardware. While a lot of recent research has been dedicated to network architecture search, relatively little attention has been devoted to basic design principles. Neural networks are often treated as a black box with their architectures derived from the search algorithms without much understanding about the reason for their final configurations. Our goal here is to build a solid foundation

DOI: 10.1201/9781003162810-12

and demystify the reasoning about neural networks architectures from a practical perspective. Based on our experience, such foundation is indispensable for both designing new architecture search spaces, as well as adaptation of existing models for new hardware and/or specific problems. Because of this, we mostly try to avoid tying our analysis to specific hardware and instead concentrate on broad principles.

12.1 INTRODUCTION

The main requirement for designing efficient architectures is to establish target metrics. The ultimate "real" metrics, such as latency and energy consumption are often hardware and implementation dependent. Since our goal here is to develop some basic principles that can be applied to a broad spectrum of hardware, thus we limit our attention to synthetic metrics that are good proxy measures of model efficiency.

The most basic metric that reflects the complexity of a model is the number of Multiply-Adds (MAdds)[1] during inference. Basic operators such as matrix multiplication and convolution are based on a combination of primitives that multiply two values and add their result to an accumulator and each call to this primitive counts as one MAdd. This metric is good approximation for latency and power consumption. However, depending on the shape of the tensor, the MAdd throughput can vary greatly due to cache and arithmetic unit utilization. In fact, the actual energy consumption is dominated by the data transfer cost to get the data from DRAM to the local fast cache and then to the registers (See for instance Figure 1.1.9 in [402]). Therefore, not all operators will have the same efficiency. For example, depthwise convolutions [403] generally have lower throughput than regular convolutions since they require a higher ratio of memory reads to MAdds than full convolutions, but are still a valuable building block due to the small percentage of total MAdds that depthwise convolutions take up in modern architectures. Nevertheless, as a first approximation, MAdds provide a very useful metric that strongly correlates with performance.

The second metric is the total parameter count. There has been a lot of progress of reducing the parameter count in neural architecture design. From 120 million parameters for early VGG [404] models, to less than 5

---

[1]Note that, FLOPs floating point operations, is also used in the literature, however its definition varies. For instance, [400, 401] assume that FLOPs and MAdds are interchangeable, while others [268] assume 1 MAdd equals to 2 Flops. To avoid confusion we will be using MAdds, throughout this chapter.

million for most of the modern mobile architectures [403, 405, 390, 406, 407]. Even with the current model sizes, this metric is often critical for many embedded applications. Somewhat regrettably, in the literature the model size is often used interchangeably with MAdds. While they often go hand-in hand, we show in the Section 12.2, that these metrics can be adjusted in opposing directions, while keeping the accuracy of the model essentially unchanged.

The final metric that we consider is the peak activation memory. This is the amount of volatile memory needed to run a single inference on a single image. While this metric is also implementation specific, a good approximation that we employ here is the maximum memory required to store all inputs and outputs for any single operator. For instance, for classical sequential models such as MobileNets [403, 390, 407], a good rule of thumb is the maximum size all inputs and outputs among all layers.

Our goal for this chapter is to show multiple ways of customizing and adapting existing architectures such as MobileNets [403, 390, 407] to improve their suitability for particular hardware and use cases. However, we leave outside of the scope specific designs of convolutional blocks, such as depthwise-convolutional layers [403, 408], inverted residual bottleneck layers [390], group convolutions [405] and others. We refer the reader to these and other works for in-depth convolutional block design.

The rest of this chapter is organized as follows. In Section 12.2 we describe simple parameterizations that can be used to adjust model size while keeping different target metrics under the budget. The goal of the section is to demonstrate how these different multipliers can be adjusted against each other. In Section 12.3, we pay special attention to the first few layers as they often are on the critical path when trying to run a model on hardware with limited memory. In the Section 12.4, we describe two techniques on how to optimize the last layers, as those are typically responsible for a large portion of the model size. In the Section 12.5, we describe two quantization friendly non-linearities, as well as a technique to introduce them even if fully optimized implementation is not available. Finally, in the Section 12.6, we propose a specific action plan for practical model optimization in order to meet the requirements for a specific use case and hardware deployment.

12.2 ARCHITECTURE PARAMETERIZATIONS

In this section we describe several ways to adjust all target metrics using simple parametric transformations. Imagine you are given a deep neural network architecture. What would be the most natural way to change its size? One way is to scale the size of its hidden layers by some constant γ. That is each convolutional layer with c channels becomes a convolutional layer with γc channels, thus increasing or decreasing both the model size, the number of MAdds and activation memory requirement. This scaling is called the network width multiplier, or simply width multiplier. Despite its simplicity, this method provides a strong baseline that are commonly used when evaluating compression methods [403, 400]. Another alternative is to change the input resolution which we term the input resolution multiplier. Finally, the network depth multiplier proposed in EfficientNet architectures [400], but also used in earlier work such as in ResNets [370, 410], applies a multiplier to the number of layers in the network. We illustrate all these different parameterizations in Figure 12.1 and we explore them in greater detail below.

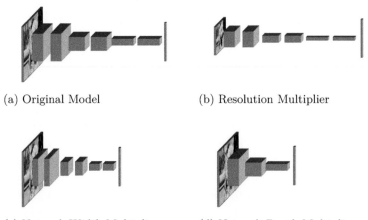

(a) Original Model　　　　　　　　(b) Resolution Multiplier

(c) Network Width Multiplier　　　(d) Network Depth Multiplier

Figure 12.1 Different types of multipliers. Figure 12.1a is the original model and Figures 12.1b, 12.1c, and 12.1d show how the different multipliers affect the model. Adapted from [409].

12.2.1 Network Width Multiplier

Network width multiplier, is a method where the number of neurons in each layers is scaled uniformly by a constant factor to slim down the model or scale it up. For example, if a layer had 512 neurons initially, applying a multiplier of 2 yields a new layer with 1024 neurons. For a convolutional layer that transforms $h \times w \times c$ into $h' \times w' \times d$, multiplier γ changes the sizes of each block to $h \times w \times c\gamma$, and $h' \times w' \times d\gamma$. Note that the amount of memory needed to store the input and the output changes by a factor of γ. While the total compute cost changes by a factor of γ^2: from $h'w'dck^2$ to $h'w'dck^2\gamma^2$, where k is the convolution kernel size. Similarly the total number of parameters scales by factor of γ^2 from k^2cd to $k^2cd\gamma^2$. Thus, applying width multiplier to every layer changes the total number of operations and the number of parameters by γ^2 and the peak memory requirement by a factor of γ.

Variants of width multiplier have been utilized in [411] to arrive at new architectures, as well as proposed as a meta-parameter to arrive at a family of architectures in MobileNetV1 [403]. Since then it became a common baseline method used to compare with other model slimming methods such as architecture search [406]. While many more advanced methods have been proposed, width multiplier remains a simple and strong technique that can be applied to reduce the size and compute requirements of virtually any neural network.

One important practical detail when applying a width multiplier (especially in the early layers), is to keep the number of channels a multiple of the parallelism supported by the target hardware system. For instance, most modern mobile CPU implementations of neural network layers are most efficient when the number of channels is divisible by 8 or 16 and some accelerators prefer multiples of 32 or 64. In some extreme cases, a layer with 3 channels could be slower than layer with 16 or even 32 channels.

12.2.2 Input Resolution Multiplier

Input resolution multiplier, or resolution multiplier for short, is another simple technique that reduces the computational complexity of a model. The main idea is to feed the network with a lower resolution image thereby processing fewer pixels and subsequently scaling the hidden layers by the corresponding amount. The model size stays the same, but as can be seen in Table 12.1, both the number of arithmetic operations and the memory requirements both change proportionally to the area of the

TABLE 12.1 Impact of Different Scaling techniques on Model Size, Activation Footprint and MAdds as Function of the Scale Multiplier γ.

Transformation	Activation	Model size	MAdds
Network width multiplier	γ	γ^2	γ^2
Data resolution multiplier	γ^2	1	γ^2
Network depth multiplier	1	γ	γ

image. Thus, a greater than 1 resolution multiplier can be used as a way to increase the model accuracy without increasing the number of parameters in the model. One important implicit assumption that we discuss below is the trade-off between the resolution of the input and the resolution of the hidden layers whose capacity is increased by using more fine-grained intermediate representations.

12.2.3 Data and Internal Resolution

One often overlooked property of a resolution multiplier, is that it changes not only the input resolution, but also the spatial resolution in each subsequent hidden layer. It is important to note that those are distinct. Specifically, the data resolution controls how discriminative the data is— intuitively there is not much a model can do if it needs to differentiate using 2x2 images. The hidden layer resolution on the other hand, controls the capacity of the model. A natural question is how can these two mechanisms be disentangled? A simple experiment, proposed in [409], uses low-resolution images and up-samples them before feeding them to the model in order control the discriminative aspects of image resolution while studying the capacity of the model. It turns out, that for ImageNet [412], the model capacity plays a much bigger role than image resolution. For instance, as can be seen in Figure 12.2, the resolution plays a very minor role until extremely low resolutions such as 32x32 and below. These experimental models using up-scaled low resolution images are useful for experimentation, but are not very practical because their compute and memory requirements are essentially the same as full-resolution models. In [409], a much more efficient modification procedure was utilized called the skip-stride method. The idea is that instead of upsampling the initial input, change the stride of all early layers until the internal resolution matches the original model. A visual comparison between upsampling and the skip-stride modification is shown on Figure 12.3a and 12.3b.

Similarly, the model's internal resolution could be kept fixed while feeding high-resolution data using the space-to-depth transformation.

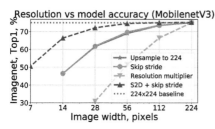
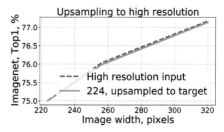

(a) Low resolution (b) High resolution

Figure 12.2 From [409]. Figure 12.2a shows the impact of the input resolution on the final accuracy for different types of internal-resolution scaling. Upsample and skip-stride methods preserve all or most of the internal resolution respectively and are much closer to the baseline model than the resolution multiplier. For example, changing resolution from 224x224 to 56x56 without changing internal resolution loses about 4% accuracy, while if we use resolution multiplier, the model loses nearly 25%. For very low resolution such as 7x7 or 14x14, the image resolution plays a major role. However, for resolutions beyond 224x224, the performance of high-res inputs vs. upsampled is identical as seen in 12.2b .

This transformation losslessly converts $n \times n$ 3-channel images into $n/k \times n/k$ image with $3k^2$ channels by flattening each $k \times k$ patch. This allows skip-stride to be applied to the new "low-resolution" image as shown in Figure 12.3b thus enabling high resolution images to be used while reducing the memory requirements of the hidden layers. We discuss this in more detail in Section 12.3. Figure 12.2 shows how different combinations of input resolution and skip-stride modifications affect the accuracy.

12.2.4 Network Depth Multiplier

The network depth multiplier or depth multiplier for short is a parametrization where the number of layers is scaled by a constant. This method was proposed in [400], though other works used such parameterization informally such as [370, 409]. Typically such multipliers are applied to the most basic building blocks of the network. For example, in the case of ResNets, it would be applied at the residual block level. In the case of EfficientNet [400], it was applied to increase the total number of inverted bottleneck blocks and scale up the model. Concretely, for a model that contains 20 blocks, it would have 25 blocks after applying

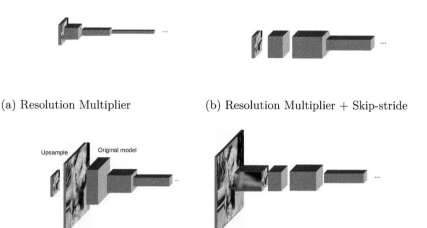

(a) Resolution Multiplier   (b) Resolution Multiplier + Skip-stride

(c) Up-sample   (d) S2D + Skip-stride

Figure 12.3 Different ways of disentangling image and model resolution as proposed in [409]. Figure adapted from [409]. Here 12.3c and 12.3b feed low-resolution input both giving comparable accuracy. Skip-stride 12.3b eliminates early operators with strides until the resolution matches the original model. Figure 12.3d shows a variant of skip stride where the input is high resolution but then fed directly into the first low-resolution hidden layer using a space-to-depth transform. Figure 12.3a shows the resolution multiplier for reference.

a multiplier of 1.25 or 15 blocks with a multiplier of 0.75. If a model utilizes multiple internal resolutions in stages, the sequence of blocks at each resolution are typically scaled independently to preserve their relative sizes. Table 12.1 shows how applying a network-depth multiplier affects the target metrics.

12.2.5 Adjusting Multipliers for Multi-criteria Optimizations

As can be seen from Table 12.1, different multipliers affect each target metric differently. Thus, given the desired budgets for all three, we can approximately compute the optimal multiplier that will utilize the budget completely. Suppose we have a baseline architecture $\mathcal{A}$, that requires a budget B_{madds} for multiply-adds, B_{params} for the model size, and B_{mem} for activation memory. Suppose we would like to produce an architecture that uses pB_{MAdds} multiply adds, qB_{params} and rB_{mem}, where p, q, and

r are budget coefficients that could be both greater or smaller than 1. Then from Table 12.1 we have

$$
\begin{aligned}
p &= \gamma_w^2 * \gamma_r^2 * \gamma_d \\
q &= \gamma_w^2 * \gamma_d \\
r &= \gamma_w * \gamma_r^2
\end{aligned}
$$

where γ_w, γ_d, and γ_r are layer width, network depths, and resolution multipliers respectively. Equivalently:

$$
\begin{aligned}
\log p &= 2 \log \gamma_w + 2 \log \gamma_r + \log \gamma_d \\
\log q &= 2 \log \gamma_w + \log \gamma_d \\
\log r &= \log \gamma_w + 2 \log \gamma_r
\end{aligned}
$$

This is a linear system of equations that we can solve explicitly resulting in:

$$
\gamma_w = \frac{qr}{p}, \quad \gamma_r = \sqrt{\frac{p}{q}}, \quad \gamma_d = \frac{p^2 q}{r^2}
$$

It is important to note that these equations are approximate because of the boundary effects such as the number of channels in the first and last layers are often constant regardless of the multiplier value. However these effects can be ignored if multiplier values stay reasonably large (e.g., > 0.2). Increasing any single multiplier to an extreme value (e.g., above 10 or below 0.1) will likely result in an architecture that is hard to train, or has poor performance. Further, the proposed multiplier values, do not guarantee optimal performance, only that the network will fully utilize the proposed budget constraints. As a rule of thumb, the formulas above will provide a reasonable baseline for the $0.2 \leq \{p, q, r\} < 5$ range. For budgets that are significantly below or above these thresholds, it is valuable to look into optimizing individual layers as described in the Sections 12.3 and 12.4.

12.3 OPTIMIZING EARLY LAYERS

The first few layers in neural networks commonly present a problem for many resource constrained mobile devices. Specifically, they often require large amounts of memory for the activations due to high initial input resolution. For instance, as can be seen in Figure 12.4, the first few layers use nearly ten times more memory than the remaining layers. Applying the width multiplier helps somewhat alleviate this problem. However, when reducing the number of features too drastically, the model

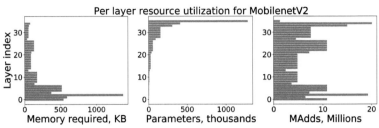

Figure 12.4 Per layer resource consumption for MobileNetV2. Note how memory requirement and parameters are counter-balanced. Earlier layers have large memory requirements but use very few parameters, while the deeper layers have a very large number of parameters but use very little memory.

performance starts to deteriorate. Furthermore, for many hardware setups using a number of channels that is not a multiple of 16 or 32 is very inefficient and padding to reach these multiples may reduce the amount of memory saved. Also note, that the memory constraint reduces linearly with the multiplier γ. On the other hand, reducing the input resolution often provides a good solution since it reduces the peak activation memory by γ^2. For many classification tasks, there is a flexibility in how much the resolution can be reduced. However, for other tasks, especially pixel-level tasks, the reduced resolution often leads to significant loss in accuracy. One way to solve this, is to convert some of the spatial dimensions into channel dimensions using the space-to-depth operation [413, 414, 415]. This operation is defined as follows. For a block size k, and image of size $n \times n \times c$ with c channels, this operation produces $n/k \times n/k \times nk^2$ tensor by flattening each $k \times k \times c$ block into $1 \times 1 \times ck^2$ tensor and arranging them in $n/k \times n/k$ grid. These operations are available in all modern ML frameworks such as TensorFlow [268] or PyTorch [416], as a built-in function. We also note that it can be efficiently implemented via a combination of two reshape and one transpose operations:

$$\text{S2D}[X, k] \equiv \underset{[b,\frac{n}{k},\frac{n}{k},k^2c]}{\text{reshape}} \left[\underset{[0,1,3,2,4]}{\text{transpose}} \left[\underset{[b,\frac{n}{k},k,\frac{n}{k},kc]}{\text{reshape}} [x] \right] \right]$$

As shown in Figure 12.2a, space-to-depth enables the spatial resolution to be reduced by a factor of 16, with only 0.5% drop in accuracy. In Figure 12.5, we show how using different space-to-depth block sizes affects the per-layer memory requirement and MAdds count.

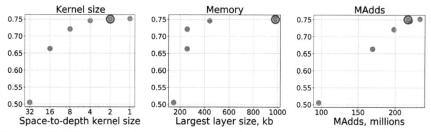

Figure 12.5 Impact of using space-to-depth on memory, MAdds, and accuracy for MobileNetV3. Circled point represents the "baseline" model. Note baseline model uses initial stride-2 convolution, while our space-to-depth implementation does not, thus k=2 results in a slightly higher multiply-add count. Space-to-depth has a small impact on parameter count but is essentially unchanged.

12.4 OPTIMIZING THE FINAL LAYERS

In this section, we explore optimizations for the last two layers independently of the rest of the archtiecture. First we consider the last layer that has spatial resolution, and then we explore of the final layer that feeds directly into embedding.

12.4.1 Adjusting the Resolution of the Final Spatial Layer

As can be seen on Figure 12.4, the last spatial layer before average pooling has a large cost in MAdds. This layer has often the highest cost in terms of overall computation. Thus a potential optimization is to reduce its spatial resolution in the final convolutional block as shown on Figure 12.6. For instance, a variation of this technique is used in MobileNetV3 and

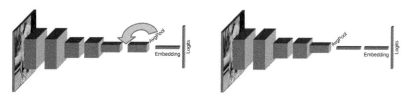

Figure 12.6 Moving average pooling deeper into the network can significantly reduce the computation cost.

reduces the compute cost by nearly 20% with negligible cost to the overall accuracy. (See Figure 9 in [407].)

12.4.2 Reducing the Size of the Embedding Layer

If we look at the size of a typical classification network, we can see that a large fraction of the model parameters are used by the last two layers that convert the hidden representation into to the final logit layer. The final logit layer size is determined by the number of classes k and thus can't be changed. But what about the size of the pre-logit *embedding* layer? In many standard models, the rule of thumb has been to use embedding size between 1024 [407] and 2048 [370, 417]. However, what is the optimal number? For example, when we apply a multiplier, should we scale the feature layer by the same factor or keep it constant?

It turns out that some architectures are much more sensitive to the size of the embedding than others. For instance, full-sized MobileNetV3-Large [407] allows the embedding size to be reduced from 1280 to 40 with only 2% accuracy loss, while reducing the number of parameters from 5M to 3M. On the other hand for MobileNetV2, the accuracy loss is nearly 7%. Yet for smaller models, as can be seen in Figure 12.7 increasing the embedding size increases the accuracy by nearly 2% without significant increase in total MAdds, and no change in memory requirements.

While it still remains an open question how to optimally select this parameter, it presents an opportunity for empirical adjustment to

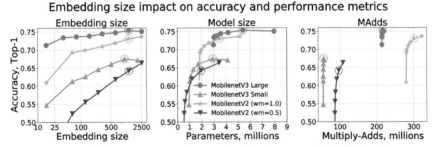

Figure 12.7 The impact of tuning embedding size on model performance characteristics. Left graph shows the direct trade-off between the number of channels in the embedding layer and the accuracy. The center graph shows the corresponding impact of the embedding on the model size. The rightmost graph shows the impact of the embedding on the number of operations. The circled dot points are the "baseline" published models.

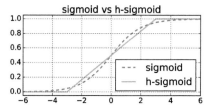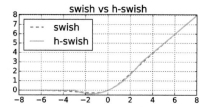

Figure 12.8 From [407]. Sigmoid and swish non-linearities and their "hard" counterparts.

significantly reduce the variable count without significant change in accuracy, or alternatively it might provide a way to increase accuracy without significant increase in MAdds. See Figure 12.7 for details.

12.5 ADJUSTING NON-LINEARITIES: H-SWISH AND H-SIGMOID

The question of what constitutes a good non-linearity for artificial neural networks has been studied extensively in the literature. Biologically inspired sigmoid and tanh dominated early literature (see for instance [418, 419]). These non-linearities suffered from vanishing gradients and eventually were replaced by simpler and more robust Rectified Linear Units (ReLU) [420, 421, 422, 423] and their bounded form such as ReLU6 (bounded at 6). Using ReLU6 eliminates the risk of overflow for low-bit floating point models such as when using float-16. We refer to [424] for an in-depth review. More complex non-linearities have also gained popularity in recent years, such as swish [383], gelu [425], silu [426], and others [424]. These new activation functions provide a meaningful gain in accuracy across wide spectrum of models without changing the architecture beyond swapping the non-linearity. In particular swish [383] has shown significant gain when used as a drop-in replacement for ReLU. It is defined as follows.

$$\text{swish}[x] = x * \sigma(x), \text{ where } \sigma(x) = \frac{1}{1 + e^{-x}}$$

While non-linearities are often treated as having negligible cost, in practice they can add up to 30% or more of extra compute time [407] for an unoptimized implementation. This introduces somewhat of a chicken and egg problem—if a non-linearity can not be used efficiently in real world scenarios, it will not be introduced in architectures, and if it is not used by the standard architectures, there will not be an efficient implementation. This can be alleviated by introducing non-linearity only into deeper layers of the architecture. Because the size of activation map is reduced with reduced resolution, this often provides most of the gain for a tiny fraction of the original cost. This trick was used for instance in [407] to make the model efficient even without optimized implementation. The other concern is the ability to apply the non-linearity in fixed-precision (e.g., 8-bit) modes where computing exponents and ratios is often associated with significant precision loss. Here we illustrate how to convert smooth floating point non-linearities into their "hard" counterparts. Hard-swish and hard-sigmoid are quantization friendly versions of swish and sigmoid that were used in MobileNetV3 [407] with early variants also mentioned in [427]. The hard nonlinearities have the same accuracy benefits as their soft counterparts but are also implementable as a simple combination of ReLU units:

$$\text{h-sigmoid}[x] = \frac{\text{ReLU6}(x+3)}{6}, \quad \text{h-swish}[x] = x\frac{\text{ReLU6}(x+3)}{6}$$

Further optimization can be gained through operator fusion and implementation in a piecewise form. The impact of applying an optimized h-swish for MobileNetV3 is shown in Table 12.2.

Note that even though the implementation is straightforward and robust, optimization is imperative for low latency. Table 12.2 additionally includes the latency of naive vs. optimized implementation in TensorFlow Lite.

TABLE 12.2 Effect of Applying different Types of Non-linearities on MobileNetV3. The second and third column gives Pixel-1 latency for reference. MobileNetV3 uses h-swish in layers with at least 80 channels and ReLU elsewhere. If h-swish is used in all layers, latency dramatically increases for non-optimized implementations while not giving significant accuracy improvement. This table is reproduced from [407]. h-swish @112 denotes a modification where h-swish is used only in layers with at least 112 channels.

	Top-1	P-1 (ms)	P-1, no-opt (ms)
V3-Large	75.2	51.4	57.5
Use ReLU in all layers	74.5 (−.7%)	50.5 (−1%)	50.5
Use h-swish in all layers	75.4 (+.2%)	53.5 (+4%)	68.9
Use h-swish @ ≥ 112	75.0 (−.3%)	51 (−0.5%)	54.4

12.6   PUTTING IT ALL TOGETHER

We described a set of common techniques that allows a developer to modify existing architectures to fit custom hardware and deployment requirements. While these techniques can generally be used a-la carte, the basic scheme of adapting a neural network to custom deployment can be described as follows.

- Check memory requirements. If the hardware has limited peak memory requirements, use space-to-depth or downsampling techniques to reduce the memory footprint of the early layers, which have the highest peak memory.

- If custom non-linearities are sufficiently performant—use them. If they are too expensive, they can usually still be applied to the last few layers.

- Inspect the last few layers, particularly pre-embedding layer, it is commonly overlooked, but often its size can be reduced dramatically without visible accuracy loss.

- Adjust embedding size as needed.

- Apply width, depth, and resolution multipliers as needed while keeping embedding size fixed.

Here we apply multipliers last because of the well-defined impact of these multipliers on latency, model size and peak memory. Thus, they can be used to fine-tune the resource usage most easily at the end of the process.

GLOSSARY

Embedding layer: Also known as pre-logits, it the output of the final layer that connects directly to the logit layer. This layer is important because it preserves information beyond the categories of the task at hand. This vector can be used to fine-tune a model to a different task.

Input resolution multiplier: A scaling coefficient that is applied to the input size which subsequently scales the resolutions of the remaining layers. If it is greater than one, a higher-resolution image is commonly used, though a simple bi-linear upsampling often suffices. It is a commonly used technique to increase model accuracy without changing the model size.

Internal Resolution: The resolution of the final layer before average pooling of features. A typical ImageNet model with input resolution of 224x224 has 7x7 internal resolution. For Isometric models [409], the internal resolution is simply the resolution of all layers.

Isometric models: A class of neural architectures introduced in [409] that keep resolution of its hidden layers constant after initial adjustment of the input.

Logits: The non-normalized class prediction vector, typically followed by the softmax operator to produce a probability vector.

MAdd: Short for multiply-add. A single operation consisting of a multiplication and adding the result to an accumulator. The number of MAdds is often used as a proxy metric to measure the computational cost of an architecture.

Model size: Total number of the parameters used by an architecture.

Network depth multiplier: Also known as depth multiplier, is a coefficient applied to the number of layers, to adjust the depth of an architecture.

Network width multiplier: Also known as width multiplier, is a coefficient applied to the size of each layer uniformly to adjust the size of an architecture.

Number of channels: The number of features of a convolutional or fully connected layers.

Peak activation memory: The amount of RAM needed for activation storage to run a single inference on an image.

Skip-stride method: A model modification procedure described in [409] that uses low-resolution images (such as 32x32) with models that were originally designed to handle high-resolution images. This procedure operates by changing the strides of the first layers to 1 in such a way that the final layer's spatial resolution remains unchanged. For example, if the new input resolution is 56x56 while the original resolution is 224x224, this method would eliminate strides in the first two layers that have stride=2.

Space-to-depth: For a kernel size k, and image of size $n \times n \times c$ with c channels, this operation produces $n/k \times n/k \times nk^2$ tensor by flattening each $k \times k \times c$ block into $1 \times 1 \times ck^2$ tensor and arranging them in $n/k \times n/k$.

CHAPTER 13

A Survey of Quantization Methods for Efficient Neural Network Inference

Amir Gholami, Sehoon Kim, Zhen Dong, Zhewei Yao, Michael W. Mahoney, and Kurt Keutzer

University of California at Berkeley

CONTENTS

As soon as abstract mathematical computations were adapted to computation on digital computers, the problem of efficient representation, manipulation, and communication of the numerical values in those computations arose. Strongly related to the problem of numerical representation is the problem of quantization: in what manner should a set of continuous real-valued numbers be distributed over a fixed discrete set of numbers to minimize the number of bits required and also to maximize the accuracy of the attendant computations? This perennial problem of quantization is particularly relevant whenever memory and/or computational resources are severely restricted, and it has come to the forefront in recent years due to the remarkable performance of Neural Network models in computer vision, natural language processing, and related areas. Moving from floating-point representations to low-precision fixed integer values represented in four bits or less holds the potential to reduce the memory footprint and latency by a factor of 16x; and, in fact, reductions of 4x to 8x are often realized in practice in these applications. Thus, it is not surprising that quantization has emerged recently as an important and very active sub-area of research in the efficient implementation of computations associated with Neural Networks. In this article, we survey approaches to the problem of quantizing the numerical values in deep Neural Network computations, covering the advantages/disadvantages of current methods. With this survey and its organization, we hope to have presented a useful snapshot of the current research in quantization for Neural Networks and to have given an intelligent organization to ease the evaluation of future research in this area.

13.1 INTRODUCTION

Over the past decade, we have observed significant improvements in the accuracy of Neural Networks (NNs) for a wide range of problems, often achieved by highly over-parameterized models. While the accuracy of these over-parameterized (and thus very large) NN models has significantly increased, the sheer size of these models means that it is not possible to deploy them for many resource-constrained applications. This creates a problem for realizing pervasive deep learning, which requires real-time inference, with low energy consumption and high accuracy, in resource-constrained environments. This pervasive deep learning is expected to have a significant impact on a wide range of applications such as real-time intelligent healthcare monitoring, autonomous driving, audio analytics, and speech recognition.

Achieving efficient, real-time NNs with optimal accuracy requires rethinking the design, training, and deployment of NN models [428]. There is a large body of literature that has focused on addressing these issues by making NN models more efficient (in terms of latency, memory footprint, energy consumption, etc.), while still providing optimal accuracy/generalization trade-offs. These efforts can be broadly categorized as follows.

Designing efficient NN model architectures.   One line of work has focused on optimizing the NN model architecture in terms of its micro-architecture [429, 206, 430, 207, 431, 432, 433, 434] (e.g., kernel types such as depth-wise convolution or low-rank factorization) as well as its macro-architecture [288, 206, 435, 190, 212, 170] (e.g., module types such as residual, or inception). The classical techniques here mostly found new architecture modules using manual search, which is not scalable. As such, a new line of work is to design Automated machine learning (AutoML) and Neural Architecture Search (NAS) methods. These aim to find in an automated way the right NN architecture, under given constraints of model size, depth, and/or width [171, 336, 209, 169, 184, 436]. We refer interested reader to [437] for a recent survey of NAS methods.

Co-designing NN architecture and hardware together.   Another recent line of work has been to adapt (and co-design) the NN architecture for a particular target hardware platform. The importance of this is because the overhead of a NN component (in terms of latency and energy) is hardware-dependent. For example, hardware with a dedicated cache hierarchy can execute bandwidth bound operations much more efficiently than hardware without such cache hierarchy. Similar to NN architecture design, initial approaches at architecture-hardware co-design were manual, where an expert would adapt/change the NN architecture [438], followed by using automated AutoML and/or NAS techniques [183, 111, 184, 212].

Pruning. Another approach to reducing the memory footprint and computational cost of NNs is to apply pruning. In pruning, neurons with small saliency (sensitivity) are removed, resulting in a sparse computational graph. Here, neurons with small saliency are those whose removal minimally affects the model output/loss function. Pruning methods can be broadly categorized into unstructured pruning [439, 440, 441, 442, 443, 444], and structured pruning [158, 445, 446, 447, 448, 449, 450]. With unstructured pruning, one removes neurons with small saliency, wherever they occur. With this approach, one can perform aggressive pruning, removing most of the NN parameters, with very little impact on the generalization performance of the model. However, this approach leads to sparse matrix operations, which are known to be hard to accelerate, and which are typically memory-bound [451, 452]. On the other hand, with structured pruning, a group of parameters (e.g., entire convolutional filters) is removed. This has the effect of changing the input and output shapes of layers and weight matrices, thus still permitting dense matrix operations. However, aggressive structured pruning often leads to significant accuracy degradation. Training and inference with high levels of pruning/sparsity, while maintaining state-of-the-art performance, has remained an open problem [453]. We refer the interested reader to [452, 454, 455] for a thorough survey of related work in pruning/sparsity.

Knowledge distillation. Model distillation [456, 159, 457, 458, 459, 460, 461, 462] involves training a large model and then using it as a teacher to train a more compact model. Instead of using "hard" class labels during the training of the student model, the key idea of model distillation is to leverage the "soft" probabilities produced by the teacher, as these probabilities can contain more information about the input. Despite the large body of work on distillation, a major challenge here is to achieve a high compression ratio with distillation alone. Compared to quantization and pruning, which can maintain the performance with $\geq 4\times$ compression (with INT8 and lower precision), knowledge distillation methods tend to have non-negligible accuracy degradation with aggressive compression. However, the combination of knowledge distillation with prior methods (i.e., quantization and pruning) has shown great success [460].

Quantization. Finally, quantization is an approach that has shown great and consistent success in both training and inference of NN models. While the problems of numerical representation and quantization are as old as digital computing, Neural Nets offer unique opportunities for improvement. While this survey on quantization is mostly focused on inference, we should emphasize that an important success of quantization has been in NN training [463, 464, 465, 466, 467]. In particular, the breakthroughs of half-precision and mixed-precision training [468, 469, 470, 471] have been the main drivers that have enabled an order of magnitude higher throughput in AI accelerators. However, it has proven very difficult to go below half-precision without significant tuning, and most of the recent quantization research has focused on inference. This quantization for inference is the focus of this article.

Quantization and Neuroscience.   Loosely related to (and for some a motivation for) NN quantization is work in neuroscience that suggests that the human brain stores information in a discrete/quantized form, rather than in a continuous form [472, 473, 474]. A popular rationale for this idea is that information stored in continuous form will inevitably get corrupted by noise (which is always present in the physical environment, including our brains, and which can be induced by thermal, sensory, external, synaptic noise, etc.) [475, 476]. However, discrete signal representations can be more robust to such low-level noise. Other reasons, including the higher generalization power of discrete representations [477, 478, 479] and their higher efficiency under limited resources [480], have also been proposed. We refer the reader to [481] for a thorough review of related work in neuroscience literature.

The goal of this work is to introduce current methods and concepts used in quantization and to discuss the current challenges and opportunities in this line of research. In doing so, we have tried to discuss most relevant work. It is not possible to discuss every work in a field as large as NN quantization in the page limit of a short survey; and there is no doubt that we have missed some relevant papers. We apologize in advance both to the readers and the authors of papers that we may have neglected.

In terms of the structure of this survey, we will first provide a brief history of quantization in Section 13.2, and then we will introduce basic concepts underlying quantization in Section 13.3. These basic concepts are shared with most of the quantization algorithms, and they are necessary for understanding and deploying existing methods. Then we discuss more advanced topics in Section 13.4. These mostly involve recent state-of-the-art methods, especially for low/mixed-precision quantization. Then we discuss the implications of quantization in hardware accelerators in Section 13.5, with a special focus on edge processors. Finally, we provide a summary and conclusions in Section 13.7.

13.2 GENERAL HISTORY OF QUANTIZATION

Gray and Neuhoff have written a very nice survey of the history of quantization up to 1998 [482]. The article is an excellent one and merits reading in its entirety; however, for the reader's convenience we will briefly summarize some of the key points here. Quantization, as a method to map from input values in a large (often continuous) set to output values in a small (often finite) set, has a long history. Rounding and truncation are typical examples. Quantization is related to the foundations of the calculus, and related methods can be seen in the early 1800s (as well as much earlier), e.g., in early work on least-squares and related techniques for large-scale (by the standards of the early 1800s) data analysis [483]. An early work on quantization dates back to 1867, where discretization was used to approximate the calculation of integrals [484]; and, subsequently, in 1897, when Shappard investigated the impact of rounding errors on the integration result [485]. More recently, quantization has been important in digital signal processing, as the process of representing a signal in digital form ordinarily involves rounding, as well as in numerical analysis and the implementation of numerical algorithms, where computations on real-valued numbers are implemented with finite-precision arithmetic.

It was not until 1948, around the advent of the digital computer, when Shannon wrote his seminal paper on the mathematical theory of communication [486], that the effect of quantization and its use in coding theory were formally presented. In particular, Shannon argued in his lossless coding theory that using the same number of bits is wasteful, when events of interest have a non-uniform probability. He argued that a more optimal approach would be to vary the number of bits based on the probability of an event, a concept that is now known as variable-rate quantization. Huffman coding in particular is motivated by this [487]. In subsequent work in 1959 [488], Shannon introduced distortion-rate functions (which provide a lower bound on the signal distortion after coding) as well as the notion of vector quantization (also briefly discussed in Section 13.4.6). This concept was extended and became practical in [489, 490, 491, 492] for real communication applications. Other important historical research on quantization in signal processing in that time period includes [493], which introduced the Pulse Code Modulation (PCM) concept (a pulsing method proposed to approximate/represent/encode sampled analog signals), as well as the classical result of high resolution quantization [494]. We refer the interested reader to [482] for a detailed discussion of these issues.

Quantization appears in a slightly different way in algorithms that use numerical approximation for problems involving continuous mathematical quantities, an area that also has a long history, but that also received renewed interest with the advent of the digital computer. In numerical analysis, an important notion was (and still is) that of a well-posed problem—roughly, a problem is well-posed if: a solution exists; that solution is unique; and that solution depends continuously on the input data in some reasonable topology. Such problems are sometimes called well-conditioned problems. It turned out that, even when working with a given well-conditioned problem, certain algorithms that solve that problem "exactly" in some idealized sense perform very poorly in the presence of "noise" introduced by the peculiarities of roundoff and truncation errors. These roundoff errors have to do with representing real numbers with only finitely-many bits—a quantization specified, e.g., by the IEEE floating point standard; and truncation errors arise since only a finite number of iterations of an iterative algorithm can actually be performed. The latter are important even in "exact arithmetic", since most problems of continuous mathematics cannot even in principle be solved by a finite sequence of elementary operations; but the former have to do with quantization. These issues led to the notion of the numerical stability of an algorithm. Let us view a numerical algorithm as a function f attempting to map the input data x to the "true" solution y; but due to roundoff and truncation errors, the output of the algorithm is actually some other y^*. In this case, the forward error of the algorithm is $\Delta y = y^* - y$; and the backward error of the algorithm is the smallest Δx such that $f(x + \Delta x) = y^*$. Thus, the forward error tells us the difference between the exact or true answer and what was output by the algorithm; and the backward error tells us what input data the algorithm we ran actually solved exactly. The forward error and backward error for an algorithm are related by the condition number of the problem. We refer the interested reader to [495] for a detailed discussion of these issues.

Quantization in Neural Nets No doubt thousands of papers have been written on these topics, and one might wonder: how is recent work on NN quantization different from these earlier works? Certainly, many of the recently proposed "novel algorithms" have strong connections with (and in some cases are essentially rediscoveries of) past work in the literature. However, NNs bring unique challenges and opportunities to the problem of quantization. First, inference and training of Neural Nets are both computationally intensive. So, the efficient representation of numerical values is particularly important. Second, most current Neural Net models are heavily over-parameterized, so there is ample opportunity for reducing bit precision without impacting accuracy. However, one very important difference is that NNs are very robust to aggressive quantization and extreme discretization. The new degree of freedom here has to do with the number of parameters involved, i.e., that we are working with over-parameterized models. This has direct implications for whether we are solving well-posed problems, whether we are interested in forward error or backward error, etc. In the NN applications driving recent developments in quantization, there is not a single well-posed or well-conditioned problem that is being solved. Instead, one is interested in some sort of forward error metric (based on classification quality, perplexity, etc.), but due to the over-parameterization, there are many very different models that exactly or approximately optimize this metric. Thus, it is possible to have high error/distance between a quantized model and the original non-quantized model, while still attaining very good generalization performance. This added degree of freedom was not present in many of the classical research, which mostly focused on finding compression methods that would not change the signal too much, or with numerical methods in which there was strong control on the difference between the "exact" versus the "discretized" computation. This observation that has been the main driver for researching novel techniques for NN quantization. Finally,the layered structure of Neural Net models offers an additional dimension to explore. Different layers in a Neural Net have different impact on the loss function, and this motivates a mixed-precision approach to quantization.

13.3 BASIC CONCEPTS OF QUANTIZATION

In this section, we first briefly introduce common notations and the problem setup in Section 13.3.1, and then we describe the basic quantization concepts and methods in Section 13.3.2–13.3.6. Afterward, we discuss the different fine-tuning methods in Section 13.3.7, followed by stochastic quantization in Section 13.3.8.

13.3.1 Problem Setup and Notations

Assume that the NN has L layers with learnable parameters, denoted as $\{W_1, W_2, ..., W_L\}$, with θ denoting the combination of all such parameters. Without loss of generality, we focus on the supervised learning problem, where the nominal goal is to optimize the following empirical risk minimization function:

$$\mathcal{L}(\theta) = \frac{1}{N} \sum_{i=1}^{N} l(x_i, y_i; \theta), \tag{13.1}$$

where (x, y) is the input data and the corresponding label, $l(x, y; \theta)$ is the loss function (e.g., Mean Squared Error or Cross Entropy loss), and N is the total number of data points. Let us also denote the input hidden activations of the ith layer as h_i, and the corresponding output hidden activation as a_i. We assume that we have the trained model parameters θ, stored in floating point precision. In quantization, the goal is to reduce the precision of both the parameters (θ), as well as the intermediate activation maps (i.e., h_i, a_i) to low-precision, with minimal impact on the generalization power/accuracy of the model. To do this, we need to define a quantization operator that maps a floating point value to a quantized one, which is described next.

13.3.2 Uniform Quantization

We need first to define a function that can quantize NN weights and activations to a finite set of values. This function takes real values in floating point, and it maps them to a lower precision range, as illustrated in Figure 13.1. A popular choice for a quantization function is as follows:

$$Q(r) = \text{Int}(r/S) - Z, \tag{13.2}$$

where Q is the quantization operator, r is a real valued input (activation or weight), S is a real valued scaling factor, and Z is an integer zero point. Furthermore, the Int function maps a real value to an integer value through a rounding operation (e.g., round to nearest and truncation). In essence, this function is a mapping from real values r to some integer values. This method of quantization is also known as *uniform quantization*, as the resulting quantized values (aka quantization levels) are uniformly spaced (Figure 13.1, Left). There are also *non-uniform quantization* methods whose quantized values are not necessarily uniformly spaced (Figure 13.1, Right), and these methods will be discussed in more

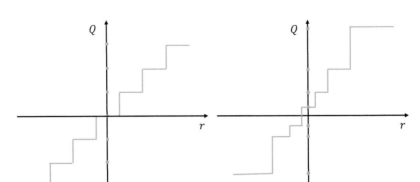

Figure 13.1 Comparison between uniform quantization (left) and non-uniform quantization (right). Real values in the continuous domain r are mapped into discrete, lower precision values in the quantized domain Q, which are marked with the orange bullets. Note that the distances between the quantized values (quantization levels) are the same in uniform quantization, whereas they can vary in non-uniform quantization.

detail in Section 13.3.6. It is possible to recover real values r from the quantized values $Q(r)$ through an operation that is often referred to as dequantization:

$$\tilde{r} = S(Q(r) + Z). \tag{13.3}$$

Note that the recovered real values $\tilde{r}$ will not exactly match r due to the rounding operation.

13.3.3 Symmetric and Asymmetric Quantization

One important factor in uniform quantization is the choice of the scaling factor S in equation 13.2. This scaling factor essentially divides a given range of real values r into a number of partitions (as discussed in [305, 496]):

$$S = \frac{\beta - \alpha}{2^b - 1}, \tag{13.4}$$

where $[\alpha, \beta]$ denotes the clipping range, a bounded range that we are clipping the real values with, and b is the quantization bit width. Therefore, in order for the scaling factor to be defined, the clipping range $[\alpha, \beta]$ should first be determined. The process of choosing the clipping range is often referred to as *calibration*. A straightforward choice is to use the min/max of the signal for the clipping range, i.e., $\alpha = r_{min}$, and $\beta = r_{max}$. This approach is an asymmetric quantization scheme,

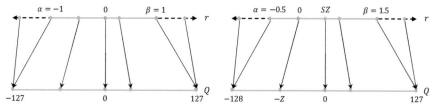

Figure 13.2 Illustration of symmetric quantization and asymmetric quantization. Symmetric quantization with restricted range maps real values to $[-127, 127]$, and full range maps to $[-128, 127]$ for 8-bit quantization.

since the clipping range is not necessarily symmetric with respect to the origin, i.e., $-\alpha \neq \beta$, as illustrated in Figure 13.2 (Right). It is also possible to use a symmetric quantization scheme by choosing a symmetric clipping range of $\alpha = -\beta$. A popular choice is to choose these based on the min/max values of the signal: $-\alpha = \beta = \max(|r|)$. Asymmetric quantization often results in a tighter clipping range as compared to symmetric quantization. This is especially important when the target weights or activations are imbalanced, e.g., the activation after ReLU that always has non-negative values. Using symmetric quantization, however, simplifies the quantization function in Eq. 13.2 by replacing the zero point with $Z = 0$:

$$Q(r) = \text{Int}\left(\frac{r}{S}\right). \tag{13.5}$$

Here, there are two choices for the scaling factor. In "full range" symmetric quantization S is chosen as $\frac{2\max(|r|)}{2^n - 1}$ (with floor rounding mode), to use the full INT8 range of $[-128, 127]$. However, in "restricted range" S is chosen as $\frac{\max(|r|)}{2^{n-1} - 1}$, which only uses the range of $[-127, 127]$. As expected, the full range approach is more accurate. Symmetric quantization is widely adopted in practice for quantizing weights because zeroing out the zero point can lead to reduction in computational cost during inference [497], and also makes the implementation more straightforward. However, note that for activation the cross terms occupying due to the offset in the asymmetric activations are a static data independent term and can be absorbed in the bias (or used to initialize the accumulator) [498].

Using the min/max of the signal for both symmetric and asymmetric quantization is a popular method. However, this approach is susceptible to outlier data in the activations. These could unnecessarily increase the range and, as a result, reduce the resolution of quantization. One approach to address this is to use percentile instead of min/max of the signal [499]. That is to say, instead of the largest/smallest value, the i-th largest/smallest values are used as β/α. Another approach is to select α and β to minimize KL divergence (i.e., information loss) between the real values and the quantized values [500]. We refer the interested readers to [497] where the different calibration methods are evaluated on various models.

Summary (Symmetric vs. Asymmetric Quantization). Symmetric quantization partitions the clipping using a symmetric range. This has the advantage of easier implementation, as it leads to $Z = 0$ in equation 13.2. However, it is sub-optimal for cases where the range could be skewed and not symmetric. For such cases, asymmetric quantization is preferred.

13.3.4 Range Calibration Algorithms: Static vs. Dynamic Quantization

So far, we discussed different calibration methods for determining the clipping range of $[\alpha, \beta]$. Another important differentiator of quantization methods is *when* the clipping range is determined. This range can be computed statically for weights, as in most cases the parameters are fixed during inference. However, the activation maps differ for each input sample (x in equation 13.1). As such, there are two approaches to quantizing activations: *dynamic quantization* and *static quantization.*

In dynamic quantization, this range is dynamically calculated for each activation map during runtime. This approach requires real-time computation of the signal statistics (min, max, percentile, etc.) which can have a very high overhead. However, dynamic quantization often results in higher accuracy as the signal range is exactly calculated for each input.

Another quantization approach is static quantization, in which the clipping range is pre-calculated and static during inference. This approach does not add any computational overhead, but it typically results in lower accuracy as compared to dynamic quantization. One popular method for the pre-calculation is to run a series of calibration inputs to compute the typical range of activations [496, 501]. Multiple different metrics have been proposed to find the best range, including minimizing Mean Squared Error (MSE) between original unquantized weight distribution and the corresponding quantized values [502, 503, 504, 505]. One could also consider using other metrics such as entropy [354], although MSE is the most common method used. Another approach is to learn/impose this clipping range during NN training [506, 507, 508, 509]. Notable work here are LQNets [509], PACT [507], LSQ [510], and LSQ+ [498] which jointly optimizes the clipping range and the weights in NN during training.

Summary (Dynamic vs. Static Quantization). Dynamic quantization dynamically computes the clipping range of each activation and often achieves the highest accuracy. However, calculating the range of a signal dynamically is very expensive, and as such, practitioners most often use static quantization where the clipping range is fixed for all inputs.

13.3.5   Quantization Granularity

In most computer vision tasks, the activation input to a layer is convolved with many different convolutional filters, as illustrated in Figure 13.3. Each of these convolutional filters can have a different range of values. As such, one differentiator for quantization methods is the granularity of how the clipping range $[\alpha, \beta]$ is calculated for the weights. We categorized them as follows.

Layerwise Quantization. In this approach, the clipping range is determined by considering all of the weights in convolutional filters of a layer [305], as shown in the third column of Figure 13.3. Here one examines the statistics of the entire parameters in that layer (e.g., min, max, percentile, etc.), and then uses the same clipping range for all the convolutional filters. While this approach is very simple to implement, it often results in sub-optimal accuracy, as the range of each convolutional filter can vary a lot. For example, a convolutional kernel that has relatively narrower range of parameters may lose its quantization resolution due to another kernel in the same layer with a wider range.

Groupwise Quantization. One could group multiple different channels inside a layer to calculate the clipping range (of either activations or convolution kernels). This could be helpful for cases where the distribution of the parameters across a single convolution/activation varies a lot. For instance, this approach was found useful in Q-BERT [511] for quantizing Transformer [512] models that consist of fully-connected attention layers. However, this approach inevitably comes with the extra cost of accounting for different scaling factors.

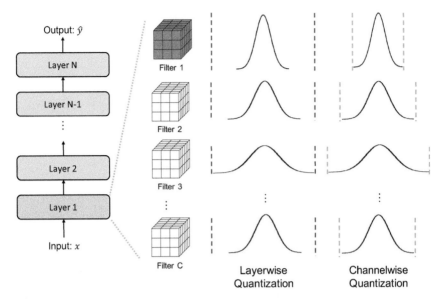

Figure 13.3 Illustration of different quantization granularities. In layerwise quantization, the same clipping range is applied to all the filters that belong to the same layer. This can result in bad quantization resolution for the channels that have narrow distributions (e.g., Filter 1 in the figure). One can achieve better quantization resolution using channelwise quantization that dedicates different clipping ranges to different channels.

Channelwise Quantization.   A popular choice of the clipping range is to use a fixed value for each convolutional filter, independent of other channels [513, 509, 496, 305, 514, 515], as shown in the last column of Figure 13.3. That is to say, each channel is assigned a dedicated scaling factor. This ensures a better quantization resolution and often results in higher accuracy.

Sub-channelwise Quantization.   The previous approach could be taken to the extreme, where the clipping range is determined with respect to any groups of parameters in a convolution or fully-connected layer. However, this approach could add considerable overhead, since the different scaling factors need to be taken into account when processing a single convolution or full-connected layer. Therefore, groupwise quantization could establish a good compromise between the quantization resolution and the computation overhead.

Summary (Quantization Granularity). Channelwise quantization is currently the standard method used for quantizing convolutional kernels. It enables the practitioner to adjust the clipping range for each individual kernel with negligible overhead. In contrast, sub-channelwise quantization may result in significant overhead and is not currently the standard choice.

13.3.6 Non-Uniform Quantization

Some work in the literature has also explored non-uniform quantization [516, 517, 469, 518, 519, 520, 521, 522, 523, 509, 524, 525, 526, 527, 528, 529, 530, 354, 531, 532], where quantization steps as well as quantization levels are allowed to be non-uniformly spaced. The formal definition of non-uniform quantization is shown in equation 13.6, where X_i represents the discrete quantization levels and Δ_i the quantization steps (thresholds):

$$Q(r) = X_i, \text{ if } r \in [\Delta_i, \Delta_{i+1}). \tag{13.6}$$

Specifically, when the value of a real number r falls in between the quantization step Δ_i and Δ_{i+1}, quantizer Q projects it to the corresponding quantization level X_i. Note that neither X_i's nor Δ_i's are uniformly spaced.

Non-uniform quantization may achieve higher accuracy for a fixed bit-width, because one could better capture the distributions by focusing more on important value regions or finding appropriate dynamic ranges. For instance, many non-uniform quantization methods have been designed for bell-shaped distributions of the weights and activations that often involve long tails [533, 522, 534, 520, 535, 536]. A typical rule-based non-uniform quantization is to use a logarithmic distribution [520, 537], where the quantization steps and levels increase exponentially instead of linearly. Another popular branch is *binary-code-based* quantization [525, 538, 509, 539, 540] where a real-number vector $\mathbf{r} \in \mathbb{R}^n$ is quantized into m binary vectors by representing $\mathbf{r} \approx \sum_{i=1}^m \alpha_i \mathbf{b}_i$, with the scaling factors $\alpha_i \in \mathbb{R}$ and the binary vectors $\mathbf{b}_i \in \{-1, +1\}^n$. Since there is no closed-form solution for minimizing the error between $\mathbf{r}$ and $\sum_{i=1}^m \alpha_i \mathbf{b}_i$, previous research relies on heuristic solutions. To further improve the quantizer, more recent work [541, 539, 538] formulates non-uniform quantization as an optimization problem. As shown in equation 13.7, the quantization steps/levels in the quantizer Q are adjusted to minimize the difference between the original tensor and the quantized counterpart.

$$\min_Q \|Q(r) - r\|^2 \tag{13.7}$$

Furthermore, the quantizer itself can also be jointly trained with the model parameters. These methods are referred to as learnable quantizers, and the quantization steps/levels are generally trained with iterative optimization [509, 538] or gradient descent [542, 526, 527].

In addition to rule-based and optimization-based non-uniform quantization, clustering can also be beneficial to alleviate the information loss due to quantization. Some works [516, 517] use k-means on different tensors to determine the quantization steps and levels, while other work [521] applies a Hessian-weighted k-means clustering on weights to minimize the performance loss. Further discussion can be found in Section 13.4.6.

Summary (Uniform vs. Non-uniform Quantization). Generally, non-uniform quantization enables us to better capture the signal information, by assigning bits and discreitizing the range of parameters non-uniformly. However, non-uniform quantization schemes are typically difficult to deploy efficiently on general computation hardware, e.g., GPU and CPU. As such, the uniform quantization is currently the de-facto method due to its simplicity and its efficient mapping to hardware.

13.3.7   Fine-tuning Methods

It is often necessary to adjust the parameters in the NN after quantization. This can either be performed by re-training the model, a process that is called Quantization-Aware Training (QAT), or done without re-training, a process that is often referred to as Post-Training Quantization (PTQ). A schematic comparison between these two approaches is illustrated in Figure 13.4, and further discussed below.

13.3.7.1   *Quantization-Aware Training*

Given a trained model, quantization may introduce a perturbation to the trained model parameters, and this can push the model away from the point to which it had converged when it was trained with floating point precision. It is possible to address this by re-training the NN model with quantized parameters so that the model can converge to a point with better loss. One popular approach is to use Quantization-Aware Training (QAT), in which the usual forward and backward pass are performed on the quantized model in floating point, but the model parameters are quantized after each gradient update (similar to projected gradient descent). In particular, it is important to do this projection after the

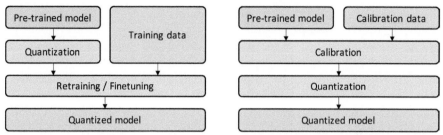

Figure 13.4 Comparison between Quantization-Aware Training (QAT, Left) and Post-Training Quantization (PTQ, Right). In QAT, a pretrained model is quantized and then finetuned using training data to adjust parameters and recover accuracy degradation. In PTQ, a pretrained model is calibrated using calibration data (e.g., a small subset of training data) to compute the clipping ranges and the scaling factors. Then, the model is quantized based on the calibration result. Note that the calibration process is often conducted in parallel with the finetuning process for QAT.

weight update is performed in floating point precision. Performing the backward pass with floating point is important, as accumulating the gradients in quantized precision can result in zero-gradient or gradients that have high error, especially in low-precision [543, 519, 540, 544, 545, 546, 547, 548].

An important subtlety in backpropagation is how the the non-differentiable quantization operator (equation 13.2) is treated. Without any approximation, the gradient of this operator is zero almost everywhere, since the rounding operation in equation 13.2 is a piece-wise flat operator. A popular approach to address this is to approximate the gradient of this operator by the so-called Straight Through Estimator (STE) [312]. STE essentially ignores the rounding operation and approximates it with an identity function, as illustrated in Figure 13.5.

Despite the coarse approximation of STE, it often works well in practice, except for ultra low-precision quantization such as binary quantization [549]. The work of [550] provides a theoretical justification for this phenomena, and it finds that the coarse gradient approximation of STE can in expectation correlate with population gradient (for a proper choice of STE). From a historical perspective, we should note that the original idea of STE can be traced back to the seminal work of [551, 552], where an identity operator was used to approximate gradient from the binary neurons.

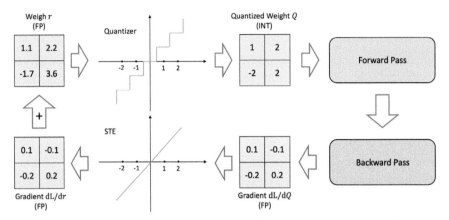

Figure 13.5 Illustration of Quantization-Aware Training procedure, including the use of Straight Through Estimator (STE).

While STE is the mainstream approach [553, 554], other approaches have also been explored in the literature [555, 556, 522, 557, 558, 559]. We should first mention that [312] also proposes a stochastic neuron approach as an alternative to STE (this is briefly discussed in Section 13.3.8). Other approaches using combinatorial optimization [560], target propagation [561], or Gumbel-softmax [562] have also been proposed. Another different class of alternative methods tries to use regularization operators to enforce the weight to be quantized. This removes the need to use the non-differentiable quantization operator in equation 13.2. These are often referred to as Non-STE methods [563, 549, 564, 558, 518, 537, 565]. Recent research in this area includes ProxQuant [549] which removes the rounding operation in the quantization formula equation 13.2, and instead uses the so-called W-shape, non-smooth regularization function to enforce the weights to quantized values. Other notable research includes using pulse training to approximate the derivative of discontinuous points [566], or replacing the quantized weights with an affine combination of floating point and quantized parameters [567]. The recent work of [568] also suggests AdaRound, which is an adaptive rounding method as an alternative to round-to-nearest method. Despite interesting works in this area, these methods often require a lot of tuning and so far STE approach is the most commonly used method.

In addition to adjusting model parameters, some prior work found it effective to learn quantization parameters during QAT as well. PACT [507] learns the clipping ranges of activations under uniform quantization, while QIT [526] also learns quantization steps and levels as an extension to a non-uniform quantization setting. LSQ [510] introduces a new gradient estimate to learn scaling factors for non-negative activations (e.g., ReLU) during QAT, and LSQ+ [498] further extends this idea to general activation functions such as swish [203] and h-swish [212] that produce negative values.

Summary (QAT). QAT has been shown to work despite the coarse approximation of STE. However, the main disadvantage of QAT is the computational cost of re-training the NN model. This re-training may need to be performed for several hundred epochs to recover accuracy, especially for low-bit precision quantization. If a quantized model is going to be deployed for an extended period, and if efficiency and accuracy are especially important, then this investment in re-training is likely to be worth it. However, this is not always the case, as some models have a relatively short lifetime. Next, we next discuss an alternative approach that does not have this overhead.

13.3.7.2 Post-Training Quantization

An alternative to the expensive QAT method is Post-Training Quantization (PTQ) which performs the quantization and the adjustments of the weights, without any fine-tuning [569, 570, 504, 505, 536, 571, 572, 573, 574, 575, 576, 577, 578, 579]. As such, the overhead of PTQ is very low and often negligible. Unlike QAT, which requires a sufficient amount of training data for retraining, PTQ has an additional advantage that it can be applied in situations where data is limited or unlabeled. However, this often comes at the cost of lower accuracy as compared to QAT, especially for low-precision quantization.

For this reason, multiple approaches have been proposed to mitigate the accuracy degradation of PTQ. For example, [569, 580] observe inherent bias in the mean and variance of the weight values following their quantization and propose bias correction methods and [570, 573] show that equalizing the weight ranges (and implicitly activation ranges) between different layers or channels can reduce quantization errors. ACIQ [569] analytically computes the optimal clipping range and the channel-wise bitwidth setting for PTQ. Although ACIQ can achieve low accuracy degradation, the channel-wise activation quantization used in ACIQ is hard to efficiently deploy on hardware. In order to address this, the OMSE method [504] removes channel-wise quantization on activation and proposes to conduct PTQ by optimizing the L2 distance between the quantized tensor and the corresponding floating point tensor. Furthermore, to better alleviate the adverse impact of outliers on PTQ, an outlier channel splitting (OCS) method is proposed in [505] which duplicates and halves the channels containing outlier values. Another notable work is AdaRound [568] which shows that the naive round-to-nearest method for quantization can counter-intuitively results in sub-optimal solutions, and it proposes an adaptive rounding method that better reduces the loss. While AdaRound restricts the changes of the quantized weights to be within ± 1 from their full-precision counterparts, AdaQuant [579] proposes a more general method that allows the quantized weights to change as needed. PTQ schemes can be taken to the extreme, where neither training nor testing data are utilized during quantization (aka zero-shot scenarios), which is discussed next.

Summary (PTQ). In PTQ, all the weights and activations quantization parameters are determined without any re-training of the NN model. As such, PTQ is a very fast method for quantizing NN models. However, this often comes at the cost of lower accuracy as compared to QAT.

13.3.7.3 Zero-shot Quantization

As discussed so far, in order to achieve minimal accuracy degradation after quantization, we need access to the entire of a fraction of training data. First, we need to know the range of activations so that we can clip the values and determine the proper scaling factors (which is usually referred to as calibration in the literature). Second, quantized models often require fine-tuning to adjust the model parameters and recover the accuracy degradation. In many cases, however, access to the original training data is not possible during the quantization procedure. This is because the training dataset is either too large to be distributed, proprietary (e.g., Google's JFT-300M), or sensitive due to security or privacy concerns (e.g., medical data). Several different methods have been proposed to address this challenge, which we refer to as zero-shot quantization (ZSQ). Inspired by [573], here we first describe two different levels of zero-shot quantization:

- **Level 1:** No data and no finetuning (aka ZSQ + PTQ).

- **Level 2:** No data but requires finetuning (aka ZSQ + QAT).

Level 1 allows faster and easier quantization without any finetuning.

Finetuning is in general time-consuming and often requires additional hyperparamenter search. However, Level 2 usually results in higher accuracy, as finetuning helps the quantized model to recover the accuracy degradation, particularly in ultra-low bit precision settings [581]. The work of [573] uses a Level 1 approach that relies on equalizing the weight ranges and correcting bias errors to make a given NN model more amenable to quantization without any data or finetuning. However, as this method is based on the scale-equivariance property of (piece-wise) linear activation functions, it can be sub-optimal for NNs with non-linear activations, such as BERT [582] with GELU [583] activation or MobileNetV3 [212] with swish activation [584].

A popular branch of research in ZSQ is to generate synthetic data similar to the real data from which the target pre-trained model is trained. The synthetic data is then used for calibrating and/or finetuning the quantized model. An early work in this area [585] exploits Generative Adversarial Networks (GANs) [586] for synthetic data generation. Using the pre-trained model as a discriminator, it trains the generator so that its outputs can be well classified by the discriminator. Then, using the synthetic data samples collected from the generator, the quantized model can be finetuned with knowledge distillation from the full-precision counterpart (see Section 13.4.4 for more details). However, this method fails to capture the internal statistics (e.g., distributions of the intermediate layer activations) of the real data, as it is generated only using the final outputs of the model. Synthetic data which does not take the internal statistics into account may not properly represent the real data distribution [581]. To address this, a number of subsequent efforts use the statistics stored in Batch Normalization (BatchNorm) [587], i.e., channel-wise mean and variance, to generate more realistic synthetic data. In particular, [581] generates data by directly minimizing the KL divergence of the internal statistics, and it uses the synthetic data to calibrate and finetune the quantized models. Furthermore, ZeroQ [574] shows that the synthetic data can be used for sensitivity measurement as well as calibration, thereby enabling mixed-precision post-training quantization without any access to the training/validation data. ZeroQ also extends ZSQ to the object detection tasks, as it does not rely on the output labels when generating data. Both [581] and [574] set the input images as trainable parameters and directly perform backpropagation on them until their internal statistics become similar to those of the real data. To take a step further, recent research [588, 589, 590] finds it effective to train and exploit generative models that can better capture the real data distribution and generate more realistic synthetic data.

Summary (ZSQ). Zero Shot (aka data free) quantization performs the entire quantization without any access to the training/validation data. This is particularly important for Machine Learning as a Service (MLaaS) providers who want to accelerate the deployment of a customer's workload, without the need to access their dataset. Moreover, this is important for cases where security or privacy concerns may limit access to the training data.

13.3.8 Stochastic Quantization

During inference, the quantization scheme is usually deterministic. However, this is not the only possibility, and some works have explored stochastic quantization for quantization aware training as well as reduced precision training [312, 469]. The high level intuition has been that the stochastic quantization may allow a NN to explore more, as compared to deterministic quantization. One popular supporting argument has been that small weight updates may not lead to any weight change, as the rounding operation may always return the same weights. However, enabling a stochastic rounding may provide the NN an opportunity to escape, thereby updating its parameters.

More formally, stochastic quantization maps the floating number up or down with a probability associated to the magnitude of the weight update. For instance, in [469, 591], the Int operator in equation 13.2 is defined as

$$\text{Int}(x) = \begin{cases} \lfloor x \rfloor & \text{with probability } \lceil x \rceil - x, \\ \lceil x \rceil & \text{with probability } x - \lfloor x \rfloor. \end{cases} \tag{13.8}$$

However, this definition cannot be used for binary quantization. Hence, [543] extends this to

$$\text{Binary}(x) = \begin{cases} -1 & \text{with probability } 1 - \sigma(x), \\ +1 & \text{with probability } \sigma(x), \end{cases} \tag{13.9}$$

where Binary is a function to binarize the real value x, and $\sigma(\cdot)$ is the sigmoid function.

Recently, another stochastic quantization method is introduced in QuantNoise [556]. QuantNoise quantizes a different random subset of weights during each forward pass and trains the model with unbiased gradients. This allows lower-bit precision quantization without significant accuracy drop in many computer vision and natural language processing models. However, a major challenge with stochastic quantization methods is the overhead of creating random numbers for every single weight update, and as such they are not yet adopted widely in practice.

13.4 ADVANCED CONCEPTS: QUANTIZATION BELOW 8 BITS

In this section, we will discuss more advanced topics in quantization which are mostly used for sub-INT8 quantization. We will first discuss simulated quantization and its difference with integer-only quantization in Section 13.4.1. Afterward, we will discuss different methods for mixed-precision quantization in Section 13.4.2, followed by hardware-aware quantization in Section 13.4.3. Then we will describe how distillation can be used to boost the quantization accuracy in Section 13.4.4, and then we will discuss extremely low bit precision quantization in Section 13.4.5. Finally, we will briefly describe the different methods for vector quantization in Section 13.4.6.

13.4.1 Simulated and Integer-only Quantization

There are two common approaches to deploy a quantized NN model, *simulated quantization* (aka fake quantization) and *integer-only quantization* (aka fixed-point quantization). In simulated quantization, the quantized model parameters are stored in low-precision, but the operations (e.g. matrix multiplications and convolutions) are carried out with floating point arithmetic. Therefore, the quantized parameters need to be dequantized before the floating point operations as schematically shown in Figure 13.6

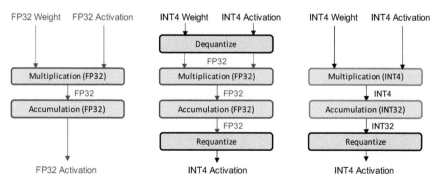

Figure 13.6 Comparison between full-precision inference (Left), inference with simulated quantization (Middle), and inference with integer-only quantization (Right).

(Middle). As such, one cannot fully benefit from fast and efficient low-precision logic with simulated quantization. However, in integer-only quantization, all the operations are performed using low-precision integer arithmetic [496, 501, 592, 593], as illustrated in Figure 13.6 (Right). This permits the entire inference to be carried out with efficient integer arithmetic, without any floating point dequantization of any parameters or activations.

In general, performing the inference in full-precision with floating point arithmetic may help the final quantization accuracy, but this comes at the cost of not being able to benefit from the low-precision logic. Low-precision logic has multiple benefits over the full-precision counterpart in terms of latency, power consumption, and area efficiency. As shown in Figure 13.7 (left), many hardware processors, including NVIDIA V100 and Titan RTX, support fast processing of low-precision arithmetic that can boost the inference throughput and latency. Moreover, as illustrated in Figure 13.7 (right) for a 45nm technology [594], low-precision logic is significantly more efficient in terms of energy and area. For example, performing INT8 addition is $30\times$ more energy efficient and $116\times$ more area efficient as compared to FP32 addition [594].

Notable integer-only quantization works include [593], which fuses Batch Normalization into the previous convolution layer, and [496], which proposes an integer-only computation method for residual networks with batch normalization. However, both methods are limited to ReLU activation. The recent work of [592] addresses this limitation by approximating

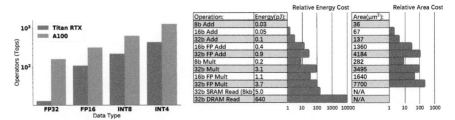

Figure 13.7 (Left) Comparison between peak throughput for different bit-precision logic on Titan RTX and A100 GPU. (Right) Comparison of the corresponding energy cost and relative area cost for different precision for 45nm technology [594]. As one can see, lower precision provides exponentially better energy efficiency and higher throughput. This trend actually increases in later/smaller process generations.

GELU [583], Softmax, and Layer Normalization [595] with integer arithmetic and further extends integer-only quantization to Transformer [512] architectures.

Dyadic quantization is another class of integer-only quantization, where all the scaling is performed with dyadic numbers, which are rational numbers with integer values in their numerator and a power of 2 in the denominator [501]. This results in a computational graph that only requires integer addition, multiplication, bit shifting, but no integer division. Importantly, in this approach, all the additions (e.g. residual connections) are enforced to have the same dyadic scale, which can make the addition logic simpler with higher efficiency.

Summary (Simulated vs Integer-only Quantization). In general integer-only and dyadic quantization are more desirable as compared to simulated/fake quantization. This is because integer-only uses lower precision logic for the arithmetic, whereas simulated quantization uses floating point logic to perform the operations. However, this does not mean that fake quantization is never useful. In fact, fake quantization methods can be beneficial for problems that are bandwidth-bound rather than compute-bound, such as in recommendation systems [596]. For these tasks, the bottleneck is the memory footprint and the cost of loading parameters from memory. Therefore, performing fake quantization can be acceptable for these cases.

13.4.2   Mixed-Precision Quantization

It is easy to see that the hardware performance improves as we use lower precision quantization. However, uniformly quantizing a model to ultra low-precision can cause significant accuracy degradation. It is possible to address this with mixed-precision quantization [597, 598, 599, 600, 601, 602, 603, 604, 605, 606, 607]. In this approach, each layer is quantized with different bit precision, as illustrated in Figure 13.8. One challenge with this approach is that the search space for choosing this bit setting is exponential in the number of layers. Different approaches have been proposed to address this huge search space.

Selecting this mixed-precision for each layer is essentially a searching problem, and many different methods have been proposed for it. The recent work of [598] proposed a reinforcement learning (RL)-based method to determine automatically the quantization policy, and the authors used a hardware simulator to take the hardware accelerator's feedback in

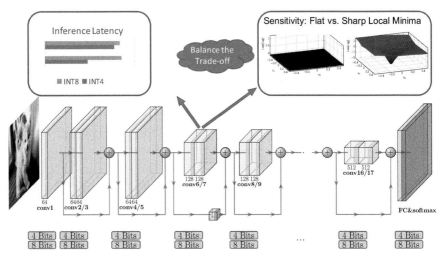

Figure 13.8 Illustration of mixed-precision quantization. In mixed-precision quantization the goal is to keep sensitive and efficient layers in higher precision, and only apply low-precision quantization to insensitive and inefficient layers. The efficiency metric is hardware dependant, and it could be latency or energy consumption.

the RL agent feedback. The paper [608] formulated the mixed-precision configuration searching problem as a Neural Architecture Search (NAS) problem and used the Differentiable NAS (DNAS) method to efficiently explore the search space. One disadvantage of these exploration-based methods [598, 608] is that they often require large computational resources, and their performance is typically sensitive to hyperparameters and even initialization.

Another class of mixed-precision methods uses periodic function regularization to train mixed-precision models by automatically distinguishing different layers and their varying importance with respect to accuracy while learning their respective bitwidths [564].

Different than these exploration and regularization-based approaches, HAWQ [599] introduces an automatic way to find the mixed-precision settings based on second-order sensitivity of the model. It was theoretically shown that the trace of the second-order operator (i.e., the Hessian) can be used to measure the sensitivity of a layer to quantization [609], similar to results for pruning in the seminal work of Optimal Brain Damage [439]. In HAWQv2, this method was extended to mixed-precision activation quantization [609], and was shown to be more than 100x faster than RL-based mixed-precision methods [598]. Recently, in HAWQv3, an integer-only, hardware-aware quantization was introduced [501] that proposed a fast Integer Linear Programming method to find the optimal bit precision for a given application-specific constraint (e.g., model size or latency). This work also addressed the common question about hardware efficiency of mixed-precision quantization by directly deploying them on T4 GPUs, showing up to 50% speed up with mixed-precision (INT4/INT8) quantization as compared to INT8 quantization.

Summary (Mixed-precision Quantization). Mixed-precision quantization has proved to be an effective and hardware-efficient method for low-precision quantization of different NN models. In this approach, the layers of a NN are grouped into sensitive/insensitive to quantization, and higher/lower bits are used for each layer. As such, one can minimize accuracy degradation and still benefit from reduced memory footprint and faster speed up with low precision quantization. Recent work [501] has also shown that this approach is hardware-efficient as mixed-precision is only used across operations/layers.

13.4.3 Hardware Aware Quantization

One of the goals of quantization is to improve the inference latency. However, not all hardware provide the same speed up after a certain layer/operation is quantized. In fact, the benefits from quantization is hardware-dependant, with many factors such as on-chip memory, bandwidth, and cache hierarchy affecting the quantization speed up.

It is important to consider this fact for achieving optimal benefits through hardware-aware quantization [517, 598, 445, 608, 610, 501, 611, 612]. In particular, the work [598] uses a reinforcement learning agent to determine the hardware-aware mixed-precision setting for quantization, based on a look-up table of latency with respect to different layers with different bitwidth. However, this approach uses simulated hardware latency. To address this the recent work of [501] directly deploys quantized operations in hardware, and measures the actual deployment latency of each layer for different quantization bit precisions.

13.4.4 Distillation-Assisted Quantization

An interesting line of work in quantization is to incorporate model distillation to boost quantization accuracy [460, 457, 501, 613]. Model distillation [456, 159, 457, 458, 460, 461, 462, 614, 553] is a method in which a large model with higher accuracy is used as a teacher to help the training of a compact student model. During the training of the student model, instead of using just the ground-truth class labels, model distillation proposes to leverage the soft probabilities produced by the teacher, which may contain more information of the input. That is the overall loss function incorporates both the student loss and the distillation loss, which is typically formulated as follows:

$$\mathcal{L} = \alpha \mathcal{H}(y, \sigma(z_s)) + \beta \mathcal{H}(\sigma(z_t, T), \sigma(z_s, T)) \tag{13.10}$$

In Eq. 13.10, α and β are weighting coefficients to tune the amount of loss from the student model and the distillation loss, y is the ground-truth class label, $\mathcal{H}$ is the cross-entropy loss function, z_s/z_t are logits generated by the student/teacher model, σ is the softmax function, and T is its temperature defined as follows:

$$p_i = \frac{\exp \frac{z_i}{T}}{\sum_j \exp \frac{z_j}{T}} \tag{13.11}$$

Previous methods of knowledge distillation focus on exploring different knowledge sources. [159, 458, 615] use logits (the soft probabilities) as the source of knowledge, while [456, 459, 461] try to leverage the knowledge from intermediate layers. The choices of teacher models are also well studied, where [616, 617] use multiple teacher models to jointly supervise the student model, while [618, 619] apply self-distillation without an extra teacher model.

13.4.5 Extreme Quantization

Binarization, where the quantized values are constrained to a 1-bit representation, thereby drastically reducing the memory requirement by 32×, is the most extreme quantization method. Besides the memory advantages, binary (1-bit) and ternary (2-bit) operations can often be computed efficiently with bit-wise arithmetic and can achieve significant acceleration over higher precisions, such as FP32 and INT8. For instance, the peak binary arithmetic on NVIDIA V100 GPUs is 8x higher than INT8. However, a naive binarization method would lead to significant accuracy degradation. As such, there is a large body of work that has proposed different solutions to address this [620, 621, 622, 603, 623, 624, 625, 626, 627, 628, 629, 630, 631, 522, 539, 632, 633, 634, 635, 636, 637, 638, 639, 640, 641, 642, 643, 644, 645, 646, 647, 648, 649, 650].

An important work here is BinaryConnect [543] which constrains the weights to either $+1$ or -1. In this approach, the weights are kept as real values and are only binarized during the forward and backward passes to simulate the binarization effect. During the forward pass, the real-value weights are converted into $+1$ or -1 based on the sign function. Then the network can be trained using the standard training method with STE to propagate the gradients through the non-differentiable sign function. Binarized NN [540] (BNN) extends this idea by binarizing the activations as well as the weights. Jointly binarizing weights and activations has the additional benefit of improved latency, since the costly floating-point matrix multiplications can be replaced with lightweight XNOR operations followed by bit-counting. Another interesting work is Binary Weight Network (BWN) and XNOR-Net proposed in [566], which achieve higher accuracy by incorporating a scaling factor to the weights and using $+\alpha$ or $-\alpha$ instead of $+1$ or -1. Here, α is the scaling factor chosen to minimize the distance between the real-valued weights and the resulting binarized weights. In other words, a real-valued weight matrix W can be formulated as $W \approx \alpha B$, where B is a binary weight matrix that satisfies the following optimization problem:

$$\alpha, B = \operatorname{argmin} \|W - \alpha B\|^2. \tag{13.12}$$

Furthermore, inspired by the observation that many learned weights are close to zero, there have been attempts to ternarize network by constraining the weights/activations with ternary values, e.g., $+1$, 0, and -1, thereby explicitly permitting the quantized values to be zero [519, 651]. Ternarization also drastically reduces the inference latency by eliminating the costly matrix multiplications as binarization does. Later, Ternary-Binary Network (TBN) [652] shows that combining binary network weights and ternary activations can achieve an optimal tradeoff between the accuracy and computational efficiency.

Since the naive binarization and ternarization methods generally result in severe accuracy degradation, especially for complex tasks such as ImageNet classification, a number of solutions have been proposed to reduce the accuracy degradation in extreme quantization. The work of [653] broadly categorizes these solutions into three branches. Here, we briefly discuss each branch, and we refer the interested readers to [653] for more details.

Quantization Error Minimization. The first branch of solutions aims to minimize the quantization error, i.e., the gap between the real values and the quantized values [654, 655, 542, 557, 656, 657, 658, 659,

660, 528, 524]. Instead of using a single binary matrix to represent real-value weights/activations, HORQ [654] and ABC-Net [542] use a linear combination of multiple binary matrices, i.e., $W \approx \alpha_1 B_1 + \cdots + \alpha_M B_M$, to reduce the quantization error. Inspired by the fact that binarizing the activations reduces their representational capability for the succeeding convolution block, [656] and [657] show that binarization of wider networks (i.e., networks with larger number of filters) can achieve a good trade-off between the accuracy and the model size.

Improved Loss function. Another branch of works focuses on the choice of loss function [518, 661, 530, 662, 627]. Important works here are loss-aware binarization and ternarization [518, 661] that directly minimize the loss with respect to the binarized/ternatized weights. This is different from other approaches that only approximate the weights and do not consider the final loss. Knowledge distillation from full-precision teacher models has also been shown as a promising method to recover the accuracy degradation after binarization/ternarization [663, 460, 630, 457].

Improved Training Method. Another interesting branch of work aims for better training methods for binary/ternary models [664, 628, 665, 666, 557, 513, 629, 650]. A number of efforts point out the limitation of STE in backpropagating gradients through the sign function: STE only propagate the gradients for the weights and/or activations that are in the range of [-1, 1]. To address this, BNN+ [664] introduces a continuous approximation for the derivative of the sign function, while [667, 625, 668] replace the sign function with smooth, differentiable functions that gradually sharpens and approaches the sign function. Bi-Real Net [557] introduces identity shortcuts connecting activations to activations in consecutive blocks, through which 32-bit activations can be propagated. While most research focuses on reducing the inference time latency, DoReFa-Net [513] quantizes the gradients in addition to the weights and activations, in order to accelerate the training as well.

Extreme quantization has been successful in drastically reducing the inference/training latency as well as the model size for many CNN models on computer vision tasks. Recently, there have been attempts to extend this idea to Natural Language Processing (NLP) tasks [669, 670, 671]. Considering the prohibitive model size and inference latency of state-of-the-art NLP models (e.g., BERT [582], RoBERTa [672], and the GPT family [673, 674, 675]) that are pre-trained on a large amount of unlabeled data, extreme quantization is emerging as a powerful tool for bringing NLP inference tasks to the edge.

Summary (Extreme Quantization). Extreme low-bit precision quantization is a very promising line of research. However, existing methods often incur high accuracy degradation as compared to baseline, unless very extensive tuning and hyperparameter search is performed. But this accuracy degradation may be acceptable for less critical applications.

13.4.6 Vector Quantization

As discussed in Section 13.2, quantization has not been invented in machine learning, but has been widely studied in the past century in information theory, and particularly in digital signal processing field as a compression tool. However, the main difference between quantization methods for machine learning is that fundamentally we are not interested to compress the signal with minimum change/error as compared to the original signal. Instead, the goal is to find a reduced-precision representation that results in as small loss as possible. As such, it is completely acceptable if the quantized weights/activations are far away from the non-quantized ones.

Having said that, there are a lot of interesting ideas in the classical quantization methods in DSP that have been applied to NN quantization, and in particular vector quantization [676]. In particular, the work of [677, 516, 517, 354, 196, 678, 679, 680, 681] clusters the weights into different groups and use the centroid of each group as quantized values during inference. As shown in equation 13.13, i is the index of weights in a tensor, $c_1, ..., c_k$ are the k centroids found by the clustering, and c_j is the corresponding centroid to w_i. After clustering, weight w_i will have a cluster index j related to c_j in the codebook (look-up table).

$$\min_{c_1,...,c_k} \sum_i \|w_i - c_j\|^2 \tag{13.13}$$

It has been found that using a k-means clustering is sufficient to reduce the model size up to $8\times$ without significant accuracy degradation [516]. In addition to that, jointly applying k-means based vector quantization with pruning and Huffman coding can further reduce the model size [196].

Product quantization [516, 517, 682] is an extension of vector quantization, where the weight matrix is divided into submatrices and vector quantization is applied to each submatrix. Besides basic product quantization method, more fine-grained usage of clustering can further improve the accuracy. For example, in [516] the residuals after k-means product quantization are further recursively quantized. And in [354], the authors apply more clusters for more important quantization ranges to better preserve the information.

13.5 QUANTIZATION AND HARDWARE PROCESSORS

We have said that quantization not only reduces the model size, but it also enables faster speed and requires less power, in particular for hardware that has low-precision logic. As such, quantization has been particularly crucial for edge deployment in IoT and mobile applications. Edge devices often have tight resource constraints including compute, memory, and importantly power budget. These are often too costly to meet for many deep NN models. In addition, many edge processors do not have any support for floating point operations, especially in micro-controllers.

Here, we briefly discuss different hardware platforms in the context of quantization. ARM Cortex-M is a group of 32-bit RISC ARM processor cores that are designed for low-cost and power-efficient embedded devices. For instance, the STM32 family are the microcontrollers based on the ARM Cortex-M cores that are also used for NN inference at the edge. Because some of the ARM Cortex-M cores do not include dedicated floating-point units, the models should first be quantized before deployment. CMSIS-NN [683] is a library from ARM that helps quantizing and deploying NN models onto the ARM Cortex-M cores. Specifically, the library leverages fixed-point quantization [593, 496, 501] with power-of-two scaling factors so that quantization and dequantization processes can be carried out efficiently with bit shifting operations. GAP-8 [684], a RISC-V SoC (System on Chip) for edge inference with a dedicated CNN accelerator, is another example of an edge processor that only supports integer arithmetic. While programmable general-purpose processors are widely adopted due to their flexibility, Google Edge TPU, a purpose-built ASIC chip, is another emerging solution for running inference at the edge. Unlike Cloud TPUs that run in Google data centers with a large amount of computing resources, the Edge TPU is designed for small and low-power devices, and thereby it only supports 8-bit arithmetic. NN models must be quantized using either quantization-aware training or post-training quantization of TensorFlow.

Figure 13.9 plots the throughput of different commercial edge processors that are widely used for NN inference at the edge. In the past few years, there has been a significant improvement in the computing power of the edge processors, and this allows deployment and inference of costly NN models that were previously available only on servers. Quantization,

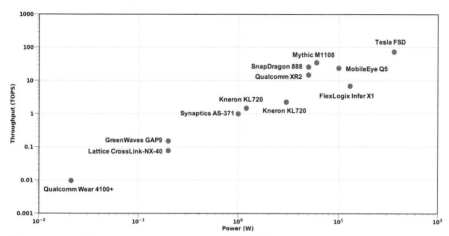

Figure 13.9 Throughput comparison of different commercial edge processors for NN inference at the edge.

combined with efficient low-precision logic and dedicated deep learning accelerators, has been one important driving force for the evolution of such edge processors.

While quantization is an indispensable technique for a lot of edge processors, it can also bring a remarkable improvement for non-edge processors, e.g., to meet Service Level Agreement (SLA) requirements such as 99th percentile latency. A good example is provided by the recent NVIDIA Turing GPUs, and in particular T4 GPUs, which include the Turing Tensor Cores. Tensor Cores are specialized execution units designed for efficient low-precision matrix multiplications.

13.6   FUTURE DIRECTIONS FOR RESEARCH IN QUANTIZATION

Here, we briefly discuss several high level challenges and opportunities for future research in quantization. This is broken down into quantization software, hardware and NN architecture co-design, coupled compression methods, and quantized training.

Quantization Software: With current methods, it is straightforward to quantize and deploy different NN models to INT8, without losing accuracy. There are several software packages that can be used to deploy INT8 quantized models (e.g., Nvidia's TensorRT, TVM, etc.), each with good documentation. Furthermore, the implementations are also quite optimal and one can easily observe speed up with quantization. However, the software for lower bit-precision quantization is not widely available, and sometimes it is non-existent. For instance, Nvidia's TensorRT does not currently support sub-INT8 quantization. Moreover, support for INT4 quantization was only recently added to TVM [501]. Recent work has shown that low precision and mixed-precision quantization with INT4/INT8 works in practice [501, 579, 597, 598, 599, 600, 601, 602, 603, 604, 605, 606, 607]. Thus, developing efficient software APIs for lower precision quantization will have an important impact.

Hardware and NN Architecture Co-Design: As discussed above, an important difference between classical work in low-precision quantization and the recent work in machine learning is the fact that NN parameters may have very different quantized values but may still generalize similarly well. For example, with quantization-aware training, we might converge to a different solution, far away from the original solution with single precision parameters, but still get good accuracy. One can take advantage of this degree of freedom and also adapt the NN architecture as it is being quantized. For instance, the recent work of [657] shows that changing the width of the NN architecture could reduce/remove generalization gap after quantization. One line of future work is to adapt jointly other architecture parameters, such as depth or individual kernels, as the model is being quantized. Another line of future work is to extend this co-design to hardware architecture. This may be particularly useful for FPGA deployment, as one can explore many different possible hardware configurations (such as different micro-architectures of multiply-accumulate elements), and then couple this with the NN architecture and quantization co-design.

Coupled Compression Methods: As discussed above, quantization is only one of the methods for efficient deployment of NNs. Other methods include efficient NN architecture design, co-design of hardware and NN architecture, pruning, and knowledge distillation. Quantization can be coupled with these other approaches. However, there is currently very little work exploring what are the optimal combinations of these methods. For instance, pruning and quantization can be applied together to a model to reduce its overhead [612], and it is important to understand the best combination of structured/unstructured pruning and quantization. Similarly, another future direction is to study the coupling between these methods and other approaches described above.

Quantized Training: Perhaps the most important use of quantization has been to accelerate NN training with half-precision [468, 469, 470, 471]. This has enabled the use of much faster and more power-efficient reduced-precision logic for training. However, it has been very difficult to push this further down to INT8 precision training. While several interesting works exist in this area [685, 686, 463, 687, 688, 689], the proposed methods often require a lot of hyperparameter tuning, or they only work for a few NN models on relatively easy learning tasks. The basic problem is that, with INT8 precision, the training can become unstable and diverge. Addressing this challenge can have a high impact on several applications, especially for training at the edge.

13.7 SUMMARY AND CONCLUSIONS

As soon as abstract mathematical computations were adapted to computation on digital computers, the problem of efficient representation, manipulation, and communication of the numerical values in those computations arose. Strongly related to the problem of numerical representation is the problem of quantization: in what manner should a set of continuous real-valued numbers be distributed over a fixed discrete set of numbers to minimize the number of bits required and also to maximize the accuracy of the attendant computations? While these problems are as old as computer science, these problems are especially relevant to the design of efficient NN models. There are several reasons for this. First, NNs are computationally intensive. So, the efficient representation of numerical values is particularly important. Second, most current NN models are heavily over-parameterized. So, there is ample opportunity for reducing the bit precision without impacting accuracy. Third, the layered structure of NN models offers an additional dimension to explore. Thus, different layers in the NN have different impact on the loss function, and this motivates interesting approaches such mixed-precision quantization.

Moving from floating-point representations to low-precision fixed integer values represented in eight/four bits or less holds the potential to reduce the memory footprint and latency. [690] shows that INT8 inference of popular computer vision models, including ResNet50 [140], VGG-19 [287], and inceptionV3 [323] using TVM [272] quantization library, can achieve $3.89\times$, $3.32\times$, and $5.02\times$ speedup on NVIDIA GTX 1080, respectively. [691] further shows that INT4 inference of ResNet50 could bring an additional 50–60% speedup on NVIDIA T4 and RTX, compared to its INT8 counterpart, emphasizing the importance of using lower-bit precision to maximize efficiency. Recently, [501] leverages mix-precision quantization to achieve 23% speedup for ResNet50, as compared to INT8 inference without accuracy degradation, and [592] extends INT8-only inference to BERT model to enable up to $4.0\times$ faster inference than FP32. While the aforementioned works focus on acceleration on GPUs, [692] also obtained $2.35\times$ and $1.40\times$ latency speedup on Intel Cascade Lake CPU and Raspberry Pi4 (which are both non-GPU architectures), respectively, through INT8 quantization of various computer vision models. As a result, as our bibliography attests, the problem of quantization in NN models has been a highly active research area.

In this work, we have tried to bring some conceptual structure to these very diverse efforts. We began with a discussion of topics common to many applications of quantization, such as uniform, non-uniform, symmetric, asymmetric, static, and dynamic quantization. We then considered quantization issues that are more unique to the quantization of NNs. These include layerwise, groupwise, channelwise, and sub-channelwise quantization. We further considered the inter-relationship between training and quantization, and we discussed the advantages and disadvantages of quantization-aware training as compared to post-training quantization. Further nuancing the discussion of the relationship between quantization and training is the issue of the availability of data. The extreme case of this is one in which the data used in training are, due to a variety of sensible reasons such as privacy, no longer available. This motivates the problem of zero-shot quantization.

As we are particularly concerned about efficient NNs targeted for edge-deployment, we considered problems that are unique to this environment. These include quantization techniques that result in parameters represented by fewer than 8 bits, perhaps as low as binary values. We also considered the problem of integer-only quantization, which enables the deployment of NNs on low-end microprocessors which often lack floating-point units.

With this survey and its organization, we hope to have presented a useful snapshot of the current research in quantization for NNs and to have given an intelligent organization to ease the evaluation of future research in this area.

Bibliography

[1] Yung-Hsiang Lu, Alexander C. Berg, and Yiran Chen. Low-power image recognition challenge. *AI Magazine*, 39(2):87–88, Jul. 2018.

[2] Yung-Hsiang Lu. Low-power image recognition. *Nature Machine Intelligence*, 1(4):199–199, Apr 2019.

[3] K. Gauen, R. Rangan, A. Mohan, Y. Lu, W. Liu, and A. C. Berg. Low-power image recognition challenge. In *22nd Asia and South Pacific Design Automation Conference (ASP-DAC)*, pages 99–104, 2017.

[4] Y. Lu, A. M. Kadin, A. C. Berg, T. M. Conte, E. P. DeBenedictis, R. Garg, G. Gingade, B. Hoang, Y. Huang, B. Li, J. Liu, W. Liu, H. Mao, J. Peng, T. Tang, E. K. Track, J. Wang, T. Wang, Y. Wang, and J. Yao. Rebooting computing and low-power image recognition challenge. In *IEEE/ACM International Conference on Computer-Aided Design (ICCAD)*, pages 927–932, 2015.

[5] S. Alyamkin, M. Ardi, A. C. Berg, A. Brighton, B. Chen, Y. Chen, H. Cheng, Z. Fan, C. Feng, B. Fu, K. Gauen, A. Goel, A. Goncharenko, X. Guo, S. Ha, A. Howard, X. Hu, Y. Huang, D. Kang, J. Kim, J. G. Ko, A. Kondratyev, J. Lee, S. Lee, S. Lee, Z. Li, Z. Liang, J. Liu, X. Liu, Y. Lu, Y. Lu, D. Malik, H. H. Nguyen, E. Park, D. Repin, L. Shen, T. Sheng, F. Sun, D. Svitov, G. K. Thiruvathukal, B. Zhang, J. Zhang, X. Zhang, and S. Zhuo. Low-power computer vision: Status, challenges, and opportunities. *IEEE Journal on Emerging and Selected Topics in Circuits and Systems*, 9(2):411–421, 2019.

[6] M. Ardi, A. C. Berg, B. Chen, Y. Chen, Y. Chen, D. Kang, J. Lee, S. Lee, Y. Lu, Y. Lu, and F. Sun. Special session: 2018 low-power image recognition challenge and beyond. In *IEEE International Conference on Artificial Intelligence Circuits and Systems (AICAS)*, pages 154–157, 2019.

[7] K. Gauen, R. Dailey, Y. Lu, E. Park, W. Liu, A. C. Berg, and Y. Chen. Three years of low-power image recognition challenge: Introduction to special session. In *Design, Automation Test in Europe Conference Exhibition (DATE)*, pages 700–703, 2018.

[8] Karen Simonyan and Andrew Zisserman. Very Deep Convolutional Networks for Large-Scale Image Recognition. *arXiv:1409.1556 [cs]*, April 2015. arXiv: 1409.1556.

[9] Andrew G. Howard, Menglong Zhu, Bo Chen, Dmitry Kalenichenko, Weijun Wang, Tobias Weyand, Marco Andreetto, and Hartwig Adam. MobileNets: Efficient Convolutional Neural Networks for Mobile Vision Applications. *arXiv:1704.04861 [cs]*, April 2017. arXiv: 1704.04861.

[10] Song Han, Huizi Mao, and William J. Dally. Deep Compression: Compressing Deep Neural Networks with Pruning, Trained Quantization and Huffman Coding. *arXiv:1510.00149 [cs]*, October 2015. arXiv: 1510.00149.

[11] Joseph Redmon and Ali Farhadi. YOLOv3: An Incremental Improvement. *arXiv:1804.02767 [cs]*, April 2018. arXiv: 1804.02767.

[12] Yann LeCun, Yoshua Bengio, and Geoffrey Hinton. Deep learning. *Nature*, 521(7553):436–444, May 2015. Number: 7553 Publisher: Nature Publishing Group.

[13] Forrest N. Iandola, Song Han, Matthew W. Moskewicz, Khalid Ashraf, William J. Dally, and Kurt Keutzer. SqueezeNet: AlexNet-level accuracy with 50x fewer parameters and <0.5MB model size. *arXiv:1602.07360 [cs]*, November 2016. arXiv: 1602.07360.

[14] Anup Mohan, Kent Gauen, Yung-Hsiang Lu, Wei Wayne Li, and Xuemin Chen. Internet of video things in 2030: A world with many cameras. In *IEEE International Symposium on Circuits and Systems (ISCAS)*, pages 1–4, Baltimore, MD, USA, May. IEEE.

[15] K. Kumar and Y. Lu. Cloud Computing for Mobile Users: Can Offloading Computation Save Energy? *Computer*, 43(4):51–56, April 2010. Conference Name: Computer.

[16] S. Alyamkin, M. Ardi, A. C. Berg, A. Brighton, B. Chen, Y. Chen, H. Cheng, Z. Fan, C. Feng, B. Fu, K. Gauen, A. Goel, A. Goncharenko, X. Guo, S. Ha, A. Howard, X. Hu, Y. Huang, D. Kang,

J. Kim, J. G. Ko, A. Kondratyev, J. Lee, S. Lee, S. Lee, Z. Li, Z. Liang, J. Liu, X. Liu, Y. Lu, Y. Lu, D. Malik, H. H. Nguyen, E. Park, D. Repin, L. Shen, T. Sheng, F. Sun, D. Svitov, G. K. Thiruvathukal, B. Zhang, J. Zhang, X. Zhang, and S. Zhuo. Low-Power Computer Vision: Status, Challenges, and Opportunities. *IEEE Journal on Emerging and Selected Topics in Circuits and Systems*, 9(2):411–421, June 2019.

[17] Forrest Iandola and Kurt Keutzer. Small neural nets are beautiful: enabling embedded systems with small deep-neural-network architectures. In *Proceedings of the Twelfth IEEE/ACM/IFIP International Conference on Hardware/Software Codesign and System Synthesis Companion*, CODES, pages 1–10, New York, NY, USA, October 2017. Association for Computing Machinery.

[18] Hao Li, Asim Kadav, Igor Durdanovic, Hanan Samet, and Hans Peter Graf. Pruning Filters for Efficient ConvNets. *arXiv:1608.08710 [cs]*, August 2016. arXiv: 1608.08710.

[19] A. Goel, C. Tung, Y.-H. Lu, and G. K. Thiruvathukal. A Survey of Methods for Low-Power Deep Learning and Computer Vision. In *IEEE 6th World Forum on Internet of Things (WF-IoT)*, pages 1–6, June 2020.

[20] Isha Ghodgaonkar, Subhankar Chakraborty, Vishnu Banna, Shane Allcroft, Mohammed Metwaly, Fischer Bordwell, Kohsuke Kimura, Xinxin Zhao, Abhinav Goel, Caleb Tung, Akhil Chinnakotla, Minghao Xue, Yung-Hsiang Lu, Mark Daniel Ward, Wei Zakharov, David S. Ebert, David M. Barbarash, and George K. Thiruvathukal. Analyzing Worldwide Social Distancing through Large-Scale Computer Vision. *arXiv:2008.12363 [cs]*, August 2020. arXiv: 2008.12363.

[21] Y. LeCun, B. Boser, J. S. Denker, D. Henderson, R. E. Howard, W. Hubbard, and L. D. Jackel. Backpropagation applied to handwritten zip code recognition. *Neural Computation*, 1(4):541–551, 1989.

[22] Raghuraman Krishnamoorthi. Quantizing deep convolutional networks for efficient inference: A whitepaper. *arXiv:1806.08342 [cs, stat]*, June 2018. arXiv: 1806.08342.

[23] Abhinav Goel, Sara Aghajanzadeh, Caleb Tung, Shuo-Han Chen, George K. Thiruvathukal, and Yung-Hsiang Lu. Modular Neural Networks for Low-Power Image Classification on Embedded Devices. *ACM Transactions on Design Automation of Electronic Systems*, 26(1):1:1–1:35, October 2020.

[24] Mahdi Nazemi, Ghasem Pasandi, and Massoud Pedram. NullaNet: Training Deep Neural Networks for Reduced-Memory-Access Inference. *arXiv:1807.08716 [cs, stat]*, August 2018. arXiv: 1807.08716.

[25] Jerry Bowles. How Athena Security found a brand new market in the midst of the COVID-19 pandemic, August 2020. Section: Governing identity privacy and security.

[26] S. Aghajanzadeh, R. Naidu, S.-H. Chen, C. Tung, A. Goel, Y.-H. Lu, and G. K. Thiruvathukal. Camera Placement Meeting Restrictions of Computer Vision. In *IEEE International Conference on Image Processing (ICIP)*, pages 3254–3258, October 2020. ISSN: 2381-8549.

[27] M. Ardi, A. C. Berg, B. Chen, Y. Chen, Y. Chen, D. Kang, J. Lee, S. Lee, Y. Lu, Y. Lu, and F. Sun. Special Session: 2018 Low-Power Image Recognition Challenge and Beyond. In *IEEE International Conference on Artificial Intelligence Circuits and Systems (AICAS)*, pages 154–157, March 2019.

[28] Kent Gauen, Ryan Dailey, Yung-Hsiang Lu, Eunbyung Park, Wei Liu, Alexander C. Berg, and Yiran Chen. Three years of low-power image recognition challenge: Introduction to special session. In *Design, Automation & Test in Europe Conference & Exhibition (DATE)*, pages 700–703, Dresden, Germany, March 2018. IEEE.

[29] K. Gauen, R. Rangan, A. Mohan, Y. Lu, W. Liu, and A. C. Berg. Low-power image recognition challenge. In *22nd Asia and South Pacific Design Automation Conference (ASP-DAC)*, pages 99–104, January 2017. ISSN: 2153-697X.

[30] Y. Lu, A. M. Kadin, A. C. Berg, T. M. Conte, E. P. DeBenedictis, R. Garg, G. Gingade, B. Hoang, Y. Huang, B. Li, J. Liu, W. Liu, H. Mao, J. Peng, T. Tang, E. K. Track, J. Wang, T. Wang, Y. Wang, and J. Yao. Rebooting computing and low-power image recognition challenge. In *IEEE/ACM International Conference on Computer-Aided Design (ICCAD)*, pages 927–932, 2015.

[31] Yung-Hsiang Lu. Low-power image recognition. *Nature Machine Intelligence*, 1(4):199–199, April 2019. Number: 4 Publisher: Nature Publishing Group.

[32] Yung-Hsiang Lu, Alexander C. Berg, and Yiran Chen. Low-Power Image Recognition Challenge. *AI Magazine*, 39(2):87–88, July 2018. Number: 2.

[33] J. Wu, C. Leng, Y. Wang, Q. Hu, and J. Cheng. Quantized Convolutional Neural Networks for Mobile Devices. In *IEEE Conference on Computer Vision and Pattern Recognition (CVPR)*, pages 4820–4828, June 2016. ISSN: 1063-6919.

[34] Markus Nagel, Rana Ali Amjad, Mart van Baalen, Christos Louizos, and Tijmen Blankevoort. Up or Down? Adaptive Rounding for Post-Training Quantization. *arXiv:2004.10568 [cs, stat]*, June 2020. arXiv: 2004.10568.

[35] M. Nagel, M. V. Baalen, T. Blankevoort, and M. Welling. Data-Free Quantization Through Weight Equalization and Bias Correction. In *IEEE/CVF International Conference on Computer Vision (ICCV)*, pages 1325–1334, October 2019. ISSN: 2380-7504.

[36] M. Sandler, A. Howard, M. Zhu, A. Zhmoginov, and L. Chen. MobileNetV2: Inverted Residuals and Linear Bottlenecks. In *IEEE/CVF Conference on Computer Vision and Pattern Recognition*, pages 4510–4520, June 2018. ISSN: 2575-7075.

[37] K. He, X. Zhang, S. Ren, and J. Sun. Deep Residual Learning for Image Recognition. In *IEEE Conference on Computer Vision and Pattern Recognition (CVPR)*, pages 770–778, June 2016. ISSN: 1063-6919.

[38] Barret Zoph, Vijay Vasudevan, Jonathon Shlens, and Quoc V. Le. Learning Transferable Architectures for Scalable Image Recognition. *arXiv:1707.07012 [cs, stat]*, April 2018. arXiv: 1707.07012.

[39] Min Lin, Qiang Chen, and Shuicheng Yan. Network In Network. *arXiv:1312.4400 [cs]*, March 2014. arXiv: 1312.4400.

[40] Matthieu Courbariaux, Yoshua Bengio, and Jean-Pierre David. Training deep neural networks with low precision multiplications. *arXiv:1412.7024 [cs]*, September 2015. arXiv: 1412.7024.

[41] Naigang Wang, Jungwook Choi, Daniel Brand, Chia-Yu Chen, and Kailash Gopalakrishnan. Training Deep Neural Networks with 8-bit Floating Point Numbers. *arXiv:1812.08011 [cs, stat]*, December 2018. arXiv: 1812.08011.

[42] A. Goel, Z. Liu, and R. D. Blanton. CompactNet: High Accuracy Deep Neural Network Optimized for On-Chip Implementation. In *2018 IEEE International Conference on Big Data (Big Data)*, pages 4723–4729, December 2018.

[43] Ruizhou Ding, Zeye Liu, Ting-Wu Chin, Diana Marculescu, and R. D. (Shawn) Blanton. FLightNNs: Lightweight Quantized Deep Neural Networks for Fast and Accurate Inference. In *Proceedings of the 56th Annual Design Automation Conference*, DAC, pages 1–6, New York, NY, USA, June. Association for Computing Machinery.

[44] Matthieu Courbariaux, Itay Hubara, Daniel Soudry, Ran El-Yaniv, and Yoshua Bengio. Binarized Neural Networks: Training Deep Neural Networks with Weights and Activations Constrained to +1 or −1. *arXiv:1602.02830 [cs]*, February 2016. arXiv: 1602.02830.

[45] Mohammad Rastegari, Vicente Ordonez, Joseph Redmon, and Ali Farhadi. XNOR-Net: ImageNet Classification Using Binary Convolutional Neural Networks. *arXiv:1603.05279 [cs]*, March 2016. arXiv: 1603.05279.

[46] Christos Louizos, Matthias Reisser, Tijmen Blankevoort, Efstratios Gavves, and Max Welling. Relaxed Quantization for Discretized Neural Networks. *arXiv:1810.01875 [cs, stat]*, October 2018. arXiv: 1810.01875.

[47] Yaohui Cai, Zhewei Yao, Zhen Dong, Amir Gholami, Michael W. Mahoney, and Kurt Keutzer. ZeroQ: A Novel Zero Shot Quantization Framework. pages 13169–13178, 2020.

[48] Shuchang Zhou, Yuxin Wu, Zekun Ni, Xinyu Zhou, He Wen, and Yuheng Zou. DoReFa-Net: Training Low Bitwidth Convolutional Neural Networks with Low Bitwidth Gradients. *arXiv:1606.06160 [cs]*, February 2018. arXiv: 1606.06160.

[49] Yann LeCun, John S. Denker, and Sara A. Solla. Optimal Brain Damage. In D. S. Touretzky, editor, *Advances in Neural Information Processing Systems 2*, pages 598–605. Morgan-Kaufmann, 1990.

[50] B. Hassibi, D.G. Stork, and G.J. Wolff. Optimal Brain Surgeon and general network pruning. In *IEEE International Conference on Neural Networks*, pages 293–299, San Francisco, CA, USA, 1993. IEEE.

[51] Xiaohan Ding, Guiguang Ding, Yuchen Guo, Jungong Han, and Chenggang Yan. Approximated Oracle Filter Pruning for Destructive CNN Width Optimization. January 2019.

[52] Song Han, Jeff Pool, John Tran, and William J. Dally. Learning both weights and connections for efficient neural networks. In *Proceedings of the 28th International Conference on Neural Information Processing Systems - Volume 1*, NIPS, page 1135–1143, Cambridge, MA, USA, 2015. MIT Press.

[53] Alex Krizhevsky, Ilya Sutskever, and Geoffrey E. Hinton. ImageNet Classification with Deep Convolutional Neural Networks. *Advances in Neural Information Processing Systems*, 25:1097–1105, 2012.

[54] Ruichi Yu, Ang Li, Chun-Fu Chen, Jui-Hsin Lai, Vlad I. Morariu, Xintong Han, Mingfei Gao, Ching-Yung Lin, and Larry S. Davis. NISP: Pruning Networks Using Neuron Importance Score Propagation. In *IEEE/CVF Conference on Computer Vision and Pattern Recognition*, pages 9194–9203, Salt Lake City, UT, June 2018. IEEE.

[55] Suraj Srinivas and R. Venkatesh Babu. Data-free parameter pruning for Deep Neural Networks. *arXiv:1507.06149 [cs]*, July 2015. arXiv: 1507.06149.

[56] P. Panda, A. Ankit, P. Wijesinghe, and K. Roy. FALCON: Feature Driven Selective Classification for Energy-Efficient Image Recognition. *IEEE Transactions on Computer-Aided Design of Integrated Circuits and Systems*, 36(12):2017–2029, December 2017. Conference Name: IEEE Transactions on Computer-Aided Design of Integrated Circuits and Systems.

[57] P. Panda and K. Roy. Semantic driven hierarchical learning for energy-efficient image classification. In *Design, Automation Test in Europe Conference Exhibition (DATE)*, pages 1582–1587, March 2017. ISSN: 1558-1101.

[58] Deboleena Roy, Priyadarshini Panda, and Kaushik Roy. Tree-CNN: A hierarchical Deep Convolutional Neural Network for incremental learning. *Neural Networks*, 121:148–160, January 2020.

[59] Abhinav Goel, Caleb Tung, Sara Aghajanzadeh, Isha Ghodgaonkar, Shreya Ghosh, George K. Thiruvathukal, and Yung-Hsiang Lu. Low-power object counting with hierarchical neural networks. In *Proceedings of the ACM/IEEE International Symposium on Low Power Electronics and Design*, ISLPED, page 163–168, New York, NY, USA, 2020. Association for Computing Machinery.

[60] Yu Cheng, Duo Wang, Pan Zhou, and Tao Zhang. A Survey of Model Compression and Acceleration for Deep Neural Networks. *arXiv:1710.09282 [cs]*, June 2020. arXiv: 1710.09282.

[61] Wei Wen, Chunpeng Wu, Yandan Wang, Yiran Chen, and Hai Li. Learning Structured Sparsity in Deep Neural Networks. In *Advances in Neural Information Processing Systems 29*, pages 2074–2082. 2016.

[62] Pruning Tutorial — PyTorch Tutorials 1.7.1 documentation.

[63] Tejalal Choudhary, Vipul Mishra, Anurag Goswami, and Jagannathan Sarangapani. A comprehensive survey on model compression and acceleration. *Artificial Intelligence Review*, 53(7):5113–5155, October 2020.

[64] Hongyang Li, Wanli Ouyang, and Xiaogang Wang. Multi-Bias Non-linear Activation in Deep Neural Networks. *arXiv:1604.00676 [cs]*, April 2016. arXiv: 1604.00676.

[65] Andrew Howard, Mark Sandler, Grace Chu, Liang-Chieh Chen, Bo Chen, Mingxing Tan, Weijun Wang, Yukun Zhu, Ruoming Pang, Vijay Vasudevan, Quoc V. Le, and Hartwig Adam. Searching for MobileNetV3. *arXiv:1905.02244 [cs]*, November 2019. arXiv: 1905.02244.

[66] G. Huang, S. Liu, L. v d Maaten, and K. Q. Weinberger. CondenseNet: An Efficient DenseNet Using Learned Group Convolutions. In *IEEE/CVF Conference on Computer Vision and Pattern Recognition*, pages 2752–2761, June 2018. ISSN: 2575-7075.

[67] Xiangyu Zhang, Xinyu Zhou, Mengxiao Lin, and Jian Sun. Shuf-fleNet: An Extremely Efficient Convolutional Neural Network for Mobile Devices. *arXiv:1707.01083 [cs]*, December 2017. arXiv: 1707.01083.

[68] C. Szegedy, Wei Liu, Yangqing Jia, P. Sermanet, S. Reed, D. Anguelov, D. Erhan, V. Vanhoucke, and A. Rabinovich. Going deeper with convolutions. In *2015 IEEE Conference on Computer Vision and Pattern Recognition (CVPR)*, pages 1–9, June 2015. ISSN: 1063-6919.

[69] B. Wu, A. Wan, X. Yue, P. Jin, S. Zhao, N. Golmant, A. Gho-laminejad, J. Gonzalez, and K. Keutzer. Shift: A Zero FLOP, Zero Parameter Alternative to Spatial Convolutions. In *IEEE/CVF Conference on Computer Vision and Pattern Recognition*, pages 9127–9135, June 2018. ISSN: 2575-7075.

[70] Han Cai, Chuang Gan, Ligeng Zhu, and Song Han. TinyTL: Reduce Memory, Not Parameters for Efficient On-Device Learning. *Advances in Neural Information Processing Systems*, 33, 2020.

[71] Max Jaderberg, Andrea Vedaldi, and Andrew Zisserman. Speeding up Convolutional Neural Networks with Low Rank Expansions. In *Proceedings of the British Machine Vision Conference*, pages 88.1–88.13, Nottingham, 2014. British Machine Vision Association.

[72] Emily Denton, Wojciech Zaremba, Joan Bruna, Yann LeCun, and Rob Fergus. Exploiting linear structure within convolutional net-works for efficient evaluation. In *Proceedings of the 27th Inter-national Conference on Neural Information Processing Systems - Volume 1*, NIPS, pages 1269–1277, Cambridge, MA, USA, December 2014. MIT Press.

[73] Tamara G. Kolda and Brett W. Bader. Tensor Decompositions and Applications. *SIAM Review*, 51(3):455–500, August 2009. Publisher: Society for Industrial and Applied Mathematics.

[74] Cheng Tai, Tong Xiao, Yi Zhang, Xiaogang Wang, and Weinan E. Convolutional neural networks with low-rank regularization. *arXiv:1511.06067 [cs, stat]*, February 2016. arXiv: 1511.06067.

[75] Vadim Lebedev, Yaroslav Ganin, Maksim Rakhuba, Ivan Oseledets, and Victor Lempitsky. Speeding-up Convolutional Neural Networks

Using Fine-tuned CP-Decomposition. *arXiv:1412.6553 [cs]*, April 2015. arXiv: 1412.6553.

[76] Huanrui Yang, Minxue Tang, Wei Wen, Feng Yan, Daniel Hu, Ang Li, Hai Li, and Yiran Chen. Learning Low-Rank Deep Neural Networks via Singular Vector Orthogonality Regularization and Singular Value Sparsification. In *Proceedings of the IEEE/CVF Conference on Computer Vision and Pattern Recognition Workshops*, pages 678–679, 2020.

[77] Hongkai Xiong, Yuhui Xu, Yuxi Li, Shuai Zhang, Yiran Chen, Yingyong Qi, Botao Wang, Wei Wen, and Weiyao Lin. TRP: Trained Rank Pruning for Efficient Deep Neural Networks. In *International Joint Conferences on Artificial Intelligence*, volume 1, pages 977–983, July 2020. ISSN: 1045-0823.

[78] Jose M. Alvarez and Mathieu Salzmann. Compression-aware Training of Deep Networks. *Advances in Neural Information Processing Systems*, 30:856–867, 2017.

[79] Wei Wen, Cong Xu, Chunpeng Wu, Yandan Wang, Yiran Chen, and Hai Li. Coordinating Filters for Faster Deep Neural Networks. In *Proceedings of the IEEE International Conference on Computer Vision*, pages 658–666, 2017.

[80] Xiaohan Ding, Guiguang Ding, Yuchen Guo, and Jungong Han. Centripetal SGD for Pruning Very Deep Convolutional Networks With Complicated Structure. pages 4943–4953, 2019.

[81] Pengzhen Ren, Yun Xiao, Xiaojun Chang, Po-Yao Huang, Zhihui Li, Xiaojiang Chen, and Xin Wang. A Comprehensive Survey of Neural Architecture Search: Challenges and Solutions. *arXiv:2006.02903 [cs, stat]*, January 2021. arXiv: 2006.02903.

[82] Barret Zoph and Quoc V. Le. Neural Architecture Search with Reinforcement Learning. *arXiv:1611.01578 [cs]*, February 2017. arXiv: 1611.01578.

[83] Bowen Baker, Otkrist Gupta, Nikhil Naik, and Ramesh Raskar. Designing Neural Network Architectures using Reinforcement Learning. *arXiv:1611.02167 [cs]*, March 2017. arXiv: 1611.02167.

[84] Esteban Real, Alok Aggarwal, Yanping Huang, and Quoc V. Le. Regularized Evolution for Image Classifier Architecture Search. *arXiv:1802.01548 [cs]*, February 2019. arXiv: 1802.01548.

[85] Mingxing Tan, Bo Chen, Ruoming Pang, Vijay Vasudevan, Mark Sandler, Andrew Howard, and Quoc V. Le. MnasNet: Platform-Aware Neural Architecture Search for Mobile. *arXiv:1807.11626 [cs]*, May 2019. arXiv: 1807.11626.

[86] Bichen Wu, Xiaoliang Dai, Peizhao Zhang, Yanghan Wang, Fei Sun, Yiming Wu, Yuandong Tian, Peter Vajda, Yangqing Jia, and Kurt Keutzer. FBNet: Hardware-Aware Efficient ConvNet Design via Differentiable Neural Architecture Search. *arXiv:1812.03443 [cs]*, May 2019. arXiv: 1812.03443.

[87] Han Cai, Ligeng Zhu, and Song Han. ProxylessNAS: Direct Neural Architecture Search on Target Task and Hardware. *arXiv:1812.00332 [cs, stat]*, February 2019. arXiv: 1812.00332.

[88] Y. Chen, G. Meng, Q. Zhang, S. Xiang, C. Huang, L. Mu, and X. Wang. RENAS: Reinforced Evolutionary Neural Architecture Search. In *IEEE/CVF Conference on Computer Vision and Pattern Recognition (CVPR)*, pages 4782–4791, June 2019. ISSN: 2575-7075.

[89] Hanxiao Liu, Karen Simonyan, and Yiming Yang. DARTS: Differentiable Architecture Search. *arXiv:1806.09055 [cs, stat]*, April 2019. arXiv: 1806.09055.

[90] Yuhui Xu, Lingxi Xie, Xiaopeng Zhang, Xin Chen, Guo-Jun Qi, Qi Tian, and Hongkai Xiong. PC-DARTS: Partial Channel Connections for Memory-Efficient Architecture Search. *arXiv:1907.05737 [cs]*, April 2020. arXiv: 1907.05737.

[91] Hanlin Chen, Li'an Zhuo, Baochang Zhang, Xiawu Zheng, Jianzhuang Liu, Rongrong Ji, David Doermann, and Guodong Guo. Binarized Neural Architecture Search for Efficient Object Recognition. *arXiv:2009.04247 [cs]*, September 2020. arXiv: 2009.04247.

[92] Y. Li, X. Jin, J. Mei, X. Lian, L. Yang, C. Xie, Q. Yu, Y. Zhou, S. Bai, and A. L. Yuille. Neural Architecture Search for Lightweight Non-Local Networks. In *IEEE/CVF Conference on Computer Vision and Pattern Recognition (CVPR)*, pages 10294–10303, 2020.

[93] Mohammad Loni, Ali Zoljodi, Sima Sinaei, Masoud Daneshtalab, and Mikael Sjödin. NeuroPower: Designing Energy Efficient Convolutional Neural Network Architecture for Embedded Systems. In *Artificial Neural Networks and Machine Learning – ICANN: Theoretical Neural Computation*, Lecture Notes in Computer Science, pages 208–222, Cham, 2019. Springer International Publishing.

[94] Hieu Pham, Melody Guan, Barret Zoph, Quoc Le, and Jeff Dean. Efficient neural architecture search via parameters sharing. In Jennifer Dy and Andreas Krause, editors, *International Conference on Machine Learning*, volume 80 of *Proceedings of Machine Learning Research*, pages 4095–4104. PMLR, 10–15 Jul 2018.

[95] Dimitrios Stamoulis, Ruizhou Ding, Di Wang, Dimitrios Lymberopoulos, Bodhi Priyantha, Jie Liu, and Diana Marculescu. Single-Path NAS: Designing Hardware-Efficient ConvNets in Less Than 4 Hours. In Ulf Brefeld, Elisa Fromont, Andreas Hotho, Arno Knobbe, Marloes Maathuis, and Céline Robardet, editors, *Machine Learning and Knowledge Discovery in Databases*, Lecture Notes in Computer Science, pages 481–497, Cham, 2020. Springer International Publishing.

[96] Zeyuan Allen-Zhu, Yuanzhi Li, and Yingyu Liang. Learning and Generalization in Overparameterized Neural Networks, Going Beyond Two Layers. September 2019.

[97] Alon Brutzkus and Amir Globerson. Why do Larger Models Generalize Better? A Theoretical Perspective via the XOR Problem. In *International Conference on Machine Learning*, pages 822–830. PMLR, May 2019. ISSN: 2640-3498.

[98] Zhuozhuo Tu, Fengxiang He, and Dacheng Tao. Understanding Generalization in Recurrent Neural Networks. In *International Conference on Learning Representations*, April 2020.

[99] Jianping Gou, Baosheng Yu, Stephen John Maybank, and Dacheng Tao. Knowledge Distillation: A Survey. *arXiv:2006.05525 [cs, stat]*, October 2020. arXiv: 2006.05525.

[100] Jimmy Ba and Rich Caruana. Do Deep Nets Really Need to be Deep? *Advances in Neural Information Processing Systems*, 27:2654–2662, 2014.

[101] Cristian Bucilua, Rich Caruana, and Alexandru Niculescu-Mizil. Model compression. In *Proceedings of the 12th ACM SIGKDD international conference on Knowledge discovery and data mining*, KDD '06, pages 535–541, New York, NY, USA, August 2006. Association for Computing Machinery.

[102] Geoffrey Hinton, Oriol Vinyals, and Jeff Dean. Distilling the Knowledge in a Neural Network. *arXiv:1503.02531 [cs, stat]*, March 2015. arXiv: 1503.02531.

[103] X. Liu, X. Wang, and S. Matwin. Improving the Interpretability of Deep Neural Networks with Knowledge Distillation. In *IEEE International Conference on Data Mining Workshops (ICDMW)*, pages 905–912, November 2018. ISSN: 2375-9259.

[104] Jang Hyun Cho and Bharath Hariharan. On the Efficacy of Knowledge Distillation. In *Proceedings of the IEEE/CVF International Conference on Computer Vision (ICCV)*, pages 4794–4802, October 2019.

[105] Byeongho Heo, Minsik Lee, Sangdoo Yun, and Jin Young Choi. Knowledge Transfer via Distillation of Activation Boundaries Formed by Hidden Neurons. *Proceedings of the AAAI Conference on Artificial Intelligence*, 33(01):3779–3787, July 2019. Number: 01.

[106] S. Ahn, S. X. Hu, A. Damianou, N. D. Lawrence, and Z. Dai. Variational Information Distillation for Knowledge Transfer. In *IEEE/CVF Conference on Computer Vision and Pattern Recognition (CVPR)*, pages 9155–9163, June 2019. ISSN: 2575-7075.

[107] J. Yim, D. Joo, J. Bae, and J. Kim. A Gift from Knowledge Distillation: Fast Optimization, Network Minimization and Transfer Learning. In *IEEE Conference on Computer Vision and Pattern Recognition (CVPR)*, pages 7130–7138, July 2017. ISSN: 1063-6919.

[108] Seyed Iman Mirzadeh, Mehrdad Farajtabar, Ang Li, Nir Levine, Akihiro Matsukawa, and Hassan Ghasemzadeh. Improved Knowledge Distillation via Teacher Assistant. *Proceedings of the AAAI Conference on Artificial Intelligence*, 34(04):5191–5198, April 2020. Number: 04.

[109] Theodore S. Nowak and Jason J. Corso. Deep Net Triage: Analyzing the Importance of Network Layers via Structural Compression. *arXiv:1801.04651 [cs]*, March 2018. arXiv: 1801.04651.

[110] Mingxing Tan and Quoc V. Le. EfficientNet: Rethinking Model Scaling for Convolutional Neural Networks. *arXiv:1905.11946 [cs, stat]*, September 2020. arXiv: 1905.11946.

[111] Han Cai, Chuang Gan, Tianzhe Wang, Zhekai Zhang, and Song Han. Once-for-all: Train one network and specialize it for efficient deployment. *arXiv preprint arXiv:1908.09791*, 2019.

[112] Jian-Hao Luo, Hao Zhang, Hong-Yu Zhou, Chen-Wei Xie, Jianxin Wu, and Weiyao Lin. Thinet: Pruning cnn filters for a thinner net. *IEEE Transactions on Pattern Analysis and Machine Intelligence*, 41(10):2525–2538, 2019.

[113] Yuhui Xu, Yuxi Li, Shuai Zhang, Wei Wen, Botao Wang, Yingyong Qi, Yiran Chen, Weiyao Lin, and Hongkai Xiong. Trained Rank Pruning for Efficient Deep Neural Networks. *arXiv preprint arXiv:1812.02402*, 2018.

[114] Kuan Wang, Zhijian Liu, Yujun Lin, Ji Lin, and Song Han. HAQ: Hardware-Aware Automated Quantization With Mixed Precision. In *IEEE/CVF Conference on Computer Vision and Pattern Recognition (CVPR)*, pages 8604–8612, Long Beach, CA, USA, June 2019. IEEE.

[115] Tien-Ju Yang, Yu-Hsin Chen, Joel Emer, and Vivienne Sze. A method to estimate the energy consumption of deep neural networks. In *Asilomar Conference on Signals, Systems, and Computers*, pages 1916–1920, 2017.

[116] Chenzhuo Zhu, Song Han, Huizi Mao, and William J. Dally. Trained Ternary Quantization. *arXiv:1612.01064 [cs]*, February 2017. arXiv: 1612.01064.

[117] Matthieu Courbariaux, Yoshua Bengio, and Jean-Pierre David. BinaryConnect: Training Deep Neural Networks with binary weights during propagations. *arXiv:1511.00363 [cs]*, April 2016. arXiv: 1511.00363.

[118] Kuan Wang, Zhijian Liu, Yujun Lin, Ji Lin, and Song Han. Haq: Hardware-aware automated quantization with mixed precision. In *Proceedings of the IEEE/CVF Conference on Computer Vision and Pattern Recognition*, pages 8612–8620, 2019.

[119] Luis Guerra, Bohan Zhuang, Ian Reid, and Tom Drummond. Automatic Pruning for Quantized Neural Networks. *arXiv:2002.00523 [cs]*, February 2020. arXiv: 2002.00523.

[120] F. Tung and G. Mori. CLIP-Q: Deep Network Compression Learning by In-parallel Pruning-Quantization. In *IEEE/CVF Conference on Computer Vision and Pattern Recognition*, pages 7873–7882, June 2018. ISSN: 2575-7075.

[121] Xiaofan Lin, Cong Zhao, and Wei Pan. Towards accurate binary convolutional neural network. In *International Conference on Neural Information Processing Systems*, NIPS, pages 344–352, Red Hook, NY, USA, December 2017. Curran Associates Inc.

[122] Asit Mishra and Debbie Marr. Apprentice: Using Knowledge Distillation Techniques To Improve Low-Precision Network Accuracy. *arXiv:1711.05852 [cs]*, November 2017. arXiv: 1711.05852.

[123] Jangho Kim, Yash Bhalgat, Jinwon Lee, Chirag Patel, and Nojun Kwak. QKD: Quantization-aware Knowledge Distillation. *arXiv:1911.12491 [cs]*, November 2019. arXiv: 1911.12491.

[124] Sridhar Swaminathan, Deepak Garg, Rajkumar Kannan, and Frederic Andres. Sparse low rank factorization for deep neural network compression. *Neurocomputing*, 398:185–196, July 2020.

[125] Michael Zhu and Suyog Gupta. To prune, or not to prune: exploring the efficacy of pruning for model compression. *arXiv:1710.01878 [cs, stat]*, November 2017. arXiv: 1710.01878.

[126] Mingxing Tan, Ruoming Pang, and Quoc V. Le. EfficientDet: Scalable and Efficient Object Detection. pages 10781–10790, 2020.

[127] Alex Krizhevsky, Geoffrey. Learning multiple layers of features from tiny images. Technical Report TR-2009, University of Toronto, Toronto, 2009.

[128] Yuval Netzer, Tao Wang, Adam Coates, Alessandro Bissacco, Bo Wu, and Andrew Y Ng. Reading digits in natural images with unsupervised feature learning. 2011.

[129] Gregory Cohen, Saeed Afshar, Jonathan Tapson, and Andre van Schaik. EMNIST: an extension of MNIST to handwritten letters. February 2017.

[130] Gregory Griffin, Alex Holub, and Pietro Perona. Caltech-256 object category dataset. 2007.

[131] J. Deng, W. Dong, R. Socher, L. Li, Kai Li, and Li Fei-Fei. ImageNet: A large-scale hierarchical image database. In *2009 IEEE Conference on Computer Vision and Pattern Recognition*, pages 248–255, June 2009. ISSN: 1063-6919.

[132] Tsung-Yi Lin, Michael Maire, Serge Belongie, James Hays, Pietro Perona, Deva Ramanan, Piotr Dollar, and C. Lawrence Zitnick. Microsoft COCO: Common Objects in Context. In *Computer Vision – ECCV*, Lecture Notes in Computer Science, pages 740–755, Cham, 2014. Springer International Publishing.

[133] M. Everingham, L. Van Gool, C. K. I. Williams, J. Winn, and A. Zisserman. The PASCAL Visual Object Classes Challenge 2012 (VOC) Results. http://www.pascal-network.org/challenges/VOC/voc2012/workshop/index.html.

[134] Gabriel J. Brostow, Jamie Shotton, Julien Fauqueur, and Roberto Cipolla. Segmentation and Recognition Using Structure from Motion Point Clouds. In David Forsyth, Philip Torr, and Andrew Zisserman, editors, *Computer Vision – ECCV 2008*, Lecture Notes in Computer Science, pages 44–57, Berlin, Heidelberg, 2008. Springer.

[135] Liang Zheng, Liyue Shen, Lu Tian, Shengjin Wang, Jingdong Wang, and Qi Tian. Scalable Person Re-Identification: A Benchmark. pages 1116–1124, 2015.

[136] Liang Zheng, Zhi Bie, Yifan Sun, Jingdong Wang, Chi Su, Shengjin Wang, and Qi Tian. MARS: A Video Benchmark for Large-Scale Person Re-Identification. In *Computer Vision – ECCV 2016*, pages 868–884, 2016.

[137] Peng Wang, Bingliang Jiao, Lu Yang, Yifei Yang, Shizhou Zhang, Wei Wei, and Yanning Zhang. Vehicle Re-Identification in Aerial Imagery: Dataset and Approach. pages 460–469, 2019.

[138] A Geiger, P Lenz, C Stiller, and R Urtasun. Vision meets robotics: The KITTI dataset. *International Journal of Robotics Research*, 32(11):1231–1237, September 2013.

[139] Marius Cordts, Mohamed Omran, Sebastian Ramos, Timo Rehfeld, Markus Enzweiler, Rodrigo Benenson, Uwe Franke, Stefan Roth, and Bernt Schiele. The Cityscapes Dataset for Semantic Urban Scene Understanding. pages 3213–3223, 2016.

[140] Kaiming He, Xiangyu Zhang, Shaoqing Ren, and Jian Sun. Deep residual learning for image recognition. In *Proceedings of the IEEE conference on computer vision and pattern recognition*, pages 770–778, 2016.

[141] Dario Amodei, Sundaram Ananthanarayanan, Rishita Anubhai, Jingliang Bai, Eric Battenberg, Carl Case, Jared Casper, Bryan Catanzaro, Qiang Cheng, Guoliang Chen, Jie Chen, Jingdong Chen, Zhijie Chen, Mike Chrzanowski, Adam Coates, Greg Diamos, Ke Ding, Niandong Du, Erich Elsen, Jesse Engel, Weiwei Fang, Linxi Fan, Christopher Fougner, Liang Gao, Caixia Gong, Awni Hannun, Tony Han, Lappi Vaino Johannes, Bing Jiang, Cai Ju, Billy Jun, Patrick LeGresley, Libby Lin, Junjie Liu, Yang Liu, Weigao Li, Xiangang Li, Dongpeng Ma, Sharan Narang, Andrew Ng, Sherjil Ozair, Yiping Peng, Ryan Prenger, Sheng Qian, Zongfeng Quan, Jonathan Raiman, Vinay Rao, Sanjeev Satheesh, David Seetapun, Shubho Sengupta, Kavya Srinet, Anuroop Sriram, Haiyuan Tang, Liliang Tang, Chong Wang, Jidong Wang, Kaifu Wang, Yi Wang, Zhijian Wang, Zhiqian Wang, Shuang Wu, Likai Wei, Bo Xiao, Wen Xie, Yan Xie, Dani Yogatama, Bin Yuan, Jun Zhan, and Zhenyao Zhu. Deep speech 2: end-to-end speech recognition in English and mandarin. In *Proceedings of the 33rd International Conference on International Conference on Machine Learning - Volume 48*, ICML'16, pages 173–182, New York, NY, USA, June 2016. JMLR.org.

[142] K. Guo, Shulin Zeng, J. Yu, Y. Wang, and H. Yang. [dl] a survey of fpga-based neural network inference accelerators. *ACM Trans. Reconfigurable Technol. Syst.*, 12:2:1–2:26, 2019.

[143] Christian Szegedy, Wei Liu, Yangqing Jia, Pierre Sermanet, Scott Reed, Dragomir Anguelov, Dumitru Erhan, Vincent Vanhoucke, and Andrew Rabinovich. Going deeper with convolutions. In *Proceedings of the IEEE conference on computer vision and pattern recognition*, pages 1–9, 2015.

[144] Jifeng Dai, Yi Li, Kaiming He, and Jian Sun. R-fcn: Object detection via region-based fully convolutional networks. In *Advances in neural information processing systems*, pages 379–387, 2016.

[145] Haoxiang Li, Zhe Lin, Xiaohui Shen, Jonathan Brandt, and Gang Hua. A convolutional neural network cascade for face detection. In *Proceedings of the IEEE Conference on Computer Vision and Pattern Recognition*, pages 5325–5334, 2015.

[146] Jure Zbontar and Yann LeCun. Stereo matching by training a convolutional neural network to compare image patches. *Journal of Machine Learning Research*, 17:1–32, 2016.

[147] Jiantao Qiu, Jie Wang, Song Yao, Kaiyuan Guo, Boxun Li, Erjin Zhou, Jincheng Yu, Tianqi Tang, Ningyi Xu, Sen Song, et al. Going deeper with embedded fpga platform for convolutional neural network. In *Proceedings of the 2016 ACM/SIGDA International Symposium on Field-Programmable Gate Arrays*, pages 26–35, 2016.

[148] Bernard Bosi, Guy Bois, and Yvon Savaria. Reconfigurable pipelined 2-d convolvers for fast digital signal processing. *VLSI*, 7(3):299–308, 1999.

[149] Srimat Chakradhar, Murugan Sankaradas, Venkata Jakkula, et al. A dynamically configurable coprocessor for conv olutional neural networks. In *ACM SIGARCH Computer Architecture News*, volume 38, pages 247–257. ACM, 2010.

[150] Vinayak Gokhale, Jonghoon Jin, Aysegul Dundar, et al. A 240 g-ops/s mobile coprocessor for deep neural networks. In *CVPRW*, pages 682–687, 2014.

[151] Chen Zhang, Peng Li, Guangyu Sun, Yijin Guan, Bingjun Xiao, and Jason Cong. Optimizing FPGA-based Accelerator Design for Deep Convolutional Neural Networks. In *Proceedings of the 2015 ACM/SIGDA International Symposium on Field-Programmable*

Gate Arrays, FPGA '15, pages 161–170, New York, NY, USA, February 2015. Association for Computing Machinery.

[152] Huimin Li, Xitian Fan, Li Jiao, Wei Cao, Xuegong Zhou, and Lingli Wang. A high performance fpga-based accelerator for large-scale convolutional neural networks. International Conference on Field Programmable Logic and Applications (FPL), 2016, pp. 1-9, doi: 10.1109/FPL.2016.7577308.

[153] Xiaowei Xu, Xinyi Zhang, Bei Yu, X Sharon Hu, Christopher Rowen, Jingtong Hu, and Yiyu Shi. Dac-sdc low power object detection challenge for uav applications. *IEEE transactions on pattern analysis and machine intelligence*, 2019.

[154] Song Han, Huizi Mao, and William J Dally. Deep compression: Compressing deep neural networks with pruning, trained quantization and huffman coding. In *International Conference on Learning Representations (ICLR)*, 2016.

[155] Wei Wen, Chunpeng Wu, Yandan Wang, Yiran Chen, and Hai Li. Learning structured sparsity in deep neural networks. *neural information processing systems*, pages 2074–2082, 2016.

[156] Hao Li, Asim Kadav, Igor Durdanovic, Hanan Samet, and Hans Peter Graf. Pruning filters for efficient convnets. *arXiv preprint arXiv:1608.08710*, 2016.

[157] Hengyuan Hu, Rui Peng, Yu-Wing Tai, and Chi-Keung Tang. Network trimming: A data-driven neuron pruning approach towards efficient deep architectures. *arXiv preprint arXiv:1607.03250*, 2016.

[158] Jian-Hao Luo, Jianxin Wu, and Weiyao Lin. Thinet: A filter level pruning method for deep neural network compression. In *Proceedings of the IEEE international conference on computer vision*, pages 5058–5066, 2017.

[159] Geoffrey Hinton, Oriol Vinyals, and Jeff Dean. Distilling the knowledge in a neural network. *arXiv preprint arXiv:1503.02531*, 2015.

[160] Shitao Tang, Litong Feng, Wenqi Shao, Zhanghui Kuang, Wei Zhang, and Yimin Chen. Learning efficient detector with semi-supervised adaptive distillation. *arXiv preprint arXiv:1901.00366*, 2019.

[161] Yihui He, Ji Lin, Zhijian Liu, Hanrui Wang, Li-Jia Li, and Song Han. Amc: Automl for model compression and acceleration on mobile devices. In *Proceedings of the European Conference on Computer Vision (ECCV)*, pages 784–800, 2018.

[162] Tien-Ju Yang, Andrew Howard, Bo Chen, Xiao Zhang, Alec Go, Mark Sandler, Vivienne Sze, and Hartwig Adam. Netadapt: Platform-aware neural network adaptation for mobile applications. In *Proceedings of the European Conference on Computer Vision (ECCV)*, pages 285–300, 2018.

[163] Zechun Liu, Haoyuan Mu, Xiangyu Zhang, Zichao Guo, Xin Yang, Kwang-Ting Cheng, and Jian Sun. Metapruning: Meta learning for automatic neural network channel pruning. In *Proceedings of the IEEE International Conference on Computer Vision*, pages 3296–3305, 2019.

[164] Ning Liu, Xiaolong Ma, Zhiyuan Xu, Yanzhi Wang, Jian Tang, and Jieping Ye. Autocompress: An automatic dnn structured pruning framework for ultra-high compression rates. In *Proceedings of the AAAI Conference on Artificial Intelligence*, 2020.

[165] Hieu Pham, Melody Y Guan, Barret Zoph, Quoc V Le, and Jeff Dean. Efficient neural architecture search via parameter sharing. In *International Conference on Machine Learning (ICML)*, 2018.

[166] Shulin Zeng, Hanbo Sun, Y. Xing, Xuefei Ning, Y. Shan, X. Chen, Yu Wang, and Hua zhong Yang. Black box search space profiling for accelerator-aware neural architecture search. *Asia and South Pacific Design Automation Conference (ASP-DAC)*, pages 518–523, 2020.

[167] Zichao Guo, Xiangyu Zhang, Haoyuan Mu, Wen Heng, Zechun Liu, Yichen Wei, and Jian Sun. Single path one-shot neural architecture search with uniform sampling. *arXiv preprint arXiv:1904.00420*, 2019.

[168] Chenxi Liu, Barret Zoph, Maxim Neumann, Jonathon Shlens, Wei Hua, Li-Jia Li, Li Fei-Fei, Alan Yuille, Jonathan Huang, and Kevin Murphy. Progressive neural architecture search. In *Proceedings of the European Conference on Computer Vision (ECCV)*, pages 19–34, 2018.

[169] Hanxiao Liu, Karen Simonyan, and Yiming Yang. Darts: Differentiable architecture search. In *ICLR*, 2019.

[170] Mingxing Tan and Quoc V Le. Efficientnet: Rethinking model scaling for convolutional neural networks. *arXiv preprint arXiv: 1905.11946*, 2019.

[171] Barret Zoph and Quoc V Le. Neural architecture search with reinforcement learning. *arXiv preprint arXiv:1611.01578*, 2016.

[172] Barret Zoph, Vijay Vasudevan, Jonathon Shlens, and Quoc V Le. Learning transferable architectures for scalable image recognition. In *CVPR*, 2018.

[173] Yukang Chen, Tong Yang, Xiangyu Zhang, Gaofeng Meng, Xinyu Xiao, and Jian Sun. Detnas: Backbone search for object detection. In *NeurIPS*, pages 6638–6648, 2019.

[174] Mingxing Tan, Ruoming Pang, and Quoc V Le. Efficientdet: Scalable and efficient object detection. *arXiv preprint arXiv:1911.09070*, 2019.

[175] Wuyang Chen, Xinyu Gong, Xianming Liu, Qian Zhang, Yuan Li, and Zhangyang Wang. Fasterseg: Searching for faster real-time semantic segmentation. In *ICLR*, 2020.

[176] Chenxi Liu, Liang-Chieh Chen, Florian Schroff, Hartwig Adam, Wei Hua, Alan L Yuille, and Li Fei-Fei. Auto-deeplab: Hierarchical neural architecture search for semantic image segmentation. In *CVPR*, 2019.

[177] Vladimir Nekrasov, Hao Chen, Chunhua Shen, and Ian Reid. Fast neural architecture search of compact semantic segmentation models via auxiliary cells. In *CVPR*, 2019.

[178] Ronald J Williams. Simple statistical gradient-following algorithms for connectionist reinforcement learning. *Machine learning*, 8(3-4):229–256, 1992.

[179] Esteban Real, Alok Aggarwal, Yanping Huang, and Quoc V Le. Regularized evolution for image classifier architecture search. In *Proceedings of the aaai conference on artificial intelligence*, volume 33, pages 4780–4789, 2019.

[180] Andrew Brock, Theodore Lim, James Millar Ritchie, and Nicholas J Weston. Smash: One-shot model architecture search through hypernetworks. In *ICLR*, 2018.

[181] Sirui Xie, Hehui Zheng, Chunxiao Liu, and Liang Lin. Snas: stochastic neural architecture search. *arXiv preprint arXiv:1812.09926*, 2018.

[182] Xinbang Zhang, Zehao Huang, and Naiyan Wang. You only search once: Single shot neural architecture search via direct sparse optimization. *arXiv preprint arXiv:1811.01567*, 2018.

[183] Han Cai, Ligeng Zhu, and Song Han. ProxylessNAS: Direct neural architecture search on target task and hardware. In *ICLR*, 2019.

[184] Bichen Wu, Xiaoliang Dai, Peizhao Zhang, Yanghan Wang, Fei Sun, Yiming Wu, Yuandong Tian, Peter Vajda, Yangqing Jia, and Kurt Keutzer. Fbnet: Hardware-aware efficient convnet design via differentiable neural architecture search. In *CVPR*, 2019.

[185] Gabriel Bender, Pieter-Jan Kindermans, Barret Zoph, Vijay Vasudevan, and Quoc V. Le. Understanding and simplifying one-shot architecture search. In *ICML*, 2018.

[186] Xiangxiang Chu, Bo Zhang, Ruijun Xu, and Jixiang Li. Fairnas: Rethinking evaluation fairness of weight sharing neural architecture search. *arXiv preprint arXiv:1907.01845*, 2019.

[187] George Adam and Jonathan Lorraine. Understanding neural architecture search techniques. *arXiv preprint arXiv:1904.00438*, 2019.

[188] Mark Fleischer. The measure of pareto optima applications to multi-objective metaheuristics. In *International Conference on Evolutionary Multi-Criterion Optimization*, pages 519–533. Springer, 2003.

[189] Martín Abadi, Ashish Agarwal, Paul Barham, Eugene Brevdo, Zhifeng Chen, Craig Citro, Greg S Corrado, Andy Davis, Jeffrey Dean, Matthieu Devin, et al. Tensorflow: Large-scale machine learning on heterogeneous distributed systems. *arXiv preprint arXiv:1603.04467*, 2016.

[190] Mark Sandler, Andrew Howard, Menglong Zhu, Andrey Zhmoginov, and Liang-Chieh Chen. Mobilenetv2: Inverted residuals and linear bottlenecks. In *CVPR*, 2018.

[191] Shan You, Tao Huang, Mingmin Yang, Fei Wang, Chen Qian, and Changshui Zhang. Greedynas: Towards fast one-shot nas with greedy supernet. In *CVPR*, 2020.

[192] Yiming Hu, Yuding Liang, Zichao Guo, Ruosi Wan, Xiangyu Zhang, Yichen Wei, Qingyi Gu, and Jian Sun. Angle-based search space shrinking for neural architecture search. In *ECCV*, 2020.

[193] Ilija Radosavovic, Raj Prateek Kosaraju, Ross Girshick, Kaiming He, and Piotr Dollár. Designing network design spaces. In *CVPR*, 2020.

[194] Xin Chen, Lingxi Xie, Jun Wu, and Qi Tian. Progressive differentiable architecture search: Bridging the depth gap between search and evaluation. In *ICCV*, 2019.

[195] Shen Yan, Biyi Fang, Faen Zhang, Yu Zheng, Xiao Zeng, Mi Zhang, and Hui Xu. Hm-nas: Efficient neural architecture search via hierarchical masking. In *Proceedings of the IEEE International Conference on Computer Vision Workshops*, 2019.

[196] Song Han, Huizi Mao, and William J Dally. Deep compression: Compressing deep neural networks with pruning, trained quantization and huffman coding. In *ICLR*, 2016.

[197] Yihui He, Xiangyu Zhang, and Jian Sun. Channel pruning for accelerating very deep neural networks. In *ICCV*, 2017.

[198] Zhuang Liu, Jianguo Li, Zhiqiang Shen, Gao Huang, Shoumeng Yan, and Changshui Zhang. Learning efficient convolutional networks through network slimming. In *ICCV*, 2017.

[199] Kalyanmoy Deb. A fast elitist non-dominated sorting genetic algorithm for multi-objective optimization: Nsga-2. *IEEE Trans. Evol. Comput.*, 6(2):182–197, 2002.

[200] Zhichao Lu, Ian Whalen, Vishnu Boddeti, Yashesh Dhebar, Kalyanmoy Deb, Erik Goodman, and Wolfgang Banzhaf. Nsga-net: neural

architecture search using multi-objective genetic algorithm. In *Proceedings of the Genetic and Evolutionary Computation Conference*, pages 419–427. ACM, 2019.

[201] Jia Deng, Wei Dong, Richard Socher, Li Jia Li, and Fei Fei Li. Imagenet: a large-scale hierarchical image database. In *CVPR*, 2009.

[202] Jie Hu, Li Shen, and Gang Sun. Squeeze-and-excitation networks. In *CVPR*, 2018.

[203] Prajit Ramachandran, Barret Zoph, and Quoc V Le. Searching for activation functions. *arXiv preprint arXiv:1710.05941*, 2017.

[204] Hongyi Zhang, Moustapha Cisse, Yann N Dauphin, and David Lopez-Paz. mixup: Beyond empirical risk minimization. In *ICLR*, 2018.

[205] Ekin D. Cubuk, Barret Zoph, Dandelion Mane, Vijay Vasudevan, and Quoc V. Le. Autoaugment: Learning augmentation strategies from data. In *CVPR*, 2019.

[206] Andrew G Howard, Menglong Zhu, Bo Chen, Dmitry Kalenichenko, Weijun Wang, Tobias Weyand, Marco Andreetto, and Hartwig Adam. Mobilenets: Efficient convolutional neural networks for mobile vision applications. *arXiv preprint arXiv:1704.04861*, 2017.

[207] Ningning Ma, Xiangyu Zhang, Hai-Tao Zheng, and Jian Sun. Shuflenet v2: Practical guidelines for efficient cnn architecture design. In *ECCV*, pages 116–131, 2018.

[208] Xin Chen, Lingxi Xie, Jun Wu, and Qi Tian. Progressive darts: Bridging the optimization gap for nas in the wild. *arXiv preprint arXiv:1912.10952*, 2019.

[209] Mingxing Tan, Bo Chen, Ruoming Pang, Vijay Vasudevan, Mark Sandler, Andrew Howard, and Quoc V Le. Mnasnet: Platform-aware neural architecture search for mobile. In *CVPR*, 2019.

[210] Jiemin Fang, Yuzhu Sun, Qian Zhang, Yuan Li, Wenyu Liu, and Xinggang Wang. Densely connected search space for more flexible neural architecture search. In *CVPR*, 2020.

[211] Dimitrios Stamoulis, Ruizhou Ding, Di Wang, Dimitrios Lymberopoulos, Bodhi Priyantha, Jie Liu, and Diana Marculescu. Single-path nas: Designing hardware-efficient convnets in less than 4 hours. *arXiv preprint arXiv:1904.02877*, 2019.

[212] Andrew Howard, Mark Sandler, Grace Chu, Liang-Chieh Chen, Bo Chen, Mingxing Tan, Weijun Wang, Yukun Zhu, Ruoming Pang, Vijay Vasudevan, et al. Searching for mobilenetv3. In *ICCV*, 2019.

[213] Maurice G Kendall. A new measure of rank correlation. *Biometrika*, 30(1/2):81–93, 1938.

[214] A. G. Howard, M. Zhu, B. Chen, D. Kalenichenko, W. Wang, T. Weyand, M. Andreetto, and H. Adam. Mobilenets: Efficient convolutional neural networks for mobile vision applications. *arXiv preprint arXiv:1704.04861*, 2017.

[215] M. Sandler, A.G. Howard, M. Zhu, A. Zhmoginov, and L.C. Chen. Inverted residuals and linear bottlenecks: Mobile networks for classification, detection and segmentation. In *IEEE Conference on Computer Vision and Pattern Recognition (CVPR)*, 2018.

[216] M. Tan, B. Chen, R. Pang, V. Vasudevan, and Q. V. Le. Mnasnet: Platform-aware neural architecture search for mobile. *arXiv preprint arXiv:1807.11626*, 2018.

[217] J. H. Lee, S. Ha, S. Choi, W. Lee, and S. Lee. Quantization for rapid deployment of deep neural networks. *arXiv preprint arXiv:1810.05488*, 2018.

[218] B. Jacob, S. Kligys, B. Chen, M. Zhu, M. Tang, A. Howard, H. Adam, and D. Kalenichenko. Quantization and training of neural networks for efficient integer-arithmetic only inference. In *Conference on Computer Vision and Pattern Recognition (CVPR)*, 2018.

[219] G. Hinton, O. Vinyals, and J. Dean. Distilling the knowledge in a neural network. *arXiv preprint arXiv:1503.02531*, 2015.

[220] A. Mishra and D. Marr. Apprentice: Using knowledge distillation techniques to improve low-precision network accuracy. *arXiv preprint arXiv:1711.05852*, 2017.

[221] A. Mishra, E. Nurvitadhi, J. J. Cook, and D. Marr. Wrpn: Wide reduced-precision networks. *arXiv preprint arXiv:1709.01134*, 2017.

[222] NVIDIA TensorRT SDK. https://developer.nvidia.com/tensorrt, 2018.

[223] M. Abadi, A. Agarwal, P. Barham, E. Brevdo, Z. Chen, C. Citro, G. S. Corrado, A. Davis, J. Dean, M. Devin, S. Ghemawat, I. Goodfellow, A. Harp, G. Irving, M. Isard, Y. Jia, R. Jozefowicz, L. Kaiser, M. Kudlur, J. Levenberg, D. Mane, R. Monga, S. Moore, D. Murray, C. Olah, M. Schuster, J. Shlens, B. Steiner, I. Sutskever, K. Talwar, P. Tucker, V. Vanhoucke, V. Vasudevan, F. Viegas, O. Vinyals, P. Warden, M. Wattenberg, M. Wicke, Y. Yu, and X. Zheng. Tensorflow: Large-scale machine learning on heterogeneous distributed systems. *arXiv preprint arXiv:1603.04467*, 2016.

[224] Neta Zmora, Guy Jacob, Lev Zlotnik, Bar Elharar, and Gal Novik. Neural network distiller: A python package for dnn compression research. *arXiv preprint arXiv:1910.12232*, 2019.

[225] M. Courbariaux, Y. Bengio, and J. David. Training deep neural networks with low precision multiplications. In *International Conference on Learning Representations (ICLR 2015)*, 2015.

[226] I. Hubara, M. Courbariaux, D. Soudry, R. El-Yaniv, and Y. Bengio. Binarized neural networks. In *Advances in Neural Information Processing Systems (NIPS)*, pages 4107–4115, 2016.

[227] M. Rastegari, V. Ordonez, J. Redmon, and A. Farhadi. Xnor-net: Imagenet classification using binary convolutional neural networks. In *European Conference on Computer Vision (ECCV)*, pages 525–542. Springer, 2016.

[228] Sh. Zhou, Y. Wu, Z. Ni, X. Zhou, H. Wen, and Y. Zou. Dorefanet: Training low bitwidth convolutional neural networks with low bitwidth gradients. *arXiv preprint arXiv:1606.06160*, 2016.

[229] Y. Bengio, N. Leonard, and A. C. Courville. Estimating or propagating gradients through stochastic neurons for conditional computation. *arXiv preprint arXiv:1308.3432*, 2013.

[230] M. D. McDonnell. Training wide residual networks for deployment using a single bit for each weight. In *International Conference on Learning Representations (ICLR)*, 2018.

[231] Sh. Zhu, X. Dong, and H. Su. Binary ensemble neural network: more bits per network or more networks per bit? *arXiv preprint arXiv:1806.07550*, 2018.

[232] Ch. Baskin, N. Liss, Y. Chai, E. Zheltonozhskii, E. Schwartz, R. Giryes, A. Mendelson, and A. M. Bronstein. Nice: Noise injection and clamping estimation for neural network quantization. *arXiv preprint arXiv:1810.00162*, 2018.

[233] O. Russakovsky, J. Deng, H. Su, J. Krause, S. Satheesh, S. Ma, Z. Huang, A. Karpathy, A. Khosla, M. Bernstein, A. C. Berg, and L. Fei-Fei. Imagenet large scale visual recognition challenge. *arXiv preprint arXiv:1409.0575*, 2014.

[234] S. Ioffe and C. Szegedy. Batch normalization: Accelerating deep network training by reducing internal covariate shift. In *International Conference on Machine Learning (ICML 2015)*, 2015.

[235] D. P. Kingma and J. L. Ba. Adam: A method for stochastic optimization. In *International Conference on Learning Representations (ICLR)*, 2015.

[236] T. Sheng, C. Feng, S. Zhuo, X. Zhang, L. Shen, and M. Aleksic. A quantization-friendly separable convolution for mobilenets. *arXiv preprint arXiv:1803.08607*, 2018.

[237] https://www.tensorflow.org/lite/guide/hosted_models

[238] https://github.com/agoncharenko1992/FAT-fast-adjustable-threshold

[239] Suyog Gupta and Berkin Akin. Accelerator-aware neural network design using automl. 2020.

[240] Weiwen Jiang, Xinyi Zhang, Edwin Hsing-Mean Sha, Lei Yang, Qingfeng Zhuge, Yiyu Shi, and Jingtong Hu. Accuracy vs. efficiency: Achieving both through fpga-implementation aware neural architecture search. In *Proceedings of the 56th Annual Design Automation Conference,DAC, Las Vegas, NV, USA*, page 5. ACM, 2019.

[241] Weiwei Chen, Ying Wang, Shuang Yang, Chen Liu, and Lei Zhang. Towards best-effort approximation: applying nas to general-purpose

approximate computing. In *Design, Automation & Test in Europe Conference & Exhibition (DATE)*, pages 1315–1318. IEEE, 2020.

[242] Xiandong Zhao, Ying Wang, Xuyi Cai, Cheng Liu, and Lei Zhang. Linear symmetric quantization of neural networks for low-precision integer hardware. In *International Conference on Learning Representations*, 2019.

[243] Ying Wang, Huawei Li, Long Cheng, and Xiaowei Li. A qos-qor aware cnn accelerator design approach. *IEEE Transactions on Computer-Aided Design of Integrated Circuits and Systems*, 38(11):1995–2007, 2018.

[244] Tianshi Chen, Zidong Du, Ninghui Sun, Jia Wang, Chengyong Wu, Yunji Chen, and Olivier Temam. Diannao: a small-footprint high-throughput accelerator for ubiquitous machine-learning. *architectural support for programming languages and operating systems*, 49(4):269–284, 2014.

[245] Y. Chen, J. Emer, and V. Sze. Eyeriss: A spatial architecture for energy-efficient dataflow for convolutional neural networks. *IEEE Micro*, pages 1–1, 2018.

[246] Ying Wang, Jie Xu, Yinhe Han, Huawei Li, and Xiaowei Li. Deepburning: automatic generation of fpga-based learning accelerators for the neural network family. In *Proceedings of the 53rd Annual Design Automation Conference*, page 110. ACM, 2016.

[247] Lili Song, Ying Wang, Yinhe Han, Xin Zhao, Bosheng Liu, and Xiaowei Li. C-brain: A deep learning accelerator that tames the diversity of cnns through adaptive data-level parallelization. In *Proceedings of the 53rd Annual Design Automation Conference*, pages 1–6, 2016.

[248] Weiwei Chen, Ying Wang, Shuang Yang, Chen Liu, and Lei Zhang. You only search once: a fast automation framework for single-stage dnn/accelerator co-design. In *Design, Automation & Test in Europe Conference & Exhibition (DATE)*, pages 1283–1286. IEEE, 2020.

[249] Ting Hu, Ying Wang, Lei Zhang, and Tingting He. A tightly-coupled light-weight neural network processing units with risc-v core. 2017.

[250] Erik Debenedictis, Yung-Hsiang Lu, Alan Kadin, Alexander Berg, Thomas Conte, Rachit Garg, Ganesh Gingade, Bichlien Hoang, Yongzhen Huang, Boxun Li, et al. Rebooting computing and low-power image recognition challenge. Technical report, Sandia National Lab.(SNL-NM), Albuquerque, NM (United States), 2016.

[251] Srimat T Chakradhar and Anand Raghunathan. Best-effort computing: Re-thinking parallel software and hardware. In *Design Automation Conference*, pages 865–870. IEEE, 2010.

[252] Cheng Wang, Ying Wang, Yinhe Han, Lili Song, Zhenyu Quan, Jiajun Li, and Xiaowei Li. Cnn-based object detection solutions for embedded heterogeneous multicore socs. In *Asia and South Pacific Design Automation Conference (ASP-DAC)*, pages 105–110. IEEE, 2017.

[253] Jasper RR Uijlings, Koen EA Van De Sande, Theo Gevers, and Arnold WM Smeulders. Selective search for object recognition. *International journal of computer vision*, 104(2):154–171, 2013.

[254] Ming-Ming Cheng, Ziming Zhang, Wen-Yan Lin, and Philip Torr. Bing: Binarized normed gradients for objectness estimation at 300fps. In *Proceedings of the IEEE conference on computer vision and pattern recognition*, pages 3286–3293, 2014.

[255] Ross Girshick, Jeff Donahue, Trevor Darrell, and Jitendra Malik. Rich feature hierarchies for accurate object detection and semantic segmentation. In *Proceedings of the IEEE conference on computer vision and pattern recognition*, pages 580–587, 2014.

[256] Pierre Sermanet, David Eigen, Xiang Zhang, Michaël Mathieu, Rob Fergus, and Yann LeCun. Overfeat: Integrated recognition, localization and detection using convolutional networks. *arXiv preprint arXiv:1312.6229*, 2013.

[257] Kaiming He, Xiangyu Zhang, Shaoqing Ren, and Jian Sun. Spatial pyramid pooling in deep convolutional networks for visual recognition. *IEEE transactions on pattern analysis and machine intelligence*, 37(9):1904–1916, 2015.

[258] Ross Girshick. Fast r-cnn. In *Proceedings of the IEEE International Conference on Computer Vision*, pages 1440–1448, 2015.

[259] Shaoqing Ren, Kaiming He, Ross Girshick, and Jian Sun. Faster r-cnn: Towards real-time object detection with region proposal networks. *arXiv preprint arXiv:1506.01497*, 2015.

[260] Ken Chatfield, Karen Simonyan, Andrea Vedaldi, and Andrew Zisserman. Return of the devil in the details: Delving deep into convolutional nets. *arXiv preprint arXiv:1405.3531*, 2014.

[261] Song Han, Huizi Mao, and William J Dally. Deep compression: Compressing deep neural networks with pruning, trained quantization and huffman coding. *international conference on learning representations*, 2016.

[262] Hongpeng Zhou, Minghao Yang, Jun Wang, and Wei Pan. Bayesnas: A bayesian approach for neural architecture search. In *International Conference on Machine Learning*, pages 7603–7613. PMLR, 2019.

[263] Stephen Boyd, Neal Parikh, and Eric Chu. *Distributed optimization and statistical learning via the alternating direction method of multipliers.* Now Publishers Inc, 2011.

[264] Ning Liu, Xiaolong Ma, Zhiyuan Xu, Yanzhi Wang, Jian Tang, and Jieping Ye. Autocompress: An automatic dnn structured pruning framework for ultra-high compression rates. In *Proceedings of the AAAI Conference on Artificial Intelligence*, volume 34, pages 4876–4883, 2020.

[265] Yuqiao Liu, Yanan Sun, Bing Xue, Mengjie Zhang, and Gary Yen. A survey on evolutionary neural architecture search. *arXiv preprint arXiv:2008.10937*, 2020.

[266] Songyun Qu, Bing Li, Ying Wang, Dawen Xu, Xiandong Zhao, and Lei Zhang. Raqu: An automatic high-utilization cnn quantization and mapping framework for general-purpose rram accelerator. In *ACM/IEEE Design Automation Conference (DAC)*, pages 1–6. IEEE, 2020.

[267] Ying Wang, Huawei Li, and Xiaowei Li. Real-time meets approximate computing: An elastic cnn inference accelerator with adaptive trade-off between qos and qor. In *ACM/EDAC/IEEE Design Automation Conference (DAC)*, pages 1–6. IEEE, 2017.

[268] Martín Abadi, Ashish Agarwal, Paul Barham, Eugene Brevdo, Zhifeng Chen, Craig Citro, Greg S. Corrado, Andy Davis, Jeffrey Dean, Matthieu Devin, Sanjay Ghemawat, Ian Goodfellow, Andrew Harp, Geoffrey Irving, Michael Isard, Yangqing Jia, Rafal Jozefowicz, Lukasz Kaiser, Manjunath Kudlur, Josh Levenberg, Dan Mané, Rajat Monga, Sherry Moore, Derek Murray, Chris Olah, Mike Schuster, Jonathon Shlens, Benoit Steiner, Ilya Sutskever, Kunal Talwar, Paul Tucker, Vincent Vanhoucke, Vijay Vasudevan, Fernanda Viégas, Oriol Vinyals, Pete Warden, Martin Wattenberg, Martin Wicke, Yuan Yu, and Xiaoqiang Zheng. TensorFlow: Large-scale machine learning on heterogeneous systems, 2015. Software available from tensorflow.org.

[269] Adam Paszke, Sam Gross, Francisco Massa, Adam Lerer, James Bradbury, Gregory Chanan, Trevor Killeen, Zeming Lin, Natalia Gimelshein, Luca Antiga, et al. Pytorch: An imperative style, high-performance deep learning library. *arXiv preprint arXiv:1912.01703*, 2019.

[270] Tianqi Chen, Mu Li, Yutian Li, Min Lin, Naiyan Wang, Minjie Wang, Tianjun Xiao, Bing Xu, Chiyuan Zhang, and Zheng Zhang. Mxnet: A flexible and efficient machine learning library for heterogeneous distributed systems. *arXiv preprint arXiv:1512.01274*, 2015.

[271] Frank Seide and Amit Agarwal. Cntk: Microsoft's open-source deep-learning toolkit. In *Proceedings of the 22nd ACM SIGKDD International Conference on Knowledge Discovery and Data Mining*, pages 2135–2135, 2016.

[272] Tianqi Chen, Thierry Moreau, Ziheng Jiang, Lianmin Zheng, Eddie Q. Yan, Haichen Shen, Meghan Cowan, Leyuan Wang, Yuwei Hu, Luis Ceze, Carlos Guestrin, and Arvind Krishnamurthy. TVM: an automated end-to-end optimizing compiler for deep learning. In *13th USENIX Symposium on Operating Systems Design and Implementation, OSDI, Carlsbad, CA, USA*, pages 578–594. USENIX Association, 2018.

[273] Nicolas Vasilache, Oleksandr Zinenko, Theodoros Theodoridis, Priya Goyal, Zachary DeVito, William S. Moses, Sven Verdoolaege, Andrew Adams, and Albert Cohen. Tensor comprehensions:

Framework-agnostic high-performance machine learning abstractions. *In arXiv:1802.04730*, 2018.

[274] Nadav Rotem, Jordan Fix, Saleem Abdulrasool, Garret Catron, Summer Deng, Roman Dzhabarov, Nick Gibson, James Hegeman, Meghan Lele, Roman Levenstein, et al. Glow: Graph lowering compiler techniques for neural networks. *arXiv preprint arXiv:1805.00907*, 2018.

[275] Scott Cyphers, Arjun K. Bansal, Anahita Bhiwandiwalla, Jayaram Bobba, Matthew Brookhart, Avijit Chakraborty, Will Constable, Christian Convey, Leona Cook, Omar Kanawi, Robert Kimball, Jason Knight, Nikolay Korovaiko, Varun Kumar, Yixing Lao, Christopher R. Lishka, Jaikrishnan Menon, Jennifer Myers, Sandeep Aswath Narayana, Adam Procter and Tristan J. Webb. Intel ngraph: An intermediate representation, compiler, and executor for deep learning. *arXiv preprint arXiv:1801.08058*, 2018.

[276] Chris Leary and Todd Wang. Xla: Tensorflow, compiled. *TensorFlow Dev Summit*, 2017.

[277] Yu-Hsin Chen, Tien-Ju Yang, Joel S. Emer, and Vivienne Sze. Eyeriss v2: A flexible accelerator for emerging deep neural networks on mobile devices. *IEEE J. Emerg. Sel. Topics Circuits Syst.*, pages 292–308, 2019.

[278] Manoj Alwani, Han Chen, Michael Ferdman, and Peter Milder. Fused-layer cnn accelerators. In *Microarchitecture (MICRO), Annual IEEE/ACM International Symposium on*, pages 1–12. IEEE, 2016.

[279] Qingcheng Xiao, Yun Liang, Liqiang Lu, Shengen Yan, and Yu-Wing Tai. Exploring heterogeneous algorithms for accelerating deep convolutional neural networks on fpgas. In *Proceedings of the 54th Annual Design Automation Conference 2017*, pages 1–6, 2017.

[280] Li Zhou, Hao Wen, Radu Teodorescu, and David HC Du. Distributing deep neural networks with containerized partitions at the edge. In *2nd {USENIX} Workshop on Hot Topics in Edge Computing (HotEdge 19)*, 2019.

[281] Yu Xing, Shuang Liang, Lingzhi Sui, Zhen Zhang, Jiantao Qiu, Xijie Jia, Xin Liu, Yushun Wang, Yi Shan, and Yu Wang. Dnnvm:

End-to-end compiler leveraging operation fusion on fpga-based cnn accelerators. In *Proceedings of the ACM/SIGDA International Symposium on Field-Programmable Gate Arrays*, pages 187–188, 2019.

[282] Abhinav Jangda and Uday Bondhugula. An effective fusion and tile size model for optimizing image processing pipelines. *ACM SIGPLAN Notices*, 53(1):261–275, 2018.

[283] Shixuan Zheng, Xianjue Zhang, Daoli Ou, Shibin Tang, Leibo Liu, Shaojun Wei, and Shouyi Yin. Efficient scheduling of irregular network structures on cnn accelerators. *IEEE Transactions on Computer-Aided Design of Integrated Circuits and Systems*, 39(11):3408–3419, 2020.

[284] Yu-Hsin Chen, Tushar Krishna, Joel Emer, and Vivienne Sze. Eyeriss: An energy-efficient reconfigurable accelerator for deep convolutional neural networks. In *ISSCC*. IEEE, 2016.

[285] Xuan Yang, Mingyu Gao, Qiaoyi Liu, Jeff Setter, Jing Pu, Ankita Nayak, Steven Bell, Kaidi Cao, Heonjae Ha, Priyanka Raina, Christos Kozyrakis and Mark Horowitz. Interstellar: Using Halide's Scheduling Language to Analyze DNN Accelerators. In *Proceedings of the Twenty-Fifth International Conference on Architectural Support for Programming Languages and Operating Systems*, ASPLOS '20, pages 369–383, New York, NY, USA, March 2020. Association for Computing Machinery.

[286] Alex Krizhevsky, Ilya Sutskever, and Geoffrey E Hinton. Imagenet classification with deep convolutional neural networks. In *NIPS*, pages 1097–1105, 2012.

[287] Karen Simonyan and Andrew Zisserman. Very deep convolutional networks for large-scale image recognition. *arXiv preprint arXiv:1409.1556*, 2014.

[288] Forrest N Iandola, Matthew W Moskewicz, Khalid Ashraf, et al. Squeezenet: Alexnet-level accuracy with 50x fewer parameters and < 1mb model size. *arXiv preprint arXiv:1602.07360*, 2016.

[289] Xiaoming Chen, Yinhe Han, and Yu Wang. Communication lower bound in convolution accelerators. In *IEEE International*

Symposium on High Performance Computer Architecture (HPCA), pages 529–541. IEEE, 2020.

[290] Orest Kupyn, Volodymyr Budzan, Mykola Mykhailych, Dmytro Mishkin, and Jiri Matas. Deblurgan: Blind motion deblurring using conditional adversarial networks. *ArXiv e-prints*, 2017.

[291] Casey Chu, Andrey Zhmoginov, and Mark Sandler. Cyclegan, a master of steganography. *arXiv preprint arXiv:1712.02950*, 2017.

[292] Jun-Yan Zhu, Taesung Park, Phillip Isola, and Alexei A Efros. Unpaired image-to-image translation using cycle-consistent adversarial networks. In *Proceedings of the IEEE international conference on computer vision*, pages 2223–2232, 2017.

[293] Haobo Xu, Ying Wang, Yujie Wang, Jiajun Li, Bosheng Liu, and Yinhe Han. Acg-engine: An inference accelerator for content generative neural networks. In *2019 IEEE/ACM International Conference on Computer-Aided Design (ICCAD)*, pages 1–7. IEEE, 2019.

[294] Norman P Jouppi, C S Young, Nishant Patil, David A Patterson, Gaurav Agrawal, Raminder Bajwa, Sarah Bates, Suresh K Bhatia, Nan Boden, Al Borchers, et al. In-datacenter performance analysis of a tensor processing unit. *international symposium on computer architecture*, 45(2):1–12, 2017.

[295] Jinmook Lee, Changhyeon Kim, Sang Hoon Kang, Dongjoo Shin, Sangyeob Kim, and Hoijun Yoo. Unpu: A 50.6tops/w unified deep neural network accelerator with 1b-to-16b fully-variable weight bit-precision. pages 218–220, 2018.

[296] Patrick Judd, Jorge Albericio, Tayler H Hetherington, Tor M Aamodt, and Andreas Moshovos. Stripes: bit-serial deep neural network computing. *international symposium on microarchitecture*, pages 1–12, 2016.

[297] Jorge Albericio, Alberto Delmás, Patrick Judd, Sayeh Sharify, Gerard O'Leary, Roman Genov, and Andreas Moshovos. Bit-pragmatic deep neural network computing. In *Proceedings of the 50th Annual IEEE/ACM International Symposium on Microarchitecture*, pages 382–394, 2017.

[298] Eunhyeok Park, Junwhan Ahn, and Sungjoo Yoo. Weighted-entropy-based quantization for deep neural networks. pages 7197–7205, 2017.

[299] Dongqing Zhang, Jiaolong Yang, Dongqiangzi Ye, and Gang Hua. Lq-nets: Learned quantization for highly accurate and compact deep neural networks. *european conference on computer vision*, pages 373–390, 2018.

[300] Shuchang Zhou, Zekun Ni, Xinyu Zhou, He Wen, Yuxin Wu, and Yuheng Zou. Dorefa-net: Training low bitwidth convolutional neural networks with low bitwidth gradients. *arXiv: Neural and Evolutionary Computing*, 2016.

[301] Christos Louizos, Matthias Reisser, Tijmen Blankevoort, Efstratios Gavves, and Max Welling. Relaxed quantization for discretized neural networks. *international conference on learning representations*, 2019.

[302] Asit K Mishra, Eriko Nurvitadhi, Jeffrey J Cook, and Debbie Marr. Wrpn: Wide reduced-precision networks. *international conference on learning representations*, 2018.

[303] Jungwook Choi. Pact: Parameterized clipping activation for quantized neural networks. *arXiv: Computer Vision and Pattern Recognition*, 2018.

[304] Chenzhuo Zhu, Song Han, Huizi Mao, and William J Dally. Trained ternary quantization. *international conference on learning representations*, 2017.

[305] Raghuraman Krishnamoorthi. Quantizing deep convolutional networks for efficient inference: A whitepaper. *arXiv preprint arXiv:1806.08342*, 2018.

[306] Joseph Redmon, Santosh Divvala, Ross Girshick, and Ali Farhadi. You only look once: Unified, real-time object detection. In *Proceedings of the IEEE conference on computer vision and pattern recognition*, pages 779–788, 2016.

[307] Zidong Du, Robert Fasthuber, Tianshi Chen, Paolo Ienne, Ling Fei Li, Tao Luo, Xiaobing Feng, Yunji Chen, and Olivier Temam. Shidiannao: shifting vision processing closer to the sensor. *international symposium on computer architecture*, 43(3):92–104, 2015.

[308] Y. Wang, J. Xu, Y. Han, H. Li, and X. Li. Deepburning: Automatic generation of fpga-based learning accelerators for the neural network family. In *ACM/EDAC/IEEE Design Automation Conference (DAC)*, pages 1–6, June 2016.

[309] S. Han, X. Liu, H. Mao, J. Pu, A. Pedram, M. A. Horowitz, and W. J. Dally. Eie: Efficient inference engine on compressed deep neural network. In *ACM/IEEE 43rd Annual International Symposium on Computer Architecture (ISCA)*, pages 243–254, June 2016.

[310] Aojun Zhou, Anbang Yao, Yiwen Guo, Lin Xu, and Yurong Chen. Incremental network quantization: Towards lossless cnns with low-precision weights. *CoRR*, abs/1702.03044, 2017.

[311] Zhaowei Cai, Xiaodong He, Jian Sun, and Nuno Vasconcelos. Deep learning with low precision by half-wave gaussian quantization. In *CVPR*, 2017.

[312] Yoshua Bengio, Nicholas Léonard, and Aaron Courville. Estimating or propagating gradients through stochastic neurons for conditional computation. *arXiv preprint arXiv:1308.3432*, 2013.

[313] Adam Paszke, Sam Gross, Soumith Chintala, Gregory Chanan, Edward Yang, Zachary DeVito, Zeming Lin, Alban Desmaison, Luca Antiga, and Adam Lerer. Automatic differentiation in PyTorch. In *NIPS Autodiff Workshop*, 2017.

[314] Mark Sandler, Andrew Howard, Menglong Zhu, Andrey Zhmoginov, and Liang-Chieh Chen. Mobilenetv2: Inverted residuals and linear bottlenecks, 2018.

[315] Joseph Redmon and Ali Farhadi. Yolo9000: Better, faster, stronger. pages 6517–6525, 2017.

[316] Xiandong Zhao, Ying Wang, Xuyi Cai, Cheng Liu, and Lei Zhang. Linear symmetric quantization of neural networks for low-precision integer hardware. 2020.

[317] Ying Wang, Zhenyu Quan, Jiajun Li, Yinhe Han, Huawei Li, and Xiaowei Li. A retrospective evaluation of energy-efficient object detection solutions on embedded devices. In *Design, Automation & Test in Europe Conference & Exhibition (DATE)*, pages 709–714. IEEE, 2018.

[318] Xiandong Zhao, Ying Wang, Cheng Liu, Cong Shi, Kaijie Tu, and Lei Zhang. Bitpruner: network pruning for bit-serial accelerators. In *ACM/IEEE Design Automation Conference (DAC)*, pages 1–6. IEEE, 2020.

[319] Dawen Xu, Cheng Liu, Ying Wang, Kaijie Tu, Bingsheng He, and Lei Zhang. Accelerating generative neural networks on unmodified deep learning processors—a software approach. *IEEE Transactions on Computers*, 69(8):1172–1184, 2020.

[320] Xuyi Cai, Ying Wang, and Lei Zhang. Optimus: towards optimal layer-fusion on deep learning processors. In *Proceedings of the 22nd ACM SIGPLAN/SIGBED International Conference on Languages, Compilers, and Tools for Embedded Systems*, pages 67–79, 2021.

[321] Weiwei Chen, Ying Wang, Gangliang Lin, Chengsi Gao, Cheng Liu, and Lei Zhang. Chanas: coordinated search for network architecture and scheduling policy. In *Proceedings of the 22nd ACM SIGPLAN/SIGBED International Conference on Languages, Compilers, and Tools for Embedded Systems*, pages 42–53, 2021.

[322] Xiangyu Zhang, Xinyu Zhou, Mengxiao Lin, and Jian Sun. ShuffleNet: An Extremely Efficient Convolutional Neural Network for Mobile Devices. In *CVPR*, 2018.

[323] Christian Szegedy, Vincent Vanhoucke, Sergey Ioffe, Jon Shlens, and Zbigniew Wojna. Rethinking the Inception architecture for computer vision. In *Proceedings of the IEEE conference on computer vision and pattern recognition*, pages 2818–2826, 2016.

[324] Min Lin, Qiang Chen, and Shuicheng Yan. Network in network. *arXiv preprint arXiv:1312.4400*, 2013.

[325] Dongyoon Han, Jiwhan Kim, and Junmo Kim. Deep pyramidal residual networks. In *Proceedings of the IEEE conference on computer vision and pattern recognition*, pages 5927–5935, 2017.

[326] Barret Zoph, Vijay Vasudevan, Jonathon Shlens, and Quoc V Le. Learning transferable architectures for scalable image recognition. *arXiv preprint arXiv:1707.07012*, 2017.

[327] Esteban Real, Alok Aggarwal, Yanping Huang, and Quoc V Le. Regularized evolution for image classifier architecture search. *arXiv preprint arXiv:1802.01548*, 2018.

[328] Bowen Baker, Otkrist Gupta, Nikhil Naik, and Ramesh Raskar. Designing neural network architectures using reinforcement learning. *arXiv preprint arXiv:1611.02167*, 2016.

[329] Zhao Zhong, Junjie Yan, and Cheng-Lin Liu. Practical network blocks design with q-learning. *arXiv preprint arXiv:1708.05552*, 2017.

[330] Hanxiao Liu, Karen Simonyan, Oriol Vinyals, Chrisantha Fernando, and Koray Kavukcuoglu. Hierarchical representations for efficient architecture search. In *ICLR*, 2018.

[331] Lei Deng, Guoqi Li, Song Han, Luping Shi, and Yuan Xie. Model compression and hardware acceleration for neural networks: A comprehensive survey. *Proceedings of the IEEE*, 108(4):485–532, 2020.

[332] Han Cai, Tianyao Chen, Weinan Zhang, Yong Yu, and Jun Wang. Efficient Architecture Search by Network Transformation. In *AAAI*, 2018.

[333] Han Cai, Jiacheng Yang, Weinan Zhang, Song Han, and Yong Yu. Path-Level Network Transformation for Efficient Architecture Search. In *ICML*, 2018.

[334] Thomas Elsken, Jan Hendrik Metzen, and Frank Hutter. Efficient multi-objective neural architecture search via lamarckian evolution. In *International Conference on Learning Representations*, 2018.

[335] Andrew Brock, Theodore Lim, James M Ritchie, and Nick Weston. Smash: one-shot model architecture search through hypernetworks. *arXiv preprint arXiv:1708.05344*, 2017.

[336] Hieu Pham, Melody Guan, Barret Zoph, Quoc Le, and Jeff Dean. Efficient neural architecture search via parameters sharing. In *International Conference on Machine Learning*, pages 4095–4104. PMLR, 2018.

[337] Golnaz Ghiasi, Tsung-Yi Lin, and Quoc V Le. Nas-fpn: Learning scalable feature pyramid architecture for object detection. In *Proceedings of the IEEE conference on computer vision and pattern recognition*, pages 7036–7045, 2019.

[338] Mingxing Tan, Ruoming Pang, and Quoc V Le. Efficientdet: Scalable and efficient object detection. In *Proceedings of the IEEE/CVF Conference on Computer Vision and Pattern Recognition*, pages 10781–10790, 2020.

[339] Liang-Chieh Chen, Maxwell Collins, Yukun Zhu, George Papandreou, Barret Zoph, Florian Schroff, Hartwig Adam, and Jon Shlens. Searching for efficient multi-scale architectures for dense image prediction. In *Advances in neural information processing systems*, pages 8699–8710, 2018.

[340] Han Cai, Chuang Gan, Tianzhe Wang, Zhekai Zhang, and Song Han. Once for all: Train one network and specialize it for efficient deployment. In *International Conference on Learning Representations*, 2020.

[341] Mingxing Tan, Bo Chen, Ruoming Pang, Vijay Vasudevan, and Quoc V Le. Mnasnet: Platform-aware neural architecture search for mobile. *arXiv preprint arXiv:1807.11626*, 2018.

[342] Emma Strubell, Ananya Ganesh, and Andrew McCallum. Energy and policy considerations for deep learning in nlp. In *ACL*, 2019.

[343] Jiahui Yu, Pengchong Jin, Hanxiao Liu, Gabriel Bender, Pieter-Jan Kindermans, Mingxing Tan, Thomas Huang, Xiaodan Song, Ruoming Pang, and Quoc Le. Bignas: Scaling up neural architecture search with big single-stage models. *arXiv preprint arXiv:2003.11142*, 2020.

[344] Jiahui Yu and Thomas Huang. Autoslim: Towards one-shot architecture search for channel numbers. *arXiv preprint arXiv:1903.11728*, 2019.

[345] S. Ren, K. He, R. Girshick, and J. Sun. Faster r-cnn: Towards real-time object detection with region proposal networks. In *Proceedings of NIPS*, pages 91–99, 2015.

[346] Joseph Redmon, Santosh Divvala, Ross Girshick, and Ali Farhadi. You only look once: Unified, real-time object detection. In *Proceedings of CVPR, IEEE*, pages 779–788, 2016.

[347] Joseph Redmon and Ali Farhadi. Yolo9000: better, faster, stronger. *arXiv preprint 1612.08242*, 2016.

[348] Joseph Redmon and Ali Farhadi, YOLOv3: An Incremental Improvement, https://arxiv.org/abs/1804.02767.

[349] Wei Liu, Dragomir Anguelov, Dumitru Erhan, Christian Szegedy, Scott Reed, Cheng-Yang Fu, and Alexander C Berg. Ssd: Single shot multibox detector. In *Proceedings of ECCV*, pages 21–37. Springer, 2016.

[350] Tsung-Yi Lin, Priya Goyal, Ross Girshick, Kaiming He, and Piotr Dollár. Focal loss for dense object detection. In *Proceedings of the IEEE International Conference on Computer Vision (ICCV)*, 2017.

[351] J. Hosang, R. Benenson, and B. Schiele. Learning non-maximum suppression. In *IEEE Conference on Computer Vision and Pattern Recognition (CVPR)*, pages 6469–6477, 2017.

[352] Y. D. Kim, E. Park, S. Yoo, T. Choi, L. Yang, and D. Shin. Compression of deep convolutional neural networks for fast and low power mobile applications. *arXiv preprint 1511.06530*, 2015.

[353] tim.lewis. Openmp.

[354] Eunhyeok Park, Junwhan Ahn, and Sungjoo Yoo. Weighted-entropy-based quantization for deep neural networks. In *Proceedings of the IEEE Conference on Computer Vision and Pattern Recognition*, pages 5456–5464, 2017.

[355] T. Y. Lin, M. Maire, S. Belongie, J. Hays, P. Perona, D. Ramanan, P. Dollár, and C. L. Zitnick. Microsoft coco: Common objects in context. In *Proceedings of ECCV*, pages 740–755. Springer, 2014.

[356] J. Redmon and A. Farhadi. Yolo website. https://pjreddie.com/darknet/yolo/, 2016–2018.

[357] T. Y. Lin, P. Dollár, R. Girshick, K. He, B. Hariharan, and S. Belongie. Feature pyramid networks for object detection. In *Proceedings of CVPR, IEEE*, volume 1, page 4, 2017.

[358] G. Huang, Z. Liu, K. Q. Weinberger, and L. van der Maaten. Densely connected convolutional networks. *arXiv preprint arXiv:1608.06993*, 2016.

[359] Facebook. Caffe2 detectron api. https://github.com/facebook-research/Detectron, 2018.

[360] J. Redmon. Yolo: Real-time object detection. http://pjreddie.com /darknet/yolo/, 2013–2016.

[361] *NVIDIA Jetson*, October 2015.

[362] M. Verucchi et al. A systematic assessment of embedded neural networks for object detection. *ETFA*, 2020.

[363] Alexey Bochkovskiy, Chien-Yao Wang, and Hong-Yuan Mark Liao. YOLOv4: Optimal Speed and Accuracy of Object Detection. *arXiv:2004.10934 [cs, eess]*, April 2020. arXiv: 2004.10934.

[364] Suyog Gupta and Berkin Akin. Accelerator-aware neural network design using automl. *arXiv preprint arXiv:2003.02838*, 2020.

[365] Tsung-Yi Lin, Michael Maire, Serge J. Belongie, Lubomir D. Bourdev, Ross B. Girshick, James Hays, Pietro Perona, Deva Ramanan, Piotr Dollár, and C. Lawrence Zitnick. Microsoft COCO: common objects in context. *CoRR*, abs/1405.0312, 2014.

[366] Kent Gauen, Ryan Dailey, Yung-Hsiang Lu, Eunbyung Park, Wei Liu, Alexander C. Berg, and Yiran Chen. Three years of low-power image recognition challenge: Introduction to special session. *DATE*, .2018.83420998, 2018.

[367] Yung-Hsiang Lu, Alexander C. Berg, and Yiran Chen. Low-power image recognition challenge. *AI Magazine*, Vol 39 No 2: Summer 2018, 2018.

[368] K. Simonyan et al. Very Deep Convolutional Networks for Large-Scale Image Recognition. *arXiv:1409.1556*.

[369] C. Szegedy, W. Liu, and Y. Jia. Going deeper with convolutions. *CVPR*, arXiv:1409.4842, 2015.

[370] Kaiming He, Xiangyu Zhang, Shaoqing Ren, and Jian Sun. Deep residual learning for image recognition. In *Proceedings of the IEEE Conference on Computer Vision and Pattern Recognition (CVPR)*, June 2016.

[371] Andrew G. Howard, Menglong Zhu, Bo Chen, Dmitry Kalenichenko, Weijun Wang, Tobias Weyand, Marco Andreetto, and Hartwig Adam. Mobilenets: Efficient convolutional neural networks for mobile vision applications. *arXiv*, abs/1704.04861, 2017.

[372] Andrew G. Howard, Menglong Zhu, Bo Chen, Dmitry Kalenichenko, Weijun Wang, Tobias Weyand, Marco Andreetto, and Hartwig Adam. Mobilenetv2: Inverted residuals and linear bottlenecks. *arXiv*, abs/1801.04381, 2018.

[373] Benoit Jacob, Skirmantas Kligys, Bo Chen, Menglong Zhu, Matthew Tang, Andrew G. Howard, Hartwig Adam, and Dmitry Kalenichenko. Quantization and training of neural networks for efficient integer-arithmetic-only inference. *CoRR*, abs/1712.05877, 2017.

[374] Tao Sheng, Chen Feng, Shaojie Zhuo, Xiaopeng Zhang, Liang Shen, and Mickey Aleksic. A quantization-friendly separable convolution for mobilenets. *arXiv*, arXiv:1803.08607, 2018.

[375] Ilya Loshchilov and Frank Hutter. SGDR: stochastic gradient descent with restarts. *CoRR*, abs/1608.03983, 2016.

[376] M. Horowitz. 1.1 computing's energy problem (and what we can do about it). In *IEEE International Solid-State Circuits Conference Digest of Technical Papers (ISSCC)*, pages 10–14, 2014.

[377] Benoit Jacob, Skirmantas Kligys, Bo Chen, Menglong Zhu, Matthew Tang, Andrew Howard, Hartwig Adam, and Dmitry Kalenichenko. Quantization and training of neural networks for efficient integer-arithmetic-only inference. *Conference on Computer Vision and Pattern Recognition (CVPR)*, 2018.

[378] Bita Rouhani, Daniel Lo, Ritchie Zhao, Ming Liu, Jeremy Fowers, Kalin Ovtcharov, Anna Vinogradsky, Sarah Massengill, Lita Yang, Ray Bittner, Alessandro Forin, Haishan Zhu, Taesik Na, Prerak Patel, Shuai Che, Lok Chand Koppaka, Xia Song, Subhojit Som, Kaustav Das, Saurabh Tiwary, Steve Reinhardt, Sitaram Lanka, Eric Chung, and Doug Burger. Pushing the limits of narrow precision inferencing at cloud scale with microsoft floating point. In *Neural Information Processing Systems (NeurIPS)*. ACM, November 2020.

[379] Pierre Stock, Armand Joulin, Rémi Gribonval, Benjamin Graham, and Hervé Jégou. And the bit goes down: Revisiting the quantization of neural networks. *CoRR*, abs/1907.05686, 2019.

[380] Marcelo Gennari do Nascimento, Roger Fawcett, and Victor Adrian Prisacariu. Dsconv: Efficient convolution operator. In *Proceedings of the IEEE/CVF International Conference on Computer Vision (ICCV)*, October 2019.

[381] Sergey Ioffe and Christian Szegedy. Batch normalization: Accelerating deep network training by reducing internal covariate shift. In Francis Bach and David Blei, editors, *Proceedings of the 32nd International Conference on Machine Learning*, volume 37 of *Proceedings of Machine Learning Research*, pages 448–456, Lille, France, 07–09 Jul 2015. PMLR.

[382] Raghuraman Krishnamoorthi. Quantizing deep convolutional networks for efficient inference: A whitepaper. *arXiv preprint arXiv:1806.08342*, 2018.

[383] Prajit Ramachandran, Barret Zoph, and Quoc V. Le. Searching for activation functions. *CoRR*, abs/1710.05941, 2017.

[384] Mart van Baalen, Christos Louizos, Markus Nagel, Rana Ali Amjad, Ying Wang, Tijmen Blankevoort, and Max Welling. Bayesian bits: Unifying quantization and pruning. In H. Larochelle, M. Ranzato, R. Hadsell, M. F. Balcan, and H. Lin, editors, *Advances in Neural Information Processing Systems*, volume 33, pages 5741–5752. Curran Associates, Inc., 2020.

[385] Zhen Dong, Zhewei Yao, Amir Gholami, Michael W. Mahoney, and Kurt Keutzer. HAWQ: hessian aware quantization of neural networks with mixed-precision. *International Conference on Computer Vision (ICCV)*, 2019.

[386] Stefan Uhlich, Lukas Mauch, Fabien Cardinaux, Kazuki Yoshiyama, Javier Alonso Garcia, Stephen Tiedemann, Thomas Kemp, and Akira Nakamura. Mixed precision dnns: All you need is a good parametrization. In *International Conference on Learning Representations*, 2020.

[387] Ron Banner, Yury Nahshan, and Daniel Soudry. Post training 4-bit quantization of convolutional networks for rapid-deployment. *Neural Information Processing Systems (NeuRIPS)*, 2019.

[388] Markus Nagel, Mart van Baalen, Tijmen Blankevoort, and Max Welling. Data-free quantization through weight equalization and

bias correction. *International Conference on Computer Vision (ICCV)*, 2019.

[389] Tao Sheng, Chen Feng, Shaojie Zhuo, Xiaopeng Zhang, Liang Shen, and Mickey Aleksic. A quantization-friendly separable convolution for mobilenets. In *1st Workshop on Energy Efficient Machine Learning and Cognitive Computing for Embedded Applications (EMC2)*, 2018.

[390] Mark Sandler, Andrew Howard, Menglong Zhu, Andrey Zhmoginov, and Liang-Chieh Chen. Mobilenetv2: Inverted residuals and linear bottlenecks. In *Proceedings of the IEEE conference on computer vision and pattern recognition*, pages 4510–4520, 2018.

[391] Eldad Meller, Alexander Finkelstein, Uri Almog, and Mark Grobman. Same, same but different: Recovering neural network quantization error through weight factorization. In *Proceedings of the 36th International Conference on Machine Learning, ICML 2019, 9-15 June 2019, Long Beach, California, USA*, pages 4486–4495, 2019.

[392] Alexander Finkelstein, Uri Almog, and Mark Grobman. Fighting quantization bias with bias. *arXiv preprint arxiv:1906.03193*, 2019.

[393] Markus Nagel, Rana Ali Amjad, Mart Van Baalen, Christos Louizos, and Tijmen Blankevoort. Up or down? Adaptive rounding for post-training quantization. In Hal Daumé III and Aarti Singh, editors, *Proceedings of the 37th International Conference on Machine Learning*, volume 119 of *Proceedings of Machine Learning Research*, pages 7197–7206. PMLR, 13–18 Jul 2020.

[394] Suyog Gupta, Ankur Agrawal, Kailash Gopalakrishnan, and Pritish Narayanan. Deep learning with limited numerical precision. *International Conference on Machine Learning, ICML*, 2015.

[395] Bert Moons, Parham Noorzad, Andrii Skliar, Giovanni Mariani, Dushyant Mehta, Chris Lott, and Tijmen Blankevoort. Distilling optimal neural networks: Rapid search in diverse spaces. *arXiv preprint arXiv:2012.08859*, 2020.

[396] Steven K. Esser, Jeffrey L. McKinstry, Deepika Bablani, Rathinakumar Appuswamy, and Dharmendra S. Modha. Learned step

size quantization. *International Conference on Learning Representations (ICLR)*, 2020.

[397] Sambhav R. Jain, Albert Gural, Michael Wu, and Chris Dick. Trained uniform quantization for accurate and efficient neural network inference on fixed-point hardware. *CoRR*, abs/1903.08066, 2019.

[398] Yash Bhalgat, Jinwon Lee, Markus Nagel, Tijmen Blankevoort, and Nojun Kwak. Lsq+: Improving low-bit quantization through learnable offsets and better initialization. In *Proceedings of the IEEE/CVF Conference on Computer Vision and Pattern Recognition (CVPR) Workshops*, June 2020.

[399] Jiwei Yang, Xu Shen, Jun Xing, Xinmei Tian, Houqiang Li, Bing Deng, Jianqiang Huang, and Xian-sheng Hua. Quantization networks. In *Proceedings of the IEEE/CVF Conference on Computer Vision and Pattern Recognition (CVPR)*, June 2019.

[400] Mingxing Tan and Quoc Le. EfficientNet: Rethinking model scaling for convolutional neural networks. In Kamalika Chaudhuri and Ruslan Salakhutdinov, editors, *Proceedings of the 36th International Conference on Machine Learning*, volume 97 of *Proceedings of Machine Learning Research*, pages 6105–6114. PMLR, 09–15 Jun 2019.

[401] Gao Huang, Shichen Liu, Laurens van der Maaten, and Kilian Q. Weinberger. Condensenet: An efficient densenet using learned group convolutions. In *Proceedings of the IEEE Conference on Computer Vision and Pattern Recognition (CVPR)*, June 2018.

[402] M. Horowitz. 1.1 computing's energy problem (and what we can do about it). In *IEEE International Solid-State Circuits Conference Digest of Technical Papers (ISSCC)*, pages 10–14, 2014.

[403] Andrew G. Howard, Menglong Zhu, Bo Chen, Dmitry Kalenichenko, Weijun Wang, Tobias Weyand, Marco Andreetto, and Hartwig Adam. Mobilenets: Efficient convolutional neural networks for mobile vision applications. *CoRR*, abs/1704.04861, 2017.

[404] Karen Simonyan and Andrew Zisserman. Very deep convolutional networks for large-scale image recognition. In *International Conference on Learning Representations*, 2015.

[405] Xiangyu Zhang, Xinyu Zhou, Mengxiao Lin, and Jian Sun. Shufflenet: An extremely efficient convolutional neural network for mobile devices. In *Proceedings of the IEEE Conference on Computer Vision and Pattern Recognition (CVPR)*, June 2018.

[406] Mingxing Tan and Quoc V. Le. Efficientnet: Rethinking model scaling for convolutional neural networks. *CoRR*, abs/1905.11946, 2019.

[407] Andrew Howard, Mark Sandler, Grace Chu, Liang-Chieh Chen, Bo Chen, Mingxing Tan, Weijun Wang, Yukun Zhu, Ruoming Pang, Vijay Vasudevan, et al. Searching for mobilenetv3. In *Proceedings of the IEEE/CVF International Conference on Computer Vision*, pages 1314–1324, 2019.

[408] Francois Chollet. Xception: Deep learning with depthwise separable convolutions. In *The IEEE Conference on Computer Vision and Pattern Recognition (CVPR)*, July 2017.

[409] Mark Sandler, Jonathan Baccash, Andrey Zhmoginov, and Andrew Howard. Non-discriminative data or weak model? on the relative importance of data and model resolution. In *Proceedings of the IEEE/CVF International Conference on Computer Vision (ICCV) Workshops*, Oct 2019.

[410] Saining Xie, Ross Girshick, Piotr Dollar, Zhuowen Tu, and Kaiming He. Aggregated residual transformations for deep neural networks. In *Proceedings of the IEEE Conference on Computer Vision and Pattern Recognition (CVPR)*, July 2017.

[411] Sergey Zagoruyko and Nikos Komodakis. Wide Residual Networks. In *British Machine Vision Conference 2016*, York, France, January 2016. British Machine Vision Association.

[412] Imagenet.

[413] Wenzhe Shi, Jose Caballero, Ferenc Huszar, Johannes Totz, Andrew P. Aitken, Rob Bishop, Daniel Rueckert, and Zehan Wang. Real-time single image and video super-resolution using an efficient sub-pixel convolutional neural network. In *The IEEE Conference on Computer Vision and Pattern Recognition (CVPR)*, June 2016.

[414] Mehdi SM Sajjadi, Raviteja Vemulapalli, and Matthew Brown. Frame-recurrent video super-resolution. In *Proceedings of the IEEE Conference on Computer Vision and Pattern Recognition*, pages 6626–6634, 2018.

[415] Tien-Ju Yang, Maxwell D Collins, Yukun Zhu, Jyh-Jing Hwang, Ting Liu, Xiao Zhang, Vivienne Sze, George Papandreou, and Liang-Chieh Chen. Deeperlab: Single-shot image parser. *arXiv preprint arXiv:1902.05093*, 2019.

[416] Adam Paszke, Sam Gross, Francisco Massa, Adam Lerer, James Bradbury, Gregory Chanan, Trevor Killeen, Zeming Lin, Natalia Gimelshein, Luca Antiga, Alban Desmaison, Andreas Kopf, Edward Yang, Zachary DeVito, Martin Raison, Alykhan Tejani, Sasank Chilamkurthy, Benoit Steiner, Lu Fang, Junjie Bai, and Soumith Chintala. Pytorch: An imperative style, high-performance deep learning library. In H. Wallach, H. Larochelle, A. Beygelzimer, F. Alché-Buc, E. Fox, and R. Garnett, editors, *Advances in Neural Information Processing Systems 32*, pages 8024–8035. Curran Associates, Inc., 2019.

[417] Christian Szegedy, Sergey Ioffe, Vincent Vanhoucke, and Alex A. Alemi. Inception-v4, inception-resnet and the impact of residual connections on learning. In *ICLR 2016 Workshop*, 2016.

[418] Radford M. Neal. Connectionist learning of belief networks. *Artif. Intell.*, 56(1):71–113, July 1992.

[419] Yann LeCun, Léon Bottou, Genevieve B. Orr, and Klaus-Robert Müller. Efficient backprop. In *Neural Networks: Tricks of the Trade, This Book is an Outgrowth of a 1996 NIPS Workshop*, page 9–50, Berlin, Heidelberg, 1998. Springer-Verlag.

[420] Richard Hahnloser, Rahul Sarpeshkar, Misha Mahowald, Rodney Douglas, and H. Seung. Digital selection and analogue amplification coexist in a cortex-inspired silicon circuit. *Nature*, 405:947–51, 07 2000.

[421] Richard Hahnloser and H. Sebastian Seung. Permitted and forbidden sets in symmetric threshold-linear networks. In T. Leen, T. Dietterich, and V. Tresp, editors, *Advances in Neural Information Processing Systems*, volume 13. MIT Press, 2001.

[422] K. Fukushima. Visual feature extraction by a multilayered network of analog threshold elements. *IEEE Transactions on Systems Science and Cybernetics*, 5(4):322–333, 1969.

[423] Xavier Glorot, Antoine Bordes, and Yoshua Bengio. Deep sparse rectifier neural networks. In Geoffrey Gordon, David Dunson, and Miroslav Dudík, editors, *Proceedings of the Fourteenth International Conference on Artificial Intelligence and Statistics*, volume 15 of *Proceedings of Machine Learning Research*, pages 315–323, Fort Lauderdale, FL, USA, 11–13 Apr 2011. JMLR Workshop and Conference Proceedings.

[424] Chigozie Nwankpa, Winifred Ijomah, Anthony Gachagan, and Stephen Marshall. Activation functions: Comparison of trends in practice and research for deep learning. *CoRR*, abs/1811.03378, 2018.

[425] Dan Hendrycks and Kevin Gimpel. Bridging nonlinearities and stochastic regularizers with gaussian error linear units. *CoRR*, abs/1606.08415, 2016.

[426] Elfwing Stefan, Uchibe Eiji, and Doya Kenji. Sigmoid-weighted linear units for neural network function approximation in reinforcement learning. *Neural Networks*, Jan 2018.

[427] R. Avenash and P. Vishawanth. Semantic segmentation of satellite images using a modified cnn with hard-swish activation function. In *VISIGRAPP*, 2019.

[428] Amir Gholami, Michael W Mahoney, and Kurt Keutzer. An integrated approach to neural network design, training, and inference. *Univ. California, Berkeley, Berkeley, CA, USA, Tech. Rep*, 2020.

[429] Yani Ioannou, Duncan Robertson, Roberto Cipolla, and Antonio Criminisi. Deep roots: Improving cnn efficiency with hierarchical filter groups. In *Proceedings of the IEEE conference on computer vision and pattern recognition*, pages 1231–1240, 2017.

[430] Franck Mamalet and Christophe Garcia. Simplifying convnets for fast learning. In *International Conference on Artificial Neural Networks*, pages 58–65. Springer, 2012.

[431] Bichen Wu, Alvin Wan, Xiangyu Yue, Peter Jin, Sicheng Zhao, Noah Golmant, Amir Gholaminejad, Joseph Gonzalez, and Kurt Keutzer. Shift: A zero flop, zero parameter alternative to spatial convolutions. In *Proceedings of the IEEE Conference on Computer Vision and Pattern Recognition*, pages 9127–9135, 2018.

[432] Tara N Sainath, Brian Kingsbury, Vikas Sindhwani, Ebru Arisoy, and Bhuvana Ramabhadran. Low-rank matrix factorization for deep neural network training with high-dimensional output targets. In *IEEE international conference on acoustics, speech and signal processing*, pages 6655–6659. IEEE, 2013.

[433] PP Kanjilal, PK Dey, and DN Banerjee. Reduced-size neural networks through singular value decomposition and subset selection. *Electronics Letters*, 29(17):1516–1518, 1993.

[434] Qibin Zhao, Masashi Sugiyama, Longhao Yuan, and Andrzej Cichocki. Learning efficient tensor representations with ring-structured networks. In *ICASSP 2019-2019 IEEE International Conference on Acoustics, Speech and Signal Processing (ICASSP)*, pages 8608–8612. IEEE, 2019.

[435] Gao Huang, Zhuang Liu, Laurens Van Der Maaten, and Kilian Q Weinberger. Densely connected convolutional networks. In *Proceedings of the IEEE conference on computer vision and pattern recognition*, pages 4700–4708, 2017.

[436] Dilin Wang, Meng Li, Chengyue Gong, and Vikas Chandra. Attentivenas: Improving neural architecture search via attentive sampling. *arXiv preprint arXiv:2011.09011*, 2020.

[437] Thomas Elsken, Jan Hendrik Metzen, and Frank Hutter. Neural architecture search: A survey. *J. Mach. Learn. Res.*, 20(55):1–21, 2019.

[438] Amir Gholami, Kiseok Kwon, Bichen Wu, Zizheng Tai, Xiangyu Yue, Peter Jin, Sicheng Zhao, and Kurt Keutzer. SqueezeNext: Hardware-aware neural network design. *Workshop paper in CVPR*, 2018.

[439] Yann LeCun, John S Denker, and Sara A Solla. Optimal brain damage. In *Advances in neural information processing systems*, pages 598–605, 1990.

[440] Babak Hassibi and David G Stork. *Second order derivatives for network pruning: Optimal brain surgeon.* Morgan Kaufmann, 1993.

[441] Xin Dong, Shangyu Chen, and Sinno Jialin Pan. Learning to prune deep neural networks via layer-wise optimal brain surgeon. *arXiv preprint arXiv:1705.07565*, 2017.

[442] Namhoon Lee, Thalaiyasingam Ajanthan, and Philip HS Torr. Snip: Single-shot network pruning based on connection sensitivity. *arXiv preprint arXiv:1810.02340*, 2018.

[443] Xia Xiao, Zigeng Wang, and Sanguthevar Rajasekaran. Autoprune: Automatic network pruning by regularizing auxiliary parameters. In *Advances in Neural Information Processing Systems*, pages 13681–13691, 2019.

[444] Sejun Park, Jaeho Lee, Sangwoo Mo, and Jinwoo Shin. Lookahead: a far-sighted alternative of magnitude-based pruning. *arXiv preprint arXiv:2002.04809*, 2020.

[445] Yihui He, Ji Lin, Zhijian Liu, Hanrui Wang, Li-Jia Li, and Song Han. Amc: Automl for model compression and acceleration on mobile devices. In *Proceedings of the European Conference on Computer Vision (ECCV)*, pages 784–800, 2018.

[446] Ruichi Yu, Ang Li, Chun-Fu Chen, Jui-Hsin Lai, Vlad I Morariu, Xintong Han, Mingfei Gao, Ching-Yung Lin, and Larry S Davis. Nisp: Pruning networks using neuron importance score propagation. In *Proceedings of the IEEE Conference on Computer Vision and Pattern Recognition*, pages 9194–9203, 2018.

[447] Shaohui Lin, Rongrong Ji, Yuchao Li, Yongjian Wu, Feiyue Huang, and Baochang Zhang. Accelerating convolutional networks via global & dynamic filter pruning. In *IJCAI*, pages 2425–2432, 2018.

[448] Zehao Huang and Naiyan Wang. Data-driven sparse structure selection for deep neural networks. In *Proceedings of the European conference on computer vision (ECCV)*, pages 304–320, 2018.

[449] Chenglong Zhao, Bingbing Ni, Jian Zhang, Qiwei Zhao, Wenjun Zhang, and Qi Tian. Variational convolutional neural network pruning. In *Proceedings of the IEEE Conference on Computer Vision and Pattern Recognition*, pages 2780–2789, 2019.

[450] Shixing Yu, Zhewei Yao, Amir Gholami, Zhen Dong, Michael W Mahoney, and Kurt Keutzer. Hessian-aware pruning and optimal neural implant. *arXiv preprint arXiv:2101.08940*, 2021.

[451] Aydin Buluc and John R Gilbert. Challenges and advances in parallel sparse matrix-matrix multiplication. In *2008 37th International Conference on Parallel Processing*, pages 503–510. IEEE, 2008.

[452] Trevor Gale, Erich Elsen, and Sara Hooker. The state of sparsity in deep neural networks. *arXiv preprint arXiv:1902.09574*, 2019.

[453] Davis Blalock, Jose Javier Gonzalez Ortiz, Jonathan Frankle, and John Guttag. What is the state of neural network pruning? *arXiv preprint arXiv:2003.03033*, 2020.

[454] Torsten Hoefler, Dan Alistarh, Tal Ben-Nun, Nikoli Dryden, and Alexandra Peste. Sparsity in deep learning: Pruning and growth for efficient inference and training in neural networks. *arXiv preprint arXiv:2102.00554*, 2021.

[455] Andrey Kuzmin, Markus Nagel, Saurabh Pitre, Sandeep Pendyam, Tijmen Blankevoort, and Max Welling. Taxonomy and evaluation of structured compression of convolutional neural networks. *arXiv preprint arXiv:1912.09802*, 2019.

[456] Adriana Romero, Nicolas Ballas, Samira Ebrahimi Kahou, Antoine Chassang, Carlo Gatta, and Yoshua Bengio. Fitnets: Hints for thin deep nets. *arXiv preprint arXiv:1412.6550*, 2014.

[457] Asit Mishra and Debbie Marr. Apprentice: Using knowledge distillation techniques to improve low-precision network accuracy. In *International Conference on Learning Representations*, 2018.

[458] Yuncheng Li, Jianchao Yang, Yale Song, Liangliang Cao, Jiebo Luo, and Li-Jia Li. Learning from noisy labels with distillation. In *Proceedings of the IEEE International Conference on Computer Vision*, pages 1910–1918, 2017.

[459] Junho Yim, Donggyu Joo, Jihoon Bae, and Junmo Kim. A gift from knowledge distillation: Fast optimization, network minimization and transfer learning. In *Proceedings of the IEEE Conference on Computer Vision and Pattern Recognition*, pages 4133–4141, 2017.

[460] Antonio Polino, Razvan Pascanu, and Dan Alistarh. Model compression via distillation and quantization. In *International Conference on Learning Representations*, 2018.

[461] Sungsoo Ahn, Shell Xu Hu, Andreas Damianou, Neil D Lawrence, and Zhenwen Dai. Variational information distillation for knowledge transfer. In *Proceedings of the IEEE/CVF Conference on Computer Vision and Pattern Recognition*, pages 9163–9171, 2019.

[462] Hongxu Yin, Pavlo Molchanov, Jose M Alvarez, Zhizhong Li, Arun Mallya, Derek Hoiem, Niraj K Jha, and Jan Kautz. Dreaming to distill: Data-free knowledge transfer via deepinversion. In *Proceedings of the IEEE/CVF Conference on Computer Vision and Pattern Recognition*, pages 8715–8724, 2020.

[463] Ron Banner, Itay Hubara, Elad Hoffer, and Daniel Soudry. Scalable methods for 8-bit training of neural networks. *Advances in neural information processing systems*, 2018.

[464] Naigang Wang, Jungwook Choi, Daniel Brand, Chia-Yu Chen, and Kailash Gopalakrishnan. Training deep neural networks with 8-bit floating point numbers. *Advances in neural information processing systems*, 2018.

[465] Jangho Kim, KiYoon Yoo, and Nojun Kwak. Position-based scaled gradient for model quantization and sparse training. *Advances in neural information processing systems*, 2020.

[466] Fartash Faghri, Iman Tabrizian, Ilia Markov, Dan Alistarh, Daniel Roy, and Ali Ramezani-Kebrya. Adaptive gradient quantization for data-parallel sgd. *Advances in neural information processing systems*, 2020.

[467] Brian Chmiel, Liad Ben-Uri, Moran Shkolnik, Elad Hoffer, Ron Banner, and Daniel Soudry. Neural gradients are near-lognormal: improved quantized and sparse training. In *International Conference on Learning Representations*, 2021.

[468] Matthieu Courbariaux, Yoshua Bengio, and Jean-Pierre David. Training deep neural networks with low precision multiplications. *arXiv preprint arXiv:1412.7024*, 2014.

[469] Suyog Gupta, Ankur Agrawal, Kailash Gopalakrishnan, and Pritish Narayanan. Deep learning with limited numerical precision. In *International conference on machine learning*, pages 1737–1746. PMLR, 2015.

[470] Boris Ginsburg, Sergei Nikolaev, Ahmad Kiswani, Hao Wu, Amir Gholaminejad, Slawomir Kierat, Michael Houston, and Alex Fit-Florea. Tensor processing using low precision format, December 28 2017. US Patent App. 15/624,577.

[471] Paulius Micikevicius, Sharan Narang, Jonah Alben, Gregory Diamos, Erich Elsen, David Garcia, Boris Ginsburg, Michael Houston, Oleksii Kuchaiev, Ganesh Venkatesh, et al. Mixed precision training. *arXiv preprint arXiv:1710.03740*, 2017.

[472] Warren S McCulloch and Walter Pitts. A logical calculus of the ideas immanent in nervous activity. *The bulletin of mathematical biophysics*, 5(4):115–133, 1943.

[473] Rufin VanRullen and Christof Koch. Is perception discrete or continuous? *Trends in cognitive sciences*, 7(5):207–213, 2003.

[474] James Tee and Desmond P Taylor. Is information in the brain represented in continuous or discrete form? *IEEE Transactions on Molecular, Biological and Multi-Scale Communications*, 6(3):199–209, 2020.

[475] A Aldo Faisal, Luc PJ Selen, and Daniel M Wolpert. Noise in the nervous system. *Nature reviews neuroscience*, 9(4):292–303, 2008.

[476] Rishidev Chaudhuri and Ila Fiete. Computational principles of memory. *Nature neuroscience*, 19(3):394, 2016.

[477] Kenneth W Latimer, Jacob L Yates, Miriam LR Meister, Alexander C Huk, and Jonathan W Pillow. Single-trial spike trains in parietal cortex reveal discrete steps during decision-making. *Science*, 349(6244):184–187, 2015.

[478] Lav R Varshney and Kush R Varshney. Decision making with quantized priors leads to discrimination. *Proceedings of the IEEE*, 105(2):241–255, 2016.

[479] Mel Win Khaw, Luminita Stevens, and Michael Woodford. Discrete adjustment to a changing environment: Experimental evidence. *Journal of Monetary Economics*, 91:88–103, 2017.

[480] Lav R Varshney, Per Jesper Sjöström, and Dmitri B Chklovskii. Optimal information storage in noisy synapses under resource constraints. *Neuron*, 52(3):409–423, 2006.

[481] John Z Sun, Grace I Wang, Vivek K Goyal, and Lav R Varshney. A framework for bayesian optimality of psychophysical laws. *Journal of Mathematical Psychology*, 56(6):495–501, 2012.

[482] Robert M. Gray and David L. Neuhoff. Quantization. *IEEE transactions on information theory*, 44(6):2325–2383, 1998.

[483] S. M. Stigler. *The History of Statistics: The Measurement of Uncertainty before 1900*. Harvard University Press, Cambridge, 1986.

[484] Bernhard Riemann. *Ueber die Darstellbarkeit einer Function durch eine trigonometrische Reihe*, volume 13. Dieterich, 1867.

[485] William Fleetwood Sheppard. On the calculation of the most probable values of frequency-constants, for data arranged according to equidistant division of a scale. *Proceedings of the London Mathematical Society*, 1(1):353–380, 1897.

[486] Claude E Shannon. A mathematical theory of communication. *The Bell system technical journal*, 27(3):379–423, 1948.

[487] David A Huffman. A method for the construction of minimum-redundancy codes. *Proceedings of the IRE*, 40(9):1098–1101, 1952.

[488] Claude E Shannon. Coding theorems for a discrete source with a fidelity criterion. *IRE Nat. Conv. Rec*, 4(142-163):1, 1959.

[489] JG Dunn. The performance of a class of n dimensional quantizers for a gaussian source. In *Proc. Columbia Symp. Signal Transmission Processing*, pages 76–81, 1965.

[490] AE Gamal, L Hemachandra, Itzhak Shperling, and V Wei. Using simulated annealing to design good codes. *IEEE Transactions on Information Theory*, 33(1):116–123, 1987.

[491] William H Equitz. A new vector quantization clustering algorithm. *IEEE transactions on acoustics, speech, and signal processing*, 37(10):1568–1575, 1989.

[492] Kenneth Rose, Eitan Gurewitz, and Geoffrey Fox. A deterministic annealing approach to clustering. *Pattern Recognition Letters*, 11(9):589–594, 1990.

[493] BM Oliver, JR Pierce, and Claude E Shannon. The philosophy of pcm. *Proceedings of the IRE*, 36(11):1324–1331, 1948.

[494] William Ralph Bennett. Spectra of quantized signals. *The Bell System Technical Journal*, 27(3):446–472, 1948.

[495] L.N. Trefethen and D. Bau III. *Numerical Linear Algebra*. SIAM, Philadelphia, 1997.

[496] Benoit Jacob, Skirmantas Kligys, Bo Chen, Menglong Zhu, Matthew Tang, Andrew Howard, Hartwig Adam, and Dmitry Kalenichenko. Quantization and training of neural networks for efficient integer-arithmetic-only inference. In *Proceedings of the IEEE Conference on Computer Vision and Pattern Recognition (CVPR)*, 2018.

[497] Hao Wu, Patrick Judd, Xiaojie Zhang, Mikhail Isaev, and Paulius Micikevicius. Integer quantization for deep learning inference: Principles and empirical evaluation. *arXiv preprint arXiv:2004.09602*, 2020.

[498] Yash Bhalgat, Jinwon Lee, Markus Nagel, Tijmen Blankevoort, and Nojun Kwak. Lsq+: Improving low-bit quantization through learnable offsets and better initialization. In *Proceedings of the IEEE/CVF Conference on Computer Vision and Pattern Recognition Workshops*, pages 696–697, 2020.

[499] Jeffrey L McKinstry, Steven K Esser, Rathinakumar Appuswamy, Deepika Bablani, John V Arthur, Izzet B Yildiz, and Dharmendra S Modha. Discovering low-precision networks close to full-precision networks for efficient embedded inference. *arXiv preprint arXiv:1809.04191*, 2018.

[500] Szymon Migacz. Nvidia 8-bit inference with tensorrt. *GPU Technology Conference*, 2017.

[501] Zhewei Yao, Zhen Dong, Zhangcheng Zheng, Amir Gholami, Jiali Yu, Eric Tan, Leyuan Wang, Qijing Huang, Yida Wang, Michael W Mahoney, et al. Hawqv3: Dyadic neural network quantization. *International Conference on Machine Learning*, 2021.

[502] Wonyong Sung, Sungho Shin, and Kyuyeon Hwang. Resiliency of deep neural networks under quantization. *arXiv preprint arXiv:1511.06488*, 2015.

[503] Sungho Shin, Kyuyeon Hwang, and Wonyong Sung. Fixed-point performance analysis of recurrent neural networks. In *2016 IEEE International Conference on Acoustics, Speech and Signal Processing (ICASSP)*, pages 976–980. IEEE, 2016.

[504] Yoni Choukroun, Eli Kravchik, Fan Yang, and Pavel Kisilev. Low-bit quantization of neural networks for efficient inference. In *ICCV Workshops*, pages 3009–3018, 2019.

[505] Ritchie Zhao, Yuwei Hu, Jordan Dotzel, Christopher De Sa, and Zhiru Zhang. Improving neural network quantization without retraining using outlier channel splitting. *Proceedings of Machine Learning Research*, 2019.

[506] Rundong Li, Yan Wang, Feng Liang, Hongwei Qin, Junjie Yan, and Rui Fan. Fully quantized network for object detection. In *Proceedings of the IEEE Conference on Computer Vision and Pattern Recognition (CVPR)*, 2019.

[507] Jungwook Choi, Zhuo Wang, Swagath Venkataramani, Pierce I-Jen Chuang, Vijayalakshmi Srinivasan, and Kailash Gopalakrishnan. Pact: Parameterized clipping activation for quantized neural networks. *arXiv preprint arXiv:1805.06085*, 2018.

[508] Chenzhuo Zhu, Song Han, Huizi Mao, and William J Dally. Trained ternary quantization. *arXiv preprint arXiv:1612.01064*, 2016.

[509] Dongqing Zhang, Jiaolong Yang, Dongqiangzi Ye, and Gang Hua. Lq-nets: Learned quantization for highly accurate and compact deep neural networks. In *European conference on computer vision (ECCV)*, 2018.

[510] Steven K Esser, Jeffrey L McKinstry, Deepika Bablani, Rathinakumar Appuswamy, and Dharmendra S Modha. Learned step size quantization. *arXiv preprint arXiv:1902.08153*, 2019.

[511] Sheng Shen, Zhen Dong, Jiayu Ye, Linjian Ma, Zhewei Yao, Amir Gholami, Michael W Mahoney, and Kurt Keutzer. Q-BERT: Hessian based ultra low precision quantization of bert. In *AAAI*, pages 8815–8821, 2020.

[512] Ashish Vaswani, Noam Shazeer, Niki Parmar, Jakob Uszkoreit, Llion Jones, Aidan N Gomez, Lukasz Kaiser, and Illia Polosukhin. Attention is all you need. In *Advances in neural information processing systems*, pages 5998–6008, 2017.

[513] Shuchang Zhou, Zekun Ni, Xinyu Zhou, He Wen, Yuxin Wu, and Yuheng Zou. Dorefa-net: Training low bitwidth convolutional neural networks with low bitwidth gradients. *arXiv preprint arXiv:1606.06160*, 2016.

[514] Qijing Huang, Dequan Wang, Zhen Dong, Yizhao Gao, Yaohui Cai, Tian Li, Bichen Wu, Kurt Keutzer, and John Wawrzynek. Codenet: Efficient deployment of input-adaptive object detection on embedded fpgas. In *The 2021 ACM/SIGDA International Symposium on Field-Programmable Gate Arrays*, pages 206–216, 2021.

[515] Moran Shkolnik, Brian Chmiel, Ron Banner, Gil Shomron, Yuri Nahshan, Alex Bronstein, and Uri Weiser. Robust quantization: One model to rule them all. *Advances in neural information processing systems*, 2020.

[516] Yunchao Gong, Liu Liu, Ming Yang, and Lubomir Bourdev. Compressing deep convolutional networks using vector quantization. *arXiv preprint arXiv:1412.6115*, 2014.

[517] Jiaxiang Wu, Cong Leng, Yuhang Wang, Qinghao Hu, and Jian Cheng. Quantized convolutional neural networks for mobile devices. In *Proceedings of the IEEE Conference on Computer Vision and Pattern Recognition*, pages 4820–4828, 2016.

[518] Lu Hou, Quanming Yao, and James T Kwok. Loss-aware binarization of deep networks. *arXiv preprint arXiv:1611.01600*, 2016.

[519] Zhouhan Lin, Matthieu Courbariaux, Roland Memisevic, and Yoshua Bengio. Neural networks with few multiplications. *arXiv preprint arXiv:1510.03009*, 2015.

[520] Daisuke Miyashita, Edward H Lee, and Boris Murmann. Convolutional neural networks using logarithmic data representation. *arXiv preprint arXiv:1603.01025*, 2016.

[521] Yoojin Choi, Mostafa El-Khamy, and Jungwon Lee. Towards the limit of network quantization. *arXiv preprint arXiv:1612.01543*, 2016.

[522] Zhaowei Cai, Xiaodong He, Jian Sun, and Nuno Vasconcelos. Deep learning with low precision by half-wave gaussian quantization. In *Proceedings of the IEEE Conference on Computer Vision and Pattern Recognition*, pages 5918–5926, 2017.

[523] Eunhyeok Park, Sungjoo Yoo, and Peter Vajda. Value-aware quantization for training and inference of neural networks. In *Proceedings of the European Conference on Computer Vision (ECCV)*, pages 580–595, 2018.

[524] Peisong Wang, Qinghao Hu, Yifan Zhang, Chunjie Zhang, Yang Liu, and Jian Cheng. Two-step quantization for low-bit neural networks. In *Proceedings of the IEEE Conference on computer vision and pattern recognition*, pages 4376–4384, 2018.

[525] Yongkweon Jeon, Baeseong Park, Se Jung Kwon, Byeongwook Kim, Jeongin Yun, and Dongsoo Lee. Biqgemm: matrix multiplication with lookup table for binary-coding-based quantized dnns. *arXiv preprint arXiv:2005.09904*, 2020.

[526] Sangil Jung, Changyong Son, Seohyung Lee, Jinwoo Son, Jae-Joon Han, Youngjun Kwak, Sung Ju Hwang, and Changkyu Choi. Learning to quantize deep networks by optimizing quantization intervals with task loss. In *Proceedings of the IEEE/CVF Conference on Computer Vision and Pattern Recognition*, pages 4350–4359, 2019.

[527] Jiwei Yang, Xu Shen, Jun Xing, Xinmei Tian, Houqiang Li, Bing Deng, Jianqiang Huang, and Xian-sheng Hua. Quantization networks. In *Proceedings of the IEEE/CVF Conference on Computer Vision and Pattern Recognition*, pages 7308–7316, 2019.

[528] Julian Faraone, Nicholas Fraser, Michaela Blott, and Philip HW Leong. Syq: Learning symmetric quantization for efficient deep neural networks. In *Proceedings of the IEEE Conference on Computer Vision and Pattern Recognition*, pages 4300–4309, 2018.

[529] Frederick Tung and Greg Mori. Clip-q: Deep network compression learning by in-parallel pruning-quantization. In *Proceedings of the IEEE Conference on Computer Vision and Pattern Recognition*, pages 7873–7882, 2018.

[530] Aojun Zhou, Anbang Yao, Kuan Wang, and Yurong Chen. Explicit loss-error-aware quantization for low-bit deep neural networks. In *Proceedings of the IEEE conference on computer vision and pattern recognition*, pages 9426–9435, 2018.

[531] Zhaohui Yang, Yunhe Wang, Kai Han, Chunjing Xu, Chao Xu, Dacheng Tao, and Chang Xu. Searching for low-bit weights in quantized neural networks. *Advances in neural information processing systems*, 2020.

[532] Zhenyu Liao, Romain Couillet, and Michael W Mahoney. Sparse quantized spectral clustering. *International Conference on Learning Representations*, 2021.

[533] Chaim Baskin, Eli Schwartz, Evgenii Zheltonozhskii, Natan Liss, Raja Giryes, Alex M Bronstein, and Avi Mendelson. Uniq: Uniform noise injection for non-uniform quantization of neural networks. *arXiv preprint arXiv:1804.10969*, 2018.

[534] Yuhang Li, Xin Dong, and Wei Wang. Additive powers-of-two quantization: An efficient non-uniform discretization for neural networks. *arXiv preprint arXiv:1909.13144*, 2019.

[535] Shubham Jain, Swagath Venkataramani, Vijayalakshmi Srinivasan, Jungwook Choi, Kailash Gopalakrishnan, and Leland Chang. Biscaled-dnn: Quantizing long-tailed datastructures with two scale factors for deep neural networks. In *2019 56th ACM/IEEE Design Automation Conference (DAC)*, pages 1–6. IEEE, 2019.

[536] Jun Fang, Ali Shafiee, Hamzah Abdel-Aziz, David Thorsley, Georgios Georgiadis, and Joseph H Hassoun. Post-training piecewise linear quantization for deep neural networks. In *European Conference on Computer Vision*, pages 69–86. Springer, 2020.

[537] Aojun Zhou, Anbang Yao, Yiwen Guo, Lin Xu, and Yurong Chen. Incremental network quantization: Towards lossless cnns with low-precision weights. *arXiv preprint arXiv:1702.03044*, 2017.

[538] Chen Xu, Jianqiang Yao, Zhouchen Lin, Wenwu Ou, Yuanbin Cao, Zhirong Wang, and Hongbin Zha. Alternating multi-bit quantization for recurrent neural networks. *arXiv preprint arXiv:1802.00150*, 2018.

[539] Yiwen Guo, Anbang Yao, Hao Zhao, and Yurong Chen. Network sketching: Exploiting binary structure in deep cnns. In *Proceedings of the IEEE Conference on Computer Vision and Pattern Recognition*, pages 5955–5963, 2017.

[540] Itay Hubara, Matthieu Courbariaux, Daniel Soudry, Ran Elyaniv, and Yoshua Bengio. Binarized neural networks. *neural information processing systems*, pages 4107–4115, 2016.

[541] Wei Tang, Gang Hua, and Liang Wang. How to train a compact binary neural network with high accuracy? In *Proceedings of the AAAI Conference on Artificial Intelligence*, volume 31, 2017.

[542] Xiaofan Lin, Cong Zhao, and Wei Pan. Towards accurate binary convolutional neural network. *arXiv preprint arXiv:1711.11294*, 2017.

[543] Matthieu Courbariaux, Yoshua Bengio, and Jean-Pierre David. BinaryConnect: Training deep neural networks with binary weights during propagations. In *Advances in neural information processing systems*, pages 3123–3131, 2015.

[544] Mohammad Rastegari, Vicente Ordonez, Joseph Redmon, and Ali Farhadi. Xnor-net: Imagenet classification using binary convolutional neural networks. 2016.

[545] Philipp Gysel, Mohammad Motamedi, and Soheil Ghiasi. Hardware-oriented approximation of convolutional neural networks. *arXiv preprint arXiv:1604.03168*, 2016.

[546] Philipp Gysel, Jon Pimentel, Mohammad Motamedi, and Soheil Ghiasi. Ristretto: A framework for empirical study of resource-efficient inference in convolutional neural networks. *IEEE transactions on neural networks and learning systems*, 29(11):5784–5789, 2018.

[547] Shyam A Tailor, Javier Fernandez-Marques, and Nicholas D Lane. Degree-quant: Quantization-aware training for graph neural networks. *International Conference on Learning Representations*, 2021.

[548] Renkun Ni, Hong-min Chu, Oscar Castañeda, Ping-yeh Chiang, Christoph Studer, and Tom Goldstein. Wrapnet: Neural net inference with ultra-low-resolution arithmetic. *arXiv preprint arXiv:2007.13242*, 2020.

[549] Yu Bai, Yu-Xiang Wang, and Edo Liberty. Proxquant: Quantized neural networks via proximal operators. *International Conference on Learning Representations*, 2018.

[550] Penghang Yin, Jiancheng Lyu, Shuai Zhang, Stanley Osher, Yingyong Qi, and Jack Xin. Understanding straight-through estimator in training activation quantized neural nets. *arXiv preprint arXiv:1903.05662*, 2019.

[551] Frank Rosenblatt. *The perceptron, a perceiving and recognizing automaton Project Para*. Cornell Aeronautical Laboratory, 1957.

[552] Frank Rosenblatt. Principles of neurodynamics. perceptrons and the theory of brain mechanisms. Technical report, Cornell Aeronautical Lab Inc Buffalo NY, 1961.

[553] Bohan Zhuang, Chunhua Shen, Mingkui Tan, Lingqiao Liu, and Ian D Reid. Towards effective low-bitwidth convolutional neural networks. *computer vision and pattern recognition*, pages 7920–7928, 2018.

[554] Pierre Stock, Angela Fan, Benjamin Graham, Edouard Grave, Rémi Gribonval, Herve Jegou, and Armand Joulin. Training with quantization noise for extreme model compression. In *International Conference on Learning Representations*, 2021.

[555] Shangyu Chen, Wenya Wang, and Sinno Jialin Pan. Metaquant: Learning to quantize by learning to penetrate non-differentiable quantization. In H. Wallach, H. Larochelle, A. Beygelzimer, F. Alché-Buc, E. Fox, and R. Garnett, editors, *Advances in Neural Information Processing Systems*, volume 32. Curran Associates, Inc., 2019.

[556] Angela Fan, Pierre Stock, Benjamin Graham, Edouard Grave, Rémi Gribonval, Hervé Jégou, and Armand Joulin. Training with quantization noise for extreme model compression. *arXiv e-prints*, pages arXiv–2004, 2020.

[557] Zechun Liu, Baoyuan Wu, Wenhan Luo, Xin Yang, Wei Liu, and Kwang-Ting Cheng. Bi-real net: Enhancing the performance of 1-bit cnns with improved representational capability and advanced training algorithm. In *Proceedings of the European conference on computer vision (ECCV)*, pages 722–737, 2018.

[558] Cong Leng, Zesheng Dou, Hao Li, Shenghuo Zhu, and Rong Jin. Extremely low bit neural network: Squeeze the last bit out with admm. In *Proceedings of the AAAI Conference on Artificial Intelligence*, volume 32, 2018.

[559] Eirikur Agustsson and Lucas Theis. Universally quantized neural compression. *Advances in neural information processing systems*, 2020.

[560] Abram L Friesen and Pedro Domingos. Deep learning as a mixed convex-combinatorial optimization problem. *arXiv preprint arXiv:1710.11573*, 2017.

[561] Dong-Hyun Lee, Saizheng Zhang, Asja Fischer, and Yoshua Bengio. Difference target propagation. In *Joint european conference on machine learning and knowledge discovery in databases*, pages 498–515. Springer, 2015.

[562] Eric Jang, Shixiang Gu, and Ben Poole. Categorical reparameterization with gumbel-softmax. *arXiv preprint arXiv:1611.01144*, 2016.

[563] Yoojin Choi, Mostafa El-Khamy, and Jungwon Lee. Learning low precision deep neural networks through regularization. *arXiv preprint arXiv:1809.00095*, 2, 2018.

[564] Maxim Naumov, Utku Diril, Jongsoo Park, Benjamin Ray, Jedrzej Jablonski, and Andrew Tulloch. On periodic functions as regularizers for quantization of neural networks. *arXiv preprint arXiv:1811.09862*, 2018.

[565] Milad Alizadeh, Arash Behboodi, Mart van Baalen, Christos Louizos, Tijmen Blankevoort, and Max Welling. Gradient l1 regularization for quantization robustness. *arXiv preprint arXiv:2002.07520*, 2020.

[566] Lei Deng, Peng Jiao, Jing Pei, Zhenzhi Wu, and Guoqi Li. Gxnor-net: Training deep neural networks with ternary weights and activations without full-precision memory under a unified discretization framework. *Neural Networks*, 100:49–58, 2018.

[567] Zhi-Gang Liu and Matthew Mattina. Learning low-precision neural networks without straight-through estimator (STE). *arXiv preprint arXiv:1903.01061*, 2019.

[568] Markus Nagel, Rana Ali Amjad, Mart Van Baalen, Christos Louizos, and Tijmen Blankevoort. Up or down? adaptive rounding for post-training quantization. In *International Conference on Machine Learning*, pages 7197–7206. PMLR, 2020.

[569] Ron Banner, Yury Nahshan, Elad Hoffer, and Daniel Soudry. Post-training 4-bit quantization of convolution networks for rapid-deployment. *arXiv preprint arXiv:1810.05723*, 2018.

[570] Eldad Meller, Alexander Finkelstein, Uri Almog, and Mark Grobman. Same, same but different: Recovering neural network quantization error through weight factorization. In *International Conference on Machine Learning*, pages 4486–4495. PMLR, 2019.

[571] Jun Fang, Ali Shafiee, Hamzah Abdel-Aziz, David Thorsley, Georgios Georgiadis, and Joseph Hassoun. Near-lossless post-training quantization of deep neural networks via a piecewise linear approximation. *arXiv preprint arXiv:2002.00104*, 2020.

[572] Jun Haeng Lee, Sangwon Ha, Saerom Choi, Won-Jo Lee, and Seungwon Lee. Quantization for rapid deployment of deep neural networks. *arXiv preprint arXiv:1810.05488*, 2018.

[573] Markus Nagel, Mart van Baalen, Tijmen Blankevoort, and Max Welling. Data-free quantization through weight equalization and bias correction. In *Proceedings of the IEEE/CVF International Conference on Computer Vision*, pages 1325–1334, 2019.

[574] Yaohui Cai, Zhewei Yao, Zhen Dong, Amir Gholami, Michael W Mahoney, and Kurt Keutzer. Zeroq: A novel zero shot quantization framework. In *Proceedings of the IEEE/CVF Conference on Computer Vision and Pattern Recognition*, pages 13169–13178, 2020.

[575] Yuhang Li, Ruihao Gong, Xu Tan, Yang Yang, Peng Hu, Qi Zhang, Fengwei Yu, Wei Wang, and Shi Gu. Brecq: Pushing the limit of post-training quantization by block reconstruction. *International Conference on Learning Representations*, 2021.

[576] Xiangyu He and Jian Cheng. Learning compression from limited unlabeled data. In *Proceedings of the European Conference on Computer Vision (ECCV)*, pages 752–769, 2018.

[577] Sahaj Garg, Anirudh Jain, Joe Lou, and Mitchell Nahmias. Confounding tradeoffs for neural network quantization. *arXiv preprint arXiv:2102.06366*, 2021.

[578] Sahaj Garg, Joe Lou, Anirudh Jain, and Mitchell Nahmias. Dynamic precision analog computing for neural networks. *arXiv preprint arXiv:2102.06365*, 2021.

[579] Itay Hubara, Yury Nahshan, Yair Hanani, Ron Banner, and Daniel Soudry. Improving post training neural quantization: Layer-wise calibration and integer programming. *arXiv preprint arXiv:2006.10518*, 2020.

[580] Alexander Finkelstein, Uri Almog, and Mark Grobman. Fighting quantization bias with bias. *arXiv preprint arXiv:1906.03193*, 2019.

[581] Matan Haroush, Itay Hubara, Elad Hoffer, and Daniel Soudry. The knowledge within: Methods for data-free model compression. In *Proceedings of the IEEE/CVF Conference on Computer Vision and Pattern Recognition*, pages 8494–8502, 2020.

[582] Jacob Devlin, Ming-Wei Chang, Kenton Lee, and Kristina Toutanova. Bert: Pre-training of deep bidirectional transformers for language understanding. *arXiv preprint arXiv:1810.04805*, 2018.

[583] Dan Hendrycks and Kevin Gimpel. Gaussian error linear units (GELUs). *arXiv preprint arXiv:1606.08415*, 2016.

[584] Prajit Ramachandran, Barret Zoph, and Quoc V Le. Swish: a self-gated activation function. *arXiv preprint arXiv:1710.05941*, 7:1, 2017.

[585] Hanting Chen, Yunhe Wang, Chang Xu, Zhaohui Yang, Chuanjian Liu, Boxin Shi, Chunjing Xu, Chao Xu, and Qi Tian. Data-free learning of student networks. In *Proceedings of the IEEE/CVF International Conference on Computer Vision*, pages 3514–3522, 2019.

[586] Ian J Goodfellow, Jean Pouget-Abadie, Mehdi Mirza, Bing Xu, David Warde-Farley, Sherjil Ozair, Aaron Courville, and Yoshua Bengio. Generative adversarial networks. *arXiv preprint arXiv:1406.2661*, 2014.

[587] Sergey Ioffe and Christian Szegedy. Batch normalization: Accelerating deep network training by reducing internal covariate shift. *International conference on machine learning*, pages 448–456, 2015.

[588] Yoojin Choi, Jihwan Choi, Mostafa El-Khamy, and Jungwon Lee. Data-free network quantization with adversarial knowledge distillation. In *Proceedings of the IEEE/CVF Conference on Computer Vision and Pattern Recognition Workshops*, pages 710–711, 2020.

[589] Shoukai Xu, Haokun Li, Bohan Zhuang, Jing Liu, Jiezhang Cao, Chuangrun Liang, and Mingkui Tan. Generative low-bitwidth data free quantization. In *European Conference on Computer Vision*, pages 1–17. Springer, 2020.

[590] Xiangyu He, Qinghao Hu, Peisong Wang, and Jian Cheng. Generative zero-shot network quantization. *arXiv preprint arXiv:2101.08430*, 2021.

[591] Jianfei Chen, Yu Gai, Zhewei Yao, Michael W Mahoney, and Joseph E Gonzalez. A statistical framework for low-bitwidth training of deep neural networks. *arXiv preprint arXiv:2010.14298*, 2020.

[592] Sehoon Kim, Amir Gholami, Zhewei Yao, Michael W Mahoney, and Kurt Keutzer. I-bert: Integer-only bert quantization. *International conference on machine learning*, 2021.

[593] Darryl Lin, Sachin Talathi, and Sreekanth Annapureddy. Fixed point quantization of deep convolutional networks. In *International conference on machine learning*, pages 2849–2858. PMLR, 2016.

[594] Mark Horowitz. 1.1 computing's energy problem (and what we can do about it). In *2014 IEEE International Solid-State Circuits Conference Digest of Technical Papers (ISSCC)*, pages 10–14. IEEE, 2014.

[595] Jimmy Lei Ba, Jamie Ryan Kiros, and Geoffrey E Hinton. Layer normalization. *arXiv preprint arXiv:1607.06450*, 2016.

[596] Maxim Naumov, Dheevatsa Mudigere, Hao-Jun Michael Shi, Jianyu Huang, Narayanan Sundaraman, Jongsoo Park, Xiaodong Wang, Udit Gupta, Carole-Jean Wu, Alisson G Azzolini, et al. Deep learning recommendation model for personalization and recommendation systems. *arXiv preprint arXiv:1906.00091*, 2019.

[597] Yiren Zhou, Seyed-Mohsen Moosavi-Dezfooli, Ngai-Man Cheung, and Pascal Frossard. Adaptive quantization for deep neural network. *arXiv preprint arXiv:1712.01048*, 2017.

[598] Kuan Wang, Zhijian Liu, Yujun Lin, Ji Lin, and Song Han. HAQ: Hardware-aware automated quantization. *In Proceedings of the IEEE conference on computer vision and pattern recognition*, 2019.

[599] Zhen Dong, Zhewei Yao, Amir Gholami, Michael W Mahoney, and Kurt Keutzer. Hawq: Hessian aware quantization of neural networks with mixed-precision. In *Proceedings of the IEEE/CVF International Conference on Computer Vision*, pages 293–302, 2019.

[600] Mart van Baalen, Christos Louizos, Markus Nagel, Rana Ali Amjad, Ying Wang, Tijmen Blankevoort, and Max Welling. Bayesian bits: Unifying quantization and pruning. *Advances in neural information processing systems*, 2020.

[601] Huanrui Yang, Lin Duan, Yiran Chen, and Hai Li. Bsq: Exploring bit-level sparsity for mixed-precision neural network quantization. *arXiv preprint arXiv:2102.10462*, 2021.

[602] Zhongnan Qu, Zimu Zhou, Yun Cheng, and Lothar Thiele. Adaptive loss-aware quantization for multi-bit networks. In *IEEE/CVF Conference on Computer Vision and Pattern Recognition (CVPR)*, June 2020.

[603] Tianzhe Wang, Kuan Wang, Han Cai, Ji Lin, Zhijian Liu, Hanrui Wang, Yujun Lin, and Song Han. Apq: Joint search for network

architecture, pruning and quantization policy. In *Proceedings of the IEEE/CVF Conference on Computer Vision and Pattern Recognition*, pages 2078–2087, 2020.

[604] Peng Hu, Xi Peng, Hongyuan Zhu, Mohamed M Sabry Aly, and Jie Lin. Opq: Compressing deep neural networks with one-shot pruning-quantization. 2021.

[605] Lin Ning, Guoyang Chen, Weifeng Zhang, and Xipeng Shen. Simple augmentation goes a long way: {ADRL} for {dnn} quantization. In *International Conference on Learning Representations*, 2021.

[606] Hai Victor Habi, Roy H Jennings, and Arnon Netzer. Hmq: Hardware friendly mixed precision quantization block for cnns. *arXiv preprint arXiv:2007.09952*, 2020.

[607] Manuele Rusci, Marco Fariselli, Alessandro Capotondi, and Luca Benini. Leveraging automated mixed-low-precision quantization for tiny edge microcontrollers. In *IoT Streams for Data-Driven Predictive Maintenance and IoT, Edge, and Mobile for Embedded Machine Learning*, pages 296–308. Springer, 2020.

[608] Bichen Wu, Yanghan Wang, Peizhao Zhang, Yuandong Tian, Peter Vajda, and Kurt Keutzer. Mixed precision quantization of convnets via differentiable neural architecture search. *arXiv preprint arXiv:1812.00090*, 2018.

[609] Zhen Dong, Zhewei Yao, Daiyaan Arfeen, Amir Gholami, Michael W. Mahoney, and Kurt Keutzer. HAWQ-V2: Hessian aware trace-weighted quantization of neural networks. *Advances in neural information processing systems*, 2020.

[610] Tien-Ju Yang, Andrew Howard, Bo Chen, Xiao Zhang, Alec Go, Mark Sandler, Vivienne Sze, and Hartwig Adam. Netadapt: Platform-aware neural network adaptation for mobile applications. In *Proceedings of the European Conference on Computer Vision (ECCV)*, pages 285–300, 2018.

[611] Ying Wang, Yadong Lu, and Tijmen Blankevoort. Differentiable joint pruning and quantization for hardware efficiency. In *European Conference on Computer Vision*, pages 259–277. Springer, 2020.

[612] Benjamin Hawks, Javier Duarte, Nicholas J Fraser, Alessandro Pappalardo, Nhan Tran, and Yaman Umuroglu. Ps and qs: Quantization-aware pruning for efficient low latency neural network inference. *arXiv preprint arXiv:2102.11289*, 2021.

[613] Prad Kadambi, Karthikeyan Natesan Ramamurthy, and Visar Berisha. Comparing fisher information regularization with distillation for dnn quantization. *Advances in neural information processing systems*, 2020.

[614] Jianming Ye, Shiliang Zhang, and Jingdong Wang. Distillation guided residual learning for binary convolutional neural networks. *arXiv preprint arXiv:2007.05223*, 2020.

[615] Wonpyo Park, Dongju Kim, Yan Lu, and Minsu Cho. Relational knowledge distillation. In *Proceedings of the IEEE/CVF Conference on Computer Vision and Pattern Recognition*, pages 3967–3976, 2019.

[616] Shan You, Chang Xu, Chao Xu, and Dacheng Tao. Learning from multiple teacher networks. In *Proceedings of the 23rd ACM SIGKDD International Conference on Knowledge Discovery and Data Mining*, pages 1285–1294, 2017.

[617] Antti Tarvainen and Harri Valpola. Mean teachers are better role models: Weight-averaged consistency targets improve semi-supervised deep learning results. *arXiv preprint arXiv:1703.01780*, 2017.

[618] Elliot J Crowley, Gavin Gray, and Amos J Storkey. Moonshine: Distilling with cheap convolutions. In *NeurIPS*, pages 2893–2903, 2018.

[619] Linfeng Zhang, Jiebo Song, Anni Gao, Jingwei Chen, Chenglong Bao, and Kaisheng Ma. Be your own teacher: Improve the performance of convolutional neural networks via self distillation. In *Proceedings of the IEEE/CVF International Conference on Computer Vision*, pages 3713–3722, 2019.

[620] Minje Kim and Paris Smaragdis. Bitwise neural networks. *arXiv preprint arXiv:1601.06071*, 2016.

[621] Xiangguo Zhang, Haotong Qin, Yifu Ding, Ruihao Gong, Qinghua Yan, Renshuai Tao, Yuhang Li, Fengwei Yu, and Xianglong Liu. Diversifying sample generation for accurate data-free quantization. *CVPR*, 2021.

[622] Se Jung Kwon, Dongsoo Lee, Byeongwook Kim, Parichay Kapoor, Baeseong Park, and Gu-Yeon Wei. Structured compression by weight encryption for unstructured pruning and quantization. In *Proceedings of the IEEE/CVF Conference on Computer Vision and Pattern Recognition*, pages 1909–1918, 2020.

[623] Haichuan Yang, Shupeng Gui, Yuhao Zhu, and Ji Liu. Automatic neural network compression by sparsity-quantization joint learning: A constrained optimization-based approach. In *Proceedings of the IEEE/CVF Conference on Computer Vision and Pattern Recognition*, pages 2178–2188, 2020.

[624] Qing Jin, Linjie Yang, and Zhenyu Liao. Adabits: Neural network quantization with adaptive bit-widths. In *Proceedings of the IEEE/CVF Conference on Computer Vision and Pattern Recognition*, pages 2146–2156, 2020.

[625] Haotong Qin, Ruihao Gong, Xianglong Liu, Mingzhu Shen, Ziran Wei, Fengwei Yu, and Jingkuan Song. Forward and backward information retention for accurate binary neural networks. In *Proceedings of the IEEE/CVF Conference on Computer Vision and Pattern Recognition*, pages 2250–2259, 2020.

[626] Bohan Zhuang, Chunhua Shen, Mingkui Tan, Lingqiao Liu, and Ian Reid. Structured binary neural networks for accurate image classification and semantic segmentation. In *Proceedings of the IEEE/CVF Conference on Computer Vision and Pattern Recognition*, pages 413–422, 2019.

[627] Ziwei Wang, Jiwen Lu, Chenxin Tao, Jie Zhou, and Qi Tian. Learning channel-wise interactions for binary convolutional neural networks. In *Proceedings of the IEEE/CVF Conference on Computer Vision and Pattern Recognition*, pages 568–577, 2019.

[628] Chunlei Liu, Wenrui Ding, Xin Xia, Baochang Zhang, Jiaxin Gu, Jianzhuang Liu, Rongrong Ji, and David Doermann. Circulant binary convolutional networks: Enhancing the performance of 1-bit dcnns with circulant back propagation. In *Proceedings of the*

IEEE/CVF Conference on Computer Vision and Pattern Recognition, pages 2691–2699, 2019.

[629] Shilin Zhu, Xin Dong, and Hao Su. Binary ensemble neural network: More bits per network or more networks per bit? In *Proceedings of the IEEE/CVF Conference on Computer Vision and Pattern Recognition*, pages 4923–4932, 2019.

[630] Yinghao Xu, Xin Dong, Yudian Li, and Hao Su. A main/subsidiary network framework for simplifying binary neural networks. In *Proceedings of the IEEE/CVF Conference on Computer Vision and Pattern Recognition*, pages 7154–7162, 2019.

[631] Zhezhi He and Deliang Fan. Simultaneously optimizing weight and quantizer of ternary neural network using truncated gaussian approximation. In *Proceedings of the IEEE/CVF Conference on Computer Vision and Pattern Recognition*, pages 11438–11446, 2019.

[632] Felix Juefei-Xu, Vishnu Naresh Boddeti, and Marios Savvides. Local binary convolutional neural networks. In *Proceedings of the IEEE conference on computer vision and pattern recognition*, pages 19–28, 2017.

[633] Yueqi Duan, Jiwen Lu, Ziwei Wang, Jianjiang Feng, and Jie Zhou. Learning deep binary descriptor with multi-quantization. In *Proceedings of the IEEE conference on computer vision and pattern recognition*, pages 1183–1192, 2017.

[634] Xiaodi Wang, Baochang Zhang, Ce Li, Rongrong Ji, Jungong Han, Xianbin Cao, and Jianzhuang Liu. Modulated convolutional networks. In *Proceedings of the IEEE Conference on Computer Vision and Pattern Recognition*, pages 840–848, 2018.

[635] Qinghao Hu, Gang Li, Peisong Wang, Yifan Zhang, and Jian Cheng. Training binary weight networks via semi-binary decomposition. In *Proceedings of the European Conference on Computer Vision (ECCV)*, pages 637–653, 2018.

[636] Koen Helwegen, James Widdicombe, Lukas Geiger, Zechun Liu, Kwang-Ting Cheng, and Roeland Nusselder. Latent weights do not exist: Rethinking binarized neural network optimization. *Advances in neural information processing systems*, 2019.

[637] Dongsoo Lee, Se Jung Kwon, Byeongwook Kim, Yongkweon Jeon, Baeseong Park, and Jeongin Yun. Flexor: Trainable fractional quantization. *Advances in neural information processing systems*, 2020.

[638] Alexander Shekhovtsov, Viktor Yanush, and Boris Flach. Path sample-analytic gradient estimators for stochastic binary networks. *Advances in neural information processing systems*, 2020.

[639] Kai Jia and Martin Rinard. Efficient exact verification of binarized neural networks. *Advances in neural information processing systems*, 2020.

[640] Mingbao Lin, Rongrong Ji, Zihan Xu, Baochang Zhang, Yan Wang, Yongjian Wu, Feiyue Huang, and Chia-Wen Lin. Rotated binary neural network. *Advances in neural information processing systems*, 2020.

[641] Hyungjun Kim, Kyungsu Kim, Jinseok Kim, and Jae-Joon Kim. Binaryduo: Reducing gradient mismatch in binary activation network by coupling binary activations. *International Conference on Learning Representations*, 2020.

[642] Yuhang Li, Ruihao Gong, Fengwei Yu, Xin Dong, and Xianglong Liu. Dms: Differentiable dimension search for binary neural networks. *International Conference on Learning Representations*, 2020.

[643] Kai Han, Yunhe Wang, Yixing Xu, Chunjing Xu, Enhua Wu, and Chang Xu. Training binary neural networks through learning with noisy supervision. In *International Conference on Machine Learning*, pages 4017–4026. PMLR, 2020.

[644] Haotong Qin, Zhongang Cai, Mingyuan Zhang, Yifu Ding, Haiyu Zhao, Shuai Yi, Xianglong Liu, and Hao Su. Bipointnet: Binary neural network for point clouds. *International Conference on Learning Representations*, 2021.

[645] Adrian Bulat, Brais Martinez, and Georgios Tzimiropoulos. High-capacity expert binary networks. *International Conference on Learning Representations*, 2021.

[646] James Diffenderfer and Bhavya Kailkhura. Multi-prize lottery ticket hypothesis: Finding accurate binary neural networks by pruning

a randomly weighted network. In *International Conference on Learning Representations*, 2021.

[647] Cheng Fu, Shilin Zhu, Hao Su, Ching-En Lee, and Jishen Zhao. Towards fast and energy-efficient binarized neural network inference on fpga. In *FPGA*, 2019.

[648] Jiaxin Gu, Junhe Zhao, Xiaolong Jiang, Baochang Zhang, Jianzhuang Liu, Guodong Guo, and Rongrong Ji. Bayesian optimized 1-bit cnns. In *Proceedings of the IEEE/CVF International Conference on Computer Vision*, pages 4909–4917, 2019.

[649] Mingzhu Shen, Kai Han, Chunjing Xu, and Yunhe Wang. Searching for accurate binary neural architectures. In *Proceedings of the IEEE/CVF International Conference on Computer Vision Workshops*, 2019.

[650] Milad Alizadeh, Javier Fernández-Marqués, Nicholas D Lane, and Yarin Gal. An empirical study of binary neural networks' optimisation. In *International Conference on Learning Representations*, 2018.

[651] Fengfu Li, Bo Zhang, and Bin Liu. Ternary weight networks. *arXiv preprint arXiv:1605.04711*, 2016.

[652] Diwen Wan, Fumin Shen, Li Liu, Fan Zhu, Jie Qin, Ling Shao, and Heng Tao Shen. Tbn: Convolutional neural network with ternary inputs and binary weights. In *Proceedings of the European Conference on Computer Vision (ECCV)*, pages 315–332, 2018.

[653] Haotong Qin, Ruihao Gong, Xianglong Liu, Xiao Bai, Jingkuan Song, and Nicu Sebe. Binary neural networks: A survey. *Pattern Recognition*, 105:107281, 2020.

[654] Zefan Li, Bingbing Ni, Wenjun Zhang, Xiaokang Yang, and Wen Gao. Performance guaranteed network acceleration via high-order residual quantization. In *Proceedings of the IEEE international conference on computer vision*, pages 2584–2592, 2017.

[655] Qinghao Hu, Peisong Wang, and Jian Cheng. From hashing to cnns: Training binary weight networks via hashing. In *Proceedings of the AAAI Conference on Artificial Intelligence*, volume 32, 2018.

[656] Asit Mishra, Eriko Nurvitadhi, Jeffrey J Cook, and Debbie Marr. WRPN: Wide reduced-precision networks. In *International Conference on Learning Representations*, 2018.

[657] Ting-Wu Chin, Pierce I-Jen Chuang, Vikas Chandra, and Diana Marculescu. One weight bitwidth to rule them all. *Proceedings of the European Conference on Computer Vision (ECCV)*, 2020.

[658] Mingzhu Shen, Xianglong Liu, Ruihao Gong, and Kai Han. Balanced binary neural networks with gated residual. In *ICASSP 2020-2020 IEEE International Conference on Acoustics, Speech and Signal Processing (ICASSP)*, pages 4197–4201. IEEE, 2020.

[659] Adrian Bulat and Georgios Tzimiropoulos. Xnor-net++: Improved binary neural networks. *British Machine Vision Conference*, 2019.

[660] Brais Martinez, Jing Yang, Adrian Bulat, and Georgios Tzimiropoulos. Training binary neural networks with real-to-binary convolutions. *arXiv preprint arXiv:2003.11535*, 2020.

[661] Lu Hou and James T. Kwok. Loss-aware weight quantization of deep networks. In *International Conference on Learning Representations*, 2018.

[662] Ruizhou Ding, Ting-Wu Chin, Zeye Liu, and Diana Marculescu. Regularizing activation distribution for training binarized deep networks. In *Proceedings of the IEEE/CVF Conference on Computer Vision and Pattern Recognition*, pages 11408–11417, 2019.

[663] Xiuyi Chen, Guangcan Liu, Jing Shi, Jiaming Xu, and Bo Xu. Distilled binary neural network for monaural speech separation. In *International Joint Conference on Neural Networks (IJCNN)*, pages 1–8. IEEE, 2018.

[664] Sajad Darabi, Mouloud Belbahri, Matthieu Courbariaux, and Vahid Partovi Nia. Bnn+: Improved binary network training. 2018.

[665] Ruihao Gong, Xianglong Liu, Shenghu Jiang, Tianxiang Li, Peng Hu, Jiazhen Lin, Fengwei Yu, and Junjie Yan. Differentiable soft quantization: Bridging full-precision and low-bit neural networks. In *Proceedings of the IEEE/CVF International Conference on Computer Vision*, pages 4852–4861, 2019.

[666] Adrian Bulat, Georgios Tzimiropoulos, Jean Kossaifi, and Maja Pantic. Improved training of binary networks for human pose estimation and image recognition. *arXiv preprint arXiv:1904.05868*, 2019.

[667] Zhe Xu and Ray CC Cheung. Accurate and compact convolutional neural networks with trained binarization. *British Machine Vision Conference*, 2019.

[668] Penghang Yin, Shuai Zhang, Jiancheng Lyu, Stanley Osher, Yingyong Qi, and Jack Xin. Blended coarse gradient descent for full quantization of deep neural networks. *Research in the Mathematical Sciences*, 6(1):14, 2019.

[669] Wei Zhang, Lu Hou, Yichun Yin, Lifeng Shang, Xiao Chen, Xin Jiang, and Qun Liu. Ternarybert: Distillation-aware ultra-low bit bert. *arXiv preprint arXiv:2009.12812*, 2020.

[670] Haoli Bai, Wei Zhang, Lu Hou, Lifeng Shang, Jing Jin, Xin Jiang, Qun Liu, Michael Lyu, and Irwin King. Binarybert: Pushing the limit of bert quantization. *arXiv preprint arXiv:2012.15701*, 2020.

[671] Jing Jin, Cai Liang, Tiancheng Wu, Liqin Zou, and Zhiliang Gan. Kdlsq-bert: A quantized bert combining knowledge distillation with learned step size quantization. *arXiv preprint arXiv:2101.05938*, 2021.

[672] Yinhan Liu, Myle Ott, Naman Goyal, Jingfei Du, Mandar Joshi, Danqi Chen, Omer Levy, Mike Lewis, Luke Zettlemoyer, and Veselin Stoyanov. RoBERTa: A robustly optimized bert pretraining approach. *arXiv preprint arXiv:1907.11692*, 2019.

[673] Alec Radford, Karthik Narasimhan, Tim Salimans, and Ilya Sutskever. Improving language understanding by generative pretraining, 2018.

[674] Alec Radford, Jeffrey Wu, Rewon Child, David Luan, Dario Amodei, and Ilya Sutskever. Language models are unsupervised multitask learners. *OpenAI blog*, 1(8):9, 2019.

[675] Tom B Brown, Benjamin Mann, Nick Ryder, Melanie Subbiah, Jared Kaplan, Prafulla Dhariwal, Arvind Neelakantan, Pranav Shyam, Girish Sastry, Amanda Askell, et al. Language models are few-shot learners. *arXiv preprint arXiv:2005.14165*, 2020.

[676] Dana Harry Ballard. *An introduction to natural computation*. MIT press, 1999.

[677] Herve Jegou, Matthijs Douze, and Cordelia Schmid. Product quantization for nearest neighbor search. *IEEE transactions on pattern analysis and machine intelligence*, 33(1):117–128, 2010.

[678] Eirikur Agustsson, Fabian Mentzer, Michael Tschannen, Lukas Cavigelli, Radu Timofte, Luca Benini, and Luc Van Gool. Soft-to-hard vector quantization for end-to-end learning compressible representations. *arXiv preprint arXiv:1704.00648*, 2017.

[679] Julieta Martinez, Shobhit Zakhmi, Holger H Hoos, and James J Little. Lsq++: Lower running time and higher recall in multi-codebook quantization. In *Proceedings of the European Conference on Computer Vision (ECCV)*, pages 491–506, 2018.

[680] Lopamudra Mukherjee, Sathya N Ravi, Jiming Peng, and Vikas Singh. A biresolution spectral framework for product quantization. In *Proceedings of the IEEE Conference on Computer Vision and Pattern Recognition*, pages 3329–3338, 2018.

[681] Kuilin Chen and Chi-Guhn Lee. Incremental few-shot learning via vector quantization in deep embedded space. In *International Conference on Learning Representations*, 2021.

[682] Pierre Stock, Armand Joulin, Rémi Gribonval, Benjamin Graham, and Hervé Jégou. And the bit goes down: Revisiting the quantization of neural networks. *arXiv preprint arXiv:1907.05686*, 2019.

[683] Liangzhen Lai, Naveen Suda, and Vikas Chandra. CMSIS-NN: Efficient neural network kernels for arm cortex-m cpus. *arXiv preprint arXiv:1801.06601*, 2018.

[684] Eric Flamand, Davide Rossi, Francesco Conti, Igor Loi, Antonio Pullini, Florent Rotenberg, and Luca Benini. Gap-8: A risc-v soc for ai at the edge of the iot. In *IEEE International Conference on Application-specific Systems, Architectures and Processors (ASAP)*, pages 1–4. IEEE, 2018.

[685] Jeff Johnson. Rethinking floating point for deep learning. *arXiv preprint arXiv:1811.01721*, 2018.

[686] Naveen Mellempudi, Sudarshan Srinivasan, Dipankar Das, and Bharat Kaul. Mixed precision training with 8-bit floating point. *arXiv preprint arXiv:1905.12334*, 2019.

[687] Shuang Wu. Training and inference with integers in deep neural networks. *international conference on learning representations*, 2018.

[688] Hamed F Langroudi, Zachariah Carmichael, David Pastuch, and Dhireesha Kudithipudi. Cheetah: Mixed low-precision hardware & software co-design framework for dnns on the edge. *arXiv preprint arXiv:1908.02386*, 2019.

[689] Léopold Cambier, Anahita Bhiwandiwalla, Ting Gong, Mehran Nekuii, Oguz H Elibol, and Hanlin Tang. Shifted and squeezed 8-bit floating point format for low-precision training of deep neural networks. *arXiv preprint arXiv:2001.05674*, 2020.

[690] Wuwei Lin. Automating optimization of quantized deep learning models on cuda: https://tvm.apache.org/2019/04/29/opt-cuda-quantized, 2019.

[691] Dave Salvator, Hao Wu, Milind Kulkarni, and Niall Emmart. Int4 precision for ai inference: https://developer.nvidia.com/blog/int4-for-ai-inference/, 2019.

[692] Animesh Jain, Shoubhik Bhattacharya, Masahiro Masuda, Vin Sharma, and Yida Wang. Efficient execution of quantized deep learning models: A compiler approach. *arXiv preprint arXiv:2006.10226*, 2020.

Index

Note: Locators in *italics* represent figures and **bold** indicate tables in the text.